Dynamic Modeling and Neural Network-Based Intelligent Control of Flexible Systems

IEEE Press
445 Hoes Lane
Piscataway, NJ 08854

Dynamic Modeling and Neural Network-Based Intelligent Control of Flexible Systems

Hejia Gao
Anhui University, China

Wei He
University of Science and Technology Beijing, China

Changyin Sun
Southeast University, China

IEEE Press Series on Control Systems Theory and Applications
Maria Domenica Di Benedetto, Series Editor

IEEE Press
WILEY

Published by John Wiley & Sons, Inc., Hoboken, New Jersey.
Published simultaneously in Canada.

Library of Congress Cataloging-in-Publication Data applied for:

Hardback ISBN: 9781394255276

Cover Design: Wiley
Cover Images: © Stocktrek Images/Getty Images, © Pobytov/Getty Images

Set in 9.5/12.5pt STIXTwoText by Straive, Chennai, India

Contents

About the Authors

Hejia Gao received the BEng degree in intelligence science and technology from the School of Automation and Electrical Engineering, University of Science and Technology Beijing, Beijing, China, in 2016, and the PhD degree in control science and engineering from the School of Automation and Electrical Engineering, University of Science and Technology Beijing, Beijing, China, in 2021. She has been working as a visiting researcher for fifteen months at the Department of Mechanical, Industrial and Aerospace Engineering, Concordia University, Canada, in 2019. Currently, she is working as an associate professor in the School of Artificial Intelligence, Anhui University, Hefei, China. She has published eight top international journal papers and four international conference papers, and two papers have been ranked as ESI Highly Cited Paper. She has been an associate editor of *the 2019 IEEE International Conference on Advanced Robotics and Mechatronics* and the co-chair of *IEEE 37st Youth Academic Annual Conference of Chinese Association of Automation*. At present, she is serving as a reviewer of *IEEE Transactions on Cybernetics*, *IEEE Transactions on Neural Networks and Learning Systems*, *IEEE Transactions on Systems, Man, and Cybernetics: Systems*, and *IEEE/CAA Journal of Automatica Sinica*. Her research interests include neural networks, reinforcement learning, flexible systems, and vibration control.

Wei He received the BEng and MEng degrees from the College of Automation Science and Engineering, South China University of Technology, China, in 2006 and 2008, respectively, and the PhD degree from Department of Electrical and Computer Engineering, the National University of Singapore, Singapore, in 2011. He is currently working as a full professor in School of Automation and Electrical Engineering, University of Science and Technology Beijing, Beijing, China. He has co-authored three books published in Springer and published over 100 international journal and conference papers. Professor He was awarded a Newton Advanced Fellowship from the Royal Society, UK, in 2017. He was a recipient of the IEEE SMC Society Andrew P. Sage Best Transactions Paper Award in 2017. He is serving as the chair of the IEEE SMC Society Beijing Capital Region Chapter. Since 2018, he has been the chair of Technical Committee on Autonomous Bionic Robotic Aircraft, IEEE Systems, Man and Cybernetics Society. He is serving as an associate editor of *IEEE Transactions on Robotics*, *IEEE Transactions on Neural Networks and Learning Systems*, *IEEE Transactions on Control Systems Technology*, *IEEE Transactions on Systems, Man, and Cybernetics: Systems*, *SCIENCE CHINA Information Sciences*, *IEEE/CAA Journal of Automatica Sinica*, *Neurocomputing*, and an editor of *Journal of Intelligent and Robotic Systems*.

Changyin Sun received the BEng degree in applied mathematics from the College of Mathematics, Sichuan University, Chengdu, China, in 1996, and the MS and PhD degrees in electrical engineering from Southeast University, Nanjing, China, in 2001 and 2004, respectively. He is currently a professor with the School of Automation, Southeast University, Nanjing, China. He has co-authored four books and published over 160 international journal papers. Professor Sun is a Chinese Association of Automation Fellow. He is serving as an associate editor of *IEEE Transactions on Neural Networks and Learning Systems, Neural Processing Letters, and IEEE/CAA Journal of Automatica Sinica*. His research interests include intelligent control, flight control, pattern recognition, and optimal theory.

Preface

The flexible system covers many different objects such as flexible robotic manipulators, bionic flexible flapping wing aircraft, and flexible buildings. With a large number of applications of flexible systems, its control theory and method issues have become a prospective high-tech research direction, which attracts concerns from both academic and industrial fields. At present, the control theory and method of flexible systems, such as the tracking and vibration control of multi-link flexible manipulators, the constraint control of flexible buildings under natural disasters, and the fault-tolerant control of bionic flexible flapping-wing robots, has developed into a common scientific problem, which is extremely challenging. In order to solve the technical problems of dynamic modeling and intelligent control of uncertain flexible systems with environmental adaptability, the book makes a systematic and detailed study on modeling mechanism and control strategy of several flexible systems.

Chapter 1 provides an overview of dynamic modeling and intelligent control of flexible systems, introducing several important issues in the study of flexible systems. The modeling and control methods of three typical flexible systems are discussed separately.

Chapter 2 provides the corresponding mathematical preliminaries of subsequent chapters, including the Hamilton principle, model discretization methods, Lagrange's equation method, neural networks, and Lyapunov stability theorem.

Chapter 3 develops the dynamic model of the single-link flexible robotic manipulator, which overcomes the challenge from the system dynamics being infinite dimensional. The fuzzy neural network control with uniform approximation performance is designed to solve the system uncertainties. Numerical simulations and extensive experiments have been investigated to verify the effectiveness of the proposed methods.

Chapter 4 establishes a finite-dimensional dynamic model of the two-link flexible robotic manipulator. A high-gain observer-based neural network control strategy is proposed to estimate the immeasurable states in practice. The semi-globally uniformly ultimate boundedness (SGUUB) of the closed-loop system is guaranteed via Lyapunov's stability theory. The simulation and experimental results demonstrate the effectiveness of the proposed control strategy.

Chapter 5 present the vibration control design for a string with the boundary time-varying output constraint. The dynamics of the string is a distributed parameter system described by a partial differential equation and two ordinary differential equations. A barrier Lyapunov function with a logarithmic function is adopted to prevent the time-varying constraint violations. Adaptive control is designed to handle the system parametric uncertainties. Stability analysis and the solvability of the inequality equations are provided. Numerical simulations are provided to illustrate the effectiveness of the proposed control design.

Chapter 6 focuses on a stand-alone tall building-like structure with an eccentric load. A neural network control approach is proposed to suppress vibrations caused by unknown time-varying disturbances (earthquake, strong wind, etc.). The output constraint on the angle of the eccentric load is also considered, and such angle can be ensured within the safety limit by incorporating a barrier Lyapunov function. Simulations and experiments based on MATLAB and Quanser are carried out to verify the feasibility and effectiveness of the proposed control.

Chapter 7 discusses an adaptive vibration control method for a single-floor building-like structure equipped with an active mass damper (AMD). The method uses a hybrid learning control strategy to suppress vibrations caused by unknown time-varying disturbances such as earthquakes or strong winds. The effectiveness of the proposed control approach is demonstrated through experimental investigation on a Quanser Active Mass Damper. The research results aim to bring new ideas and methods to the field of disaster reduction for engineering development.

Chapter 8 investigates a single-floor building-like structure equipped with an active mass damper (AMD). Optimal vibration control, while dealing with system uncertainties, is realized by the reinforcement learning technique. When the unexpected natural disasters occur, the proposed controller applying to the active mass damper can compensate the increase of the system vibration caused by external disturbances. The experimental results in the form of graphics and tables have shown the effectiveness of the proposed control algorithm.

Chapter 9 develops the visualization model of the rigid-flexible coupled bionic flapping wing by the advanced system-level modeling software MapleSim. A novel neural network controller based on disturbance observer technology is proposed to compensate the system uncertainties. The proposed method can successfully suppress the vibration of the flapping wing while accurately track the desired trajectory. Co-simulation results from MapleSim and Matlab/Simulink validate the effectiveness of the proposed method.

Chapter 10 focuses on the flexible wings of the aircraft, which has great advantages, such as being lightweight, having high flexibility, and offering low energy consumption. A novel adaptive finite-time controller based on the fuzzy neural network and nonsingular fast terminal slidingmode scheme are proposed for tracking control and vibration suppression of the flexible wings, while successfully addressing the issues of system uncertainties and actuator failures. Co-simulations through MapleSim and MATLAB/Simulink are carried out to verify the performance of the proposed controller.

Chapter 11 discusses the importance of vibration control for bionic flapping-wing robotic aircraft and autonomous ornithopter applications. A visualization model of the rigid-flexible coupled bionic flapping wing is established using MapleSim software. A novel adaptive vibration controller based on neural network (NN) algorithm is proposed to compensate for system uncertainties. The proposed method can successfully suppress the vibration of the flapping wing while accurately tracking the desired trajectory. The effectiveness of the proposed method is validated through co-simulation results from MapleSim and Matlab/Simulink.

Chapter 12 investigate dynamic modeling, active boundary control design, and stability analysis for a coupled floating wind turbine (FWT) system, which is connected with two flexible mooring lines. It is a coupled beam-strings structure, and we design two boundary controllers to restrain the vibrations of this flexible system caused by external disturbances based on the coupled partial differential equations and ordinary differential equations (PDEs-ODEs) model. Meanwhile, significant performance of designed boundary controllers and system's stability are theoretically analyzed, and a set of simulation results are provided to show efficacy of the proposed approach.

Chapter 13 summarizes the practical significance in the application of neural network-based intelligent control and proposes some future research directions in this field.

In summary, this book proposes high-efficiency modeling methods and novel intelligent control strategies for several representative flexible systems developed by means of neural networks. The book discusses the tracking control of multi-link flexible manipulators, the vibration control of flexible buildings under natural disasters, and the fault-tolerant control of bionic flexible flapping-wing aircraft. Expanding on its theoretical deliberations, the book includes many case studies demonstrating how the proposed approaches work in practice. The most important features of the book include:

1) a comprehensive review of modeling and control theory for flexible systems;
2) detailed presentation of the modeling methods and the neural network-based control strategies;
3) successful addressing of external disturbances, dynamic uncertainties, output constrains, and actuator faults;
4) abundance case studies illustrating the important steps in designing the neural network-based control; and
5) performance analysis of the described control approaches by a large number of figures and tables.

This book can be regarded as an authoritative reference for the field (studies) of dynamics and control of flexible systems. Interested readers will gain a systematic understanding of the flexible systems as well as the technical details involved. The material presented in this book will be useful for researchers and engineers who wish to avoid excessive time in searching high-efficiency modeling methods and neural-network-based control solutions for flexible systems. It is written for industry engineers and researchers who are interested in control theory and the applications. This book is also good for postgraduate students engaged in self-study of adaptive control for the flexible systems.

Hejia Gao
Anhui University, China

Wei He
University of Science and Technology Beijing, China

Changyin Sun
Southeast University, China

Acknowledgments

The completion of this book would not have been achieved without many help and effort from the group members of the Key Laboratory of Knowledge Automation for Industrial Processes, Ministry of Education, China. Over the course of our researching and writing this book, we hereby express our sincere thanks to all those who have helped us.

First of all, we express our sincere appreciation to our co-workers and colleagues who have contributed to the collaborative research. In particular, we thank Youmin Zhang, from Concordia University, Canada; Xiuyu He, Xinbo Yu, Zhijie Liu from University of Science and Technology Beijing; Juqi Hu, from Anhui University and their research groups for their excellent works, and helpful advice on our research.

Appreciations must be made to Zele Yu, Junjie Zhao, Chuanfeng He, Xiang Wang, Liang Zhang, Jiayi Hou, Yu Liu, Zhijia Zhao, Zhe Jing, Tingting Meng, Linghuan Kong, and Yuhua Song for the interesting discussions and constructive suggestions. Their nice comments contribute to improvements of the book in terms of both the presentation and the organization.

Last but not the least, we want to show our deep gratitude to our families for their invaluable love, support and sacrifices over the years. Their long-term support and care give us a warm mood in writing this book.

This work was supported, in part, by National Natural Science Foundation of China under Grant 62225304, 61933001, 62303010, in part, by the Anhui Natural Science Foundation under Grant 2208085QF207, 2208085QF209, and, in part, by the Anhui Provincial Key Research Program of Universities under Grant 2022AH050087.

Hejia Gao
Wei He
Changyin Sun

Acronyms

ADP	Adaptive dynamic programming
AFT	An adaptive finite-time
AMM	Assumed mode method
ARV	Acceleration response value
ASM	Absolute coordinate method
AVC	Adaptive vibration controller
DO	Disturbance observer
DOF	Degree of freedom
DC	Direct current
DiffServ	Differentiated Services
DPS	Distributed parameter system
FSM	Finite segment method
FSF	Full-state feedback
FDM	Finite difference method
FEM	Finite element method
FWMAV	Flapping wing micro aerial vehicle
FWT	Floating wind turbine
GB	Global balance
HL	Hybrid learning
LPM	Lumped parameter method
MSM	Modes synthesis method
MIMO	Multi-input and multi-output
MR	Magneto rheological
NFTSM	Nonsingular fast terminal sliding mode
NN	Neural network
ODEs	Ordinary differential equations
PC	Personal computer
PDF	Probability distribution function
PDEs	Partial differential equations
PSF	Partial state feedback
PV	Proportional-velocity
QAMD	Quanser One-Floor Active Mass Damper
RBFNN	Radial basis function neural network
SGUUB	Semi-global global uniform ultimate boundedness
TDC	Time delay control

TMD	Tuned mass damper
UDP	User Datagram Protocol
UUB	Uniform ultimate boundedness
VRV	Vibration response value

1

Introduction

1.1 Background and Motivation

Flexible systems refer to those structures that can be bent or folded without breaking [286]. Flexible systems have attracted considerable research and OpenDocument efforts due to their multi-faceted advantages such as lightweight and low energy consumption [300]. The significant advances in material science, computing, and artificial intelligence technologies, particularly during the last two decades, have further inspired the society's expectation and passion for commercially viable flexible systems [368]. However, flexible systems might generate unexpected deformation and vibration during the execution of the task. The vibration will degrade the system performance, even shorten the lifespan of the flexible systems. In addition, the complex environment and sudden failures also bring challenges to the research and application of flexible systems. Therefore, designing an effective control method for suppressing the vibration of the flexible systems is significant in practice.

In recent years, with the increasing maturity of intelligent manufacturing technology and the continuous breakthrough of scientific and technological research results, flexible systems have been widely used and developed in various fields such as aerospace, advanced manufacturing, medical health, and social services. Basically, most mechanical structures in practical engineering can be regarded as flexible systems, in other words, large-span mechanical structures can be regarded as having non-negligible deformation and vibration. The flexible systems mainly have three types of applications, which are briefly described below:

- **Flexible robotic manipulators**: Flexible robotic manipulators are the most typical representative flexible systems, which have the properties of light structure, low energy consumption, and high load/self-weight ratio. Compared with the traditional rigid manipulators, the flexible manipulators have a larger operating space, higher work efficiency, and faster response. Flexible manipulators have many potential advantages and play a very important role in industrial, national defense, and other application fields [113]. Their applications have gradually penetrated into aerospace, medical, and military fields [188], as shown in Figure 1.1. With the rapid development of the aerospace industry and the robotics industry, the traditional rigid system dynamic analysis methods and control strategies cannot meet the requirements of practical engineering [283]. In recent years, the application research of flexible robotic manipulators has received extensive attention, attracting the interest of many scholars and experts.

  The main control goal of the flexible robotic manipulators is to achieve accurate trajectory tracking, and damping vibration (also known as vibration suppression) due to low stiffness is an urgent problem to be solved. The traditional rigid manipulator has a thick base, short arm, limited operating space, and poor flexibility, which cannot meet the requirements of modern

Dynamic Modeling and Neural Network-Based Intelligent Control of Flexible Systems,
First Edition. Hejia Gao, Wei He, and Changyin Sun.

Figure 1.1 Canadarm 2, the Canadian robotic arm on the space station. Source: NASA/Public domain.

industrial automation and high-precision industries. The flexible robotic manipulator has gradually occupied an increasingly important position in the manufacturing, aerospace, and other industries due to its advantages of low energy consumption, flexible operation, and fast response. The flexible robotic manipulator is composed of flexible unit components, mainly including flexible joints and flexible links. These flexible components will produce twisting deformation, elastic deformation, and shear deformation during the movement of the manipulator. The flexible robotic manipulator is thus a rigid–flexible coupled, nonlinear infinite-dimensional distributed parameter system, and its dynamic model is more complex than that of the rigid manipulator. Moreover, the traditional dynamic modeling is too complicated and not suitable for the control design of the high-performance flexible robotic manipulator. Therefore, how to explore efficient dynamic modeling theory for flexible manipulators with dynamic uncertainties is a hot and difficult issue in current research [5].

- **Flexible building structures**: With the continuous increase of building height, high-rise, and ultra-high-rise buildings can be regarded as flexible building structures. Under external excitations such as earthquakes and strong winds, the vibration of high-rise building structures cannot be ignored. Due to the unpredictability of natural disasters such as earthquakes and strong winds, the seismic and wind-resistant design of structures faces severe challenges. Vibration and displacement control are of critical importance for both high-rise and ultra-high-rise building systems, as shown in Figure 1.2. Since the concept of structural control was first proposed in 1972, the research and application of structural vibration control have received more and more attention. At present, it has become the most cutting-edge development directions in the field of structural engineering and structural mechanics. The high-rise structure system can not only provide residents with a comfortable working and living environment can also relieve the huge pressure caused by population growth and shortage of land resources in large cities. Due to the large number of residents and the huge construction cost of the high-rise structure, when it is subjected to natural disasters such as earthquakes and strong winds, once the structure is damaged or collapsed, its influence and destructive power will be huge. Therefore, it is particularly important to effectively suppress or reduce these vibrations caused by earthquakes and winds, so that the safety, usability, and comfort of high-rise structures can be guaranteed.

  The main control goal of flexible building structures is to achieve vibration suppression under unexpected disturbances. The existing vibration control methods of flexible building

Figure 1.2 The Shanghai Tower. Source: Baycrest/Wikimedia Commons/CC BY SA 2.5.

structures mostly employed traditional passive control strategies. However, the installation and maintenance as well as the replacement of the dampers are time-consuming and non-trivial work. Compared with passive control, active control can select control objectives, such as structural response (displacement, velocity, and acceleration response) and structural displacement, improving the control effectiveness. Actually, the complex flexible building structures, whose degrees of freedom are close to infinity, are distributed parameter systems with many dynamic uncertainties [137]. Most importantly, the dynamic response cannot be predicted accurately. Therefore, how to study the vibration control method of flexible building structures under natural disasters is an urgent problem to be solved.

- **Flexible bionic flapping-wing systems**: The flexible bionic flapping-wing robotic aircraft is inspired by the flight of birds or insects, which has great advantages, such as being lightweight, having high flexibility, and offering low energy consumption. The bionic flapping-wing robotic aircraft has attracted the special attention of many researchers in recent years [316]. The aircraft can stabilize the fuselage in the horizontal position or glide in the sky through flapping wings. The bionic flapping-wing aircraft can generate lift force with high efficiency. In addition, with the advantage of low energy consumption, the bionic flapping-wing aircraft is suitable for flight missions without energy replenishment under long-distance conditions. The bionic flapping-wing aircraft is therefore widely used in both the military tasks (low-altitude surveillance, urban combat, accurate delivery, etc.) and civilian applications (disaster monitoring and relief, field exploration, etc.) [244], as shown in Figure 1.3. Compared with fixed wings and rotary wings, flapping wings have the advantage in flight efficiency [227, 363], which shows the vital importance of the study of the bionic flexible flapping wings. However, flexible wings might generate unexpected deformation and vibration during the flying process. The vibration will degrade the flight performance, even shorten the lifespan of the aircraft. Therefore, designing an effective control method for suppressing vibrations of the flexible wings is significant in practice.

The vast majority of the previous studies have shown that an accurate mathematical model of flexible systems always necessitates complex derivations and heavy computational load. Even if a few studies have employed the software modeling methods, the distinct disadvantages, such as high complexity, poor scalability, and slow response, are followed. Another finding of the comprehensive literature review is that the control theory research of the bionic flapping-wing aircraft is at its initial stage. The previous control research has mostly neglected the vibrations [116, 143], dynamic uncertainties [292], and actuator failures that are extremely unfavorable in view of the practical application. Even among the few vibration studies, most control inputs can only be applied to the wing tip, which is impractical in real flight situations. It is thus desirable to obtain a high-efficiency modeling method and design a feasible intelligent vibration control strategy while solving the system uncertainties and actuator failures.

These figures shows that the flexible system covers many different objects such as flexible robotic manipulators, flexible bionic flapping-wing aircraft, and flexible buildings. With a large number of applications of flexible systems, its control theory and method issues have become

Figure 1.3 The bionic flapping wing aircraft.

a prospective high-tech research direction, which attracts concerns from both academic and industrial fields. It is worth mentioning that only tracking control and vibration control are not sufficient to enable flexible systems moving to its desired position due to potential variations in the environment-related factors. The following issues are essential for designing effective control strategies:

- **Dynamic uncertainties**: Flexible systems are distributed parameter systems with complicated dynamics. The mass matrix, damping matrix, and stiffness matrix are usually not sufficiently accurate enough to obtain.
- **Immeasurable states**: Due to the sensor's limitations or environmental impacts, some states of the flexible systems cannot be measured in practice.
- **External disturbances**: Flexible systems can replace people to do all kinds of dangerous works, and the factors determine that the working environment of the flexible systems is unknown, complicated, and variable.
- **Actuator failures**: Due to the harsh working environment, the failures of the components related to the actuator (such as efficiency loss and actuator bias fault) will inevitably occur.
- **Composite constraints**: Due to the limitation of workspace, the output constraints need to be considered to ensure the safety of the flexible systems during the course of movement.

The issues of dynamic uncertainties, immeasurable states, unknown external disturbances, sudden failures, and composite constraints are thus also required to be addressed.

In order to solve the above technical problems of modeling and intelligent control of flexible systems, the book makes a systematic and detailed study on dynamic modeling and intelligent control of flexible systems.

1.2 Modeling and Control Strategies of Flexible Robotic Manipulators

Flexible robotic manipulators are a type of robots that have flexible links or joints, which can deform under external forces or internal actuation. Flexible robotic arms are widely used in various fields such as aerospace, manufacturing, and medical applications. They have many advantages over rigid robots, such as higher dexterity, lower weight, lower energy consumption, and better adaptability to complex environments. However, they also pose significant challenges in modeling and control, due to their nonlinear dynamics, distributed parameters, and coupling effects. Therefore, developing effective modeling and control strategies for flexible robotic manipulators is an important and active research topic in robotics.

The control of lightweight manipulators is complex based on the nature of flexibility in the system, i.e., flexible joints, flexible links, or flexible joints and links. Among them, the most difficult task is to control the flexible link manipulators because of link flexibility, underactuation, and non-minimal phase nature. Underactuation is due to a finite number of actuators to control infinite degrees of freedom that arise due to link flexibility [21]. Figure 1.4 shows a single-link flexible manipulator, and Figure 1.5 shows a two-link flexible manipulator.

Modeling of flexible robotic arms involves deriving the equations of motion that describe the coupled rigid–flexible dynamics of the system. The widely used modeling approaches are the Hamilton's principle [200], Newton–Euler equation [361], Lagrangian equation [167], Kane equation [50], Gaussian equation [88], etc. The Newton–Euler equation is based on vector mechanics, and the rest of the methods refer to analytical mechanics. The resulting models

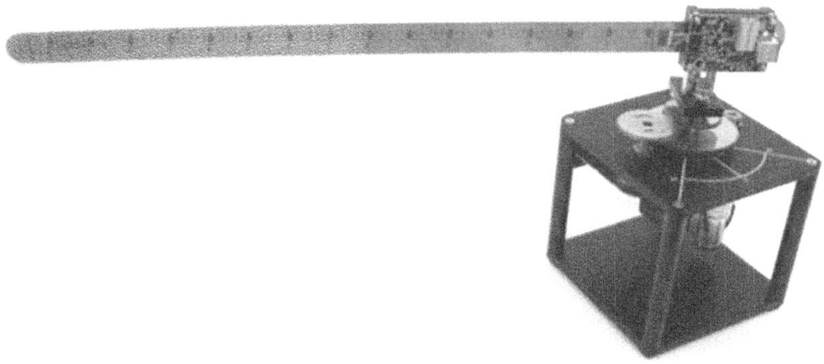

Figure 1.4 A single-link flexible manipulator system.

Figure 1.5 A two-link flexible manipulator system.

are usually partial differential equations (PDEs) or ordinary differential equations (ODEs) with boundary conditions.

Control of flexible robotic arms aims at achieving desired performance objectives such as tracking, stabilization, vibration suppression, and disturbance rejection. Various control strategies have been developed for this purpose, such as sliding mode control 1, adaptive robust boundary control 2, and fuzzy adaptive control 3. These strategies can be classified into two categories: boundary control and distributed control. Boundary control only uses actuators and sensors at the base or tip of the arm, while distributed control uses distributed actuators and sensors along the arm. Boundary control has the advantages of simplicity, robustness, and low cost. However, it may suffer from spillover effects, which occur when the unmodeled high-frequency modes are excited by the controller. Distributed control can overcome this problem by taking into account the full PDE model of the system. However, it may require more complex hardware and software implementation. In conclusion, modeling and control design of flexible robotic arms are active research topics that have attracted much attention in recent years. There are still many open problems and challenges that need to be addressed, such as model reduction, parameter estimation, sensor fusion, fault detection and isolation, and cooperative control of multiple flexible arms.

From these studies, it can be concluded that there is no single best modeling or control strategy for flexible robots, and the choice should be based on the specific application scenario and the characteristics of the robot. Future research can explore more efficient and accurate modeling methods and more intelligent and adaptive control methods to solve some of the challenges in modeling and control of flexible robots.

1.3 Vibration Control Technologies of Flexible Building-like Structures

Flexible building-like structures are civil engineering systems that have high flexibility and low damping, such as high-rise and ultra-high-rise buildings. They are susceptible to vibrations caused by natural disasters such as earthquakes and strong winds, which may endanger the safety of the structures and the occupants. Therefore, designing effective vibration control technologies for flexible building-like structures is a significant and challenging task.

Vibration control technologies can be classified into two categories: passive and active. Passive control technologies use devices that do not require external power or feedback signals, such as tuned mass dampers, base isolators, and viscoelastic dampers. Active control technologies use devices that require external power and feedback signals, such as active mass dampers (AMD), active tendon systems, and piezoelectric actuators. Figure 1.6 shows a one-floor AMD system, which consists of one flexible floor, two flexible walls, and a linear cart system. The linear cart system includes a DC motor that provides the driven force, an accelerometer that

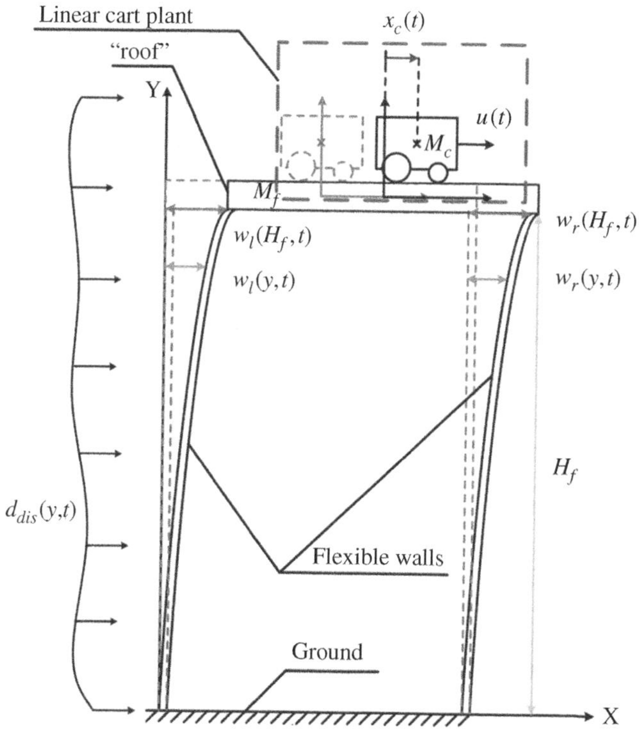

Figure 1.6 One-floor active mass damper system.

measures the acceleration of the flexible floor relative to the ground, and an encoder that measures the cart position.

Passive control technologies have the advantages of simplicity, reliability, and low maintenance cost. However, they also have some limitations, such as fixed parameters, limited performance, and possible resonance problems. Active control technologies can overcome these limitations by adjusting the control parameters according to the feedback signals and the desired performance objectives. However, they also have some drawbacks, such as complexity, high cost, and stability issues.

The main challenges in vibration control of flexible building-like structures are: (1) modeling the complex dynamics of the coupled rigid–flexible system; (2) designing robust and adaptive control strategies that can deal with system uncertainties, nonlinearities, and external disturbances; (3) implementing the control devices and sensors at appropriate locations; and (4) ensuring the stability and performance of the closed-loop system.

Various methods have been proposed to address these challenges in the literature. In a study [83], Gao et al. used Euler–Lagrange approach to model a single-floor building-like structure with an AMD and proposed a sliding mode control strategy based on reinforcement learning. In another study [120], He et al. used Hamilton's principle to model a flexible structure with a tip payload and designed an adaptive robust boundary control strategy based on Lyapunov stability theory.

In future works, the control design of the multi-floor building-like structure will be considered. Multiple constraint problems such as constraints, saturation, and backlash will be studied. In addition, excitations caused by wind and earthquake, which may have a wider range of frequency, are further included so as to more practically simulate real-world engineering problems.

1.4 Modeling and Control Approaches of Bionic Flexible Flapping-wing Aircraft

The modeling and control of bionic flexible flapping-wing aircraft is a challenging and promising research topic in the field of micro aerial vehicles (MAV). Inspired by the flight of birds or insects, bionic flapping-wing aircraft can achieve high maneuverability, low noise, and low energy consumption. However, the flexible wings of such aircraft also introduce complex aerodynamics, nonlinear dynamics, and uncertainties, which pose difficulties for the design, analysis, and control of the system. Therefore, it is important to develop effective modeling and control approaches for bionic flexible flapping-wing aircraft to improve their performance and reliability.

The modeling and control of bionic flexible flapping-wing aircraft is still an active and open research area with many challenges and opportunities. Some of the current challenges include improving the accuracy and efficiency of the modeling methods; developing more robust and adaptive control methods; incorporating sensor feedback and vision-based navigation; enhancing the flight performance and autonomy; integrating multiple flapping-wing aircraft into a swarm system; exploring new applications and scenarios for flapping-wing aircraft. Some of the future directions include using machine learning and artificial intelligence techniques to improve the modeling and control methods; exploiting bio-inspired mechanisms and materials to enhance the wing design; incorporating energy harvesting and storage devices to extend the flight endurance; developing new test beds and platforms for experimental validation; and collaborating with other disciplines such as biology, physics, chemistry, etc., to gain more insights into flapping-wing flight.

The flexible flapping wing is essentially an infinite-dimensional distributed parameter system with complex dynamic characteristics, which causes enormous challenges to system dynamic analysis and modeling [27, 176, 206]. The existing modeling methods for flexible structures include modal superposition method (MSM), finite element method (FEM), lumped mass method (LMM), finite segment method (FSM), and absolute coordinate method (ACM). In a study [71], Ferhatoglu et al. proposed a new MSM utilizing novel hybrid mode shapes for the dynamic analysis of nonlinear systems. The accuracy and computational efficiency of the proposed MSM are verified by several cases. However, obtaining the mode shape functions always requires complex derivations. FEM and assumed mode method (AMM) were compared to investigate the coupled vibrations of a flexible-disk rotor system with lacing wires [366]. Although the precision of both methods can be improved with the increasing number of elements, the issue of the heavy computational demand will be followed.

Several studies have applied different modeling and control methods to bionic flexible flapping-wing aircraft and compared their performance and effectiveness. For example, Biswal et al. [16] developed three nonlinear dynamic models with increasing complexity and designed a family of LQR controllers for each model. They tested the robustness of the controllers on different models and examined the trade-offs between performance and complexity. Shao et al. [250] designed a bionic flapping-wing aircraft model with an unfolding-bending effect and used ADAMS and XFlow co-simulation to analyze its aerodynamic characteristics. They investigated the effects of flutter frequency, airflow velocity, fuselage angle of attack, and aileron torsion angle on its aerodynamic characteristics. Gao et al. [84] proposed an adaptive finite-time fault-tolerant controller for uncertain flexible flapping-wing aircraft with actuator faults. They proved the stability of the controller using the Lyapunov theory and demonstrated its effectiveness through simulations.

1.5 Outline of the Book

Chapter 1 provides an overview of dynamic modeling and intelligent control of flexible systems, introducing several important issues in the study of flexible systems. The modeling and control methods of three typical flexible systems are discussed separately.

Chapter 2 provides the corresponding mathematical preliminaries of subsequent chapters, including the Hamilton's principle, model discretization methods, Lagrange's equation method, neural networks (NNs), and Lyapunov stability theorem.

Chapter 3 develops the dynamic model of the single-link flexible robotic manipulator, which overcomes the challenge of the system dynamics being infinite-dimensional. The fuzzy NN control with uniform approximation performance is designed to solve the system uncertainties. Numerical simulations and extensive experiments have been investigated to verify the effectiveness of the proposed methods.

Chapter 4 establishes a finite-dimensional dynamic model of the two-link flexible robotic manipulator. A high-gain observer-based NN control strategy is proposed to estimate the immeasurable states in practice. The semi-globally uniformly ultimate boundedness (SGUUB) of the closed-loop system is guaranteed via Lyapunov stability theory. The simulation and experimental results demonstrate the effectiveness of the proposed control strategy.

Chapter 5 presents the vibration control design for a string with the boundary time-varying output constraint. The dynamics of the string is a distributed parameter system described by a PDE and two ODEs. A barrier Lyapunov function with a logarithmic function is adopted to prevent the time-varying constraint violations. Adaptive control is designed to handle the system's parametric uncertainties. Stability analysis and the solvability of the inequality equations are provided. Numerical simulations are provided to illustrate the effectiveness of the proposed control design.

Chapter 6 focuses on a stand-alone tall building-like structure with an eccentric load. An NN control approach is proposed to suppress vibrations caused by unknown time-varying disturbances (earthquake, strong wind, etc.). The output constraint on the angle of the eccentric load is also considered, and such an angle can be ensured within the safety limit by incorporating a barrier Lyapunov function. Simulations and experiments based on MATLAB and Quanser are carried out to verify the feasibility and effectiveness of the proposed control.

Chapter 7 discusses an adaptive vibration control method for a single-floor building-like structure equipped with an AMD. The method uses a hybrid learning control strategy to suppress vibrations caused by unknown time-varying disturbances such as earthquakes or strong winds. The effectiveness of the proposed control approach is demonstrated through experimental investigation on a Quanser Active Mass Damper. The research results aim to bring new ideas and methods to the field of disaster reduction for engineering development.

Chapter 8 investigates a single-floor building-like structure equipped with an AMD. Optimal vibration control, while dealing with system uncertainties, is realized by the reinforcement learning technique. When the unexpected natural disasters occur, the proposed controller applying to the AMD can compensate for the increase in the system vibration caused by external disturbances. The experimental results in the form of graphics and tables have shown the effectiveness of the proposed control algorithm.

Chapter 9 develops the visualization model of the rigid–flexible coupled bionic flapping wing by the advanced system-level modeling software MapleSim. A novel NN controller based on disturbance observer technology is proposed to compensate the system uncertainties. The proposed method can successfully suppress the vibration of the flapping wing while accurately tracking the desired trajectory. Co-simulation results from MapleSim and MATLAB/Simulink validate the effectiveness of the proposed method.

Chapter 10 focuses on the flexible wings of the aircraft, which have great advantages, such as being lightweight, having high flexibility, and offering low energy consumption. A novel adaptive finite-time controller based on the fuzzy NN and nonsingular fast terminal sliding-mode scheme is proposed for tracking control and vibration suppression of the flexible wings, while successfully addressing the issues of system uncertainties and actuator failures. Co-simulations through MapleSim and MATLAB/Simulink are carried out to verify the performance of the proposed controller.

Chapter 11 discusses the importance of vibration control for bionic flapping-wing robotic aircraft and autonomous ornithopter applications. A visualization model of the rigid–flexible coupled bionic flapping wing is established using MapleSim software. A novel adaptive vibration controller based on NN algorithm is proposed to compensate for system uncertainties. The proposed method can successfully suppress the vibration of the flapping wing while accurately tracking the desired trajectory. The effectiveness of the proposed method is validated through cosimulation results from MapleSim and MATLAB/Simulink.

Chapter 12 investigates dynamic modeling, active boundary control design, and stability analysis for a coupled floating wind turbine (FWT) system, which is connected with two flexible mooring lines. It is a coupled beam-strings structure, and we design two boundary controllers to restrain the vibrations of this flexible system caused by external disturbances based on the coupled PDEs-ODEs model. Meanwhile, significant performance of designed boundary controllers and system's stability are theoretically analyzed, and a set of simulation results are provided to show the efficacy of the proposed approach.

Chapter 13 summarizes the practical significance of the application of NN-based intelligent control and proposes some future research directions in this field.

In summary, this book proposes high-efficiency modeling methods and novel intelligent control strategies for several representative flexible systems developed by means of NNs. The book discusses the tracking control of multilink flexible manipulators, the vibration control of flexible buildings under natural disasters, and the fault-tolerant control of bionic flexible flapping-wing aircraft. Expanding on its theoretical deliberations, the book includes many case studies demonstrating how the proposed approaches work in practice.

2

Mathematical Preliminaries

After tierreading this chapter, you should be able to:

- Provide a comprehensive introduction to the mathematical preliminaries required for understanding the topics discussed in the subsequent chapters
- Cover a range of essential concepts and methods, including the Hamilton's principle, model discretization methods, Lagrange's equation method, neural networks, and the Lyapunov stability theorem

2.1 Mathematical Preliminaries

2.1.1 Hamilton Principle

The Hamilton's principle is commonly used to model the dynamics of distributed parameter systems [200]. Its main characteristic is to construct the dynamic model of the system through the variation of energy. Hamiltonian principle is a commonly used and effective method for describing the dynamic characteristics of flexible structures. The specific steps are as follows: Firstly, from the perspective of energy analysis, we derive the work done by the dynamic energy, potential energy, and other non-conservative forces of the system, and calculate their variations respectively [1]. Then, combined with the Hamilton's principle [200].

$$\int_{t_1}^{t_2} \delta(E_k - E_p + W)dt = 0. \tag{2.1}$$

Thus, a set of system master control equations described by partial differential equations and system boundary conditions described by ordinary differential equations can be obtained.

The advantage of the Hamilton's Principle is that it provides a powerful and elegant way of formulating the equations of motion for a system in terms of a single function [97] by the Lagranges. This makes it possible to describe the motion of complex systems in a simple and unified way, and to derive equations of motion for systems that cannot be solved using other methods [107].

Dynamic Modeling and Neural Network-Based Intelligent Control of Flexible Systems,
First Edition. Hejia Gao, Wei He, and Changyin Sun.

2.1.2 Model Discretization

2.1.2.1 Assumed Mode Method

The hypothetical mode method is a method of separating variables [366]. In the hypothetical mode method, the elastic deflection and deformation of a flexible link can be expressed as an infinite number of separable modes [30]:

$$w(x,t) = \sum_{i=1}^{n} F(x)q(t), \tag{2.2}$$

where $F(w)$ is a modal function, and $q(t)$ is a time-dependent function.

In elastic flexural deformation, only the first few low-frequency modes dominate. With the increase in frequency, the influence of the modes on vibration will attenuate exponentially. Therefore, most research work has simplified the dynamic modeling process by reducing the modal truncation to a finite dimension [190]. Hypothetical mode method is divided into two types: constrained mode and unconstrained mode [65].

Constrained mode: Generally, we assume that the acceleration of the joint motion of a flexible link system is zero, and obtain corresponding types of boundary conditions through energy analysis to solve the Euler-Bernoulli equation [171].

Unconstrained mode: Decoupling the model for each vibration mode of a flexible link system. Due to the fact that modal functions are strictly formulated and most of them rely on control inputs, it may be difficult to obtain the analytical expression of the model [324].

It should be noted that in the hypothetical modal method, relevant definitions such as modal functions and time functions are introduced [38]. However, in the actual dynamic analysis process, the proposed modal function and time function are defined for the purpose of model discretization, and do not have actual physical meaning [46].

2.1.2.2 Finite Rigid Element Method

The finite rigid body element method has high modeling efficiency, is simple and easy to implement, and is currently widely used to solve the infinite-dimensional characteristics of flexible system dynamics[149]. Taking a flexible link as an example, the basic idea of the finite rigid body element method is to divide the flexible link into a limited number of small rigid segments. To describe the flexible characteristics, the small rigid segments are connected by springs and dampers that can undergo bending deformation [135]. Of course, their mass is very light and their dimensions are very small, which can be ignored. This achieves the transformation of dynamic characteristics from infinite to finite dimensions.

It is worth noting that in the finite rigid body element method, flexible objects of arbitrary shape are acceptable [255]. This method is used to characterize its dynamic properties by changing tiny rigid body segments and spring-dampers. The arrangement can be realized, and the method is simple and easy to operate, which greatly improves the modeling efficiency and solves the problem of flexibility. For problems with a large amount of computation in the dynamic analysis of sexual systems, the numerical model obtained by this method can also be obtained. This method can be used in various fields such as aerospace, industrial production, and rehabilitation medicine. In addition, the finite rigid body element method can also be used to solve the dynamic analysis of flexible surface systems [152], and the finite rigid body element method is used to perform on flexible surfaces. When modeling, rigid elements are described by tiny rigid body segment strips, flexible elements are described by spring-dampers description, and the rigid body section is connected by a spring-damper [163].

The main advantages of the finite rigid body element method are that it is simple and convenient, the dynamic analysis method is easy to understand, the calculation amount is small, and the modeling efficiency is high. The modeling accuracy can also be adjusted by changing the number of rigid body segments, which can be achieved in the model analysis of various flexible objects [164].

2.1.3 Lagrange's Equation Method

In order to analyze the dynamic properties of the system, from the energy point of view, mechanics and Newtonian mechanics are used, the dynamic model of the system can be directly established by generalized coordinates, generalized velocity, and generalized forces. The essence of the Lagrange equation method is actually a combination of the dynamic and static method and the virtual displacement principle, so as to avoid it. Complex vector calculations in the modeling process can be done without considering system constraints [287]. Take advantage of lager, the Lange equation method can obtain dynamic equations in a perfectly structured, closed form, and the modeling process is clear. It has been widely used in the dynamic modeling of systems [252]. The Lagrange equation is usually written as:

$$\frac{d}{dx}\frac{\partial L}{\partial \dot{q}} - \frac{\partial L}{\partial q} = \tau, \tag{2.3}$$

where $L = T - U$ is a Lagrange function, τ denotes generalized driving force, and q is the generalized coordinate variation of the system.

2.1.4 Neural Networks

In practical engineering, the dynamic model of flexible systems is often uncertain by the relevant matrix parameters [242]. As we all know, neural networks (NNs) have a strong fitting ability to uncertainty functions. Especially, for uncertain nonlinear systems, the NN can continuously adjust the input and output data weight, so as to achieve strong learning ability through training. Therefore, it is powerful through NNs. The fitting characteristics can solve the uncertainty problem of the flexible system model in actual engineering and achieve strong fault tolerance and adaptability [174].

Radial basis function neural networks (RBFNNs) have a simple structure and have good nonlinear fitting characteristics. The ability to generalize, and the fast convergence speed, have been widely used in modeling machines for nonlinear systems [130]. Aspects of control system design. The specific structure of the radial basis function neural network is shown in Figure 2.1. It is a three-layer forward network consisting of an input layer, a hidden layer, and an output layer. The processing function of the layer is a radial basis function, a nonlinear function that is radially symmetric and attenuated by the center point [353]. The idea is to continuously train the weights of the implicit layer NN through the input data composed of low-dimensional vectors. Practice to achieve consistent approximation ability.

For continuous functions $f(Z) : \mathbb{R}^k \rightarrow \mathbb{R}$, based on the uniform approximation performance of RBFNNs, there is the following relationship [35]:

$$f(Z) = \Phi B(Z), \tag{2.4}$$

where $Z = [Z_1, Z_2, ..., Z_k] \in \Omega$ represents system input data, $\Phi = [\Phi_1, \Phi_2, ..., \Phi_l]$ is the network weight vector, $l > 1$ is the number of nodes in the middle layer.

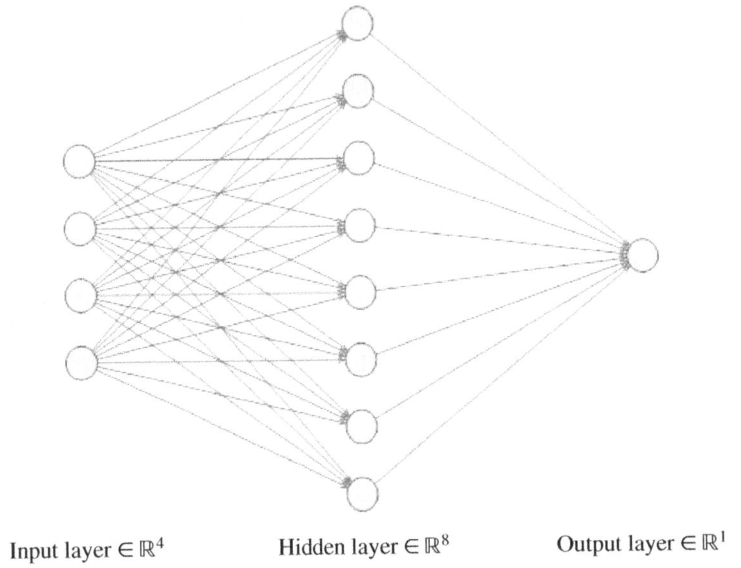

Figure 2.1 Structure diagram of the RBF neural network.

Define $B(Z) = [b_1(Z), b_2(Z)]$, where $b_i(Z)$ uses a Gaussian function in the form of:

$$b_i(Z) = \exp\left[\frac{-(Z-\mu_i)^T(Z-\mu_i)}{\mu_k^2}\right], i = 1, 2, \dots, l, \tag{2.5}$$

where $\mu_i = [\mu_{i1}, \mu_{i2}, \dots \mu_{ik}]$ is the center value of each field, and μ_i is the radial distance parameter of the Gaussian function.

Therefore, the approximation performance of RBFNNs for uncertain continuous functions can be written as [77]:

$$f(Z) = \Phi^* B(Z) + \varepsilon, \forall Z \in \Omega_Z, \tag{2.6}$$

where Φ^* represents ideal weight, and ε represents fitting error. And there are $\|\varepsilon\| \leq \|\varepsilon^*\|$. In addition, the selection rule for Φ^* is:

$$\Phi^* = \arg\min\left\{\sup_{Z\in\Omega_z} |f(Z) - \Phi^* B(Z)|\right\}. \tag{2.7}$$

2.1.5 Lyapunov Stability Theorem

The Lyapunov function is a powerful tool in the analysis of dynamical systems, and it is widely used in control theory, robotics, and other fields of engineering and mathematics [74]. It states that if a system's Lyapunov function satisfies certain conditions, then the system is stable. It is particularly useful in control theory, where it is used to design controllers for feedback systems that can maintain stability in the face of disturbances or uncertainties. The theorem was first proposed by Russian mathematician Aleksandr Lyapunov in the late 19th century [54], as a way to address the question of whether a given equilibrium point in a dynamical system is stable or unstable. The theorem establishes the conditions under which an equilibrium point is stable or asymptotically stable, meaning that the system will converge to the equilibrium point over time. The basic idea of a Lyapunov function is to prove the stability of a dynamical system by constructing a scalar function that decreases along the trajectories of the system. This function is called a Lyapunov function, and its construction relies on the properties of the system's dynamics. In other words, a Lyapunov function is a function that assigns a scalar value to each state of a dynamical system, and this value is such that it decreases over time as the system evolves. The Lyapunov function is used to establish the stability of the system by proving that it is bounded and that its trajectory converges to a fixed point, limit cycle, or other stable set.

More specifically, the Lyapunov stability theorem states that if there exists a continuously differentiable function $V(x)$ (known as a Lyapunov function) that is positive definite ($V(x) > 0$ for all $x \neq 0$), radially unbounded ($V(x) \rightarrow \infty$ as $\|x\| \rightarrow \infty$), and satisfies the condition that the derivative of V along the system trajectories is negative semidefinite ($\frac{dV}{dt} \leq 0$ for all $x \neq 0$), then the system is asymptotically stable.

In practical applications, the Lyapunov stability theorem can be used to design controllers that stabilize a system [89]. This involves selecting an appropriate Lyapunov function for the system and then using it to design a control law that stabilizes the system. Additionally, the Lyapunov stability theorem can also be used to prove the stability of flexible systems, by showing that the conditions of the theorem are satisfied for the system in question [218].

A Lyapunov function is said to be positive definite if it is greater than zero for all nonzero inputs, and positive semidefinite if it is greater than or equal to zero for all inputs [53]. The theorem states that if a Lyapunov function exists that is positive definite or positive semidefinite, and if its time derivative is negative semidefinite (i.e., it is less than or equal to zero for all inputs), then the equilibrium point is stable or asymptotically stable.

Lyapunov stability theorem is a method to judge the stability of nonlinear dynamic systems. Here is the process of using the Lyapunov stability theorem to determine the stability of a system [344]: Determine the state equation or difference equation of the system. For continuous systems, the state equation is usually represented in the form of a differential equation, while for discrete systems, the state equation is usually represented in the form of a difference equation [2]. Construct a Lyapunov function. A Lyapunov function is a real-valued function that satisfies the following conditions:

(1) the function value is non-negative;
(2) the function value is zero at only one point, the equilibrium point of the system;
(3) the first derivative of the function exists and is continuous;
(4) the second derivative of the function exists or is integrable.

Based on the derivative of the Lyapunov function, determine the stability of the system. If the derivative is less than zero, the system is stable. If the derivative is equal to zero, further analysis is required [299]. If the derivative is greater than zero, the system is unstable. It should be noted that using the Lyapunov stability theorem to determine the stability of a system is a complex method that requires a deep understanding and analysis of the characteristics of the system. In practical applications, numerical simulations and other methods are often used to verify the stability of the system [70].

In general, the Lyapunov function is complex and needs to be constructed according to the characteristics of the system. As a design tool for general nonlinear systems, Lyapunov function is unknown. The task of finding a suitable Lyapunov function can be as complex as the task of designing a stable feedback law. For flexible systems, however, these two tasks can be solved simultaneously. It should be noted that there may be many Lyapunov functions for a given nonlinear system. Specific choices of Lyapunov's functions may result in better, cleaner controllers than other functions [154].

2.1.6 Summary

In this chapter, a series of principles and methods in the modeling process of distributed parameter systems are presented, and the uniform approximation performance of NNs is analyzed. Finally, the stability verification methods in control theory are introduced in detail. The preliminary knowledge provided in this chapter provides a powerful theoretical basis for the theoretical analysis and research in the following article, and also enables readers to more clearly understand the research methods and research content used in the article.

3

Fuzzy Neural Network Control of the Single-Link Flexible Robotic Manipulator

After reading this chapter, you should be able to:

- Focuses on developing the dynamic model of a single-link flexible robotic manipulator, addressing the complexity arising from the system's infinite-dimensional dynamics
- To tackle the system uncertainties, a fuzzy neural network control strategy with uniform approximation performance is introduced and meticulously designed
- The validity and efficiency of the proposed methods are thoroughly demonstrated through a series of numerical simulations and extensive experimental investigations

3.1 Introduction

Recently, as the application of robotic technology is advancing rapidly, robot research field and research methods are constantly updated [119, 351]. The study of the traditional robot is based on rigid body dynamics theory under the assumption of its components as rigidity. However, there does not exist the absolute rigid body in practice. In general, the deformation of the object is very small and its main movement is rigid movement. In this case, the elastic deformation can be ignored, and these assumptions can be established. With the development of the robot system toward high-speed, heavy load, high precision and light components, the elastic deformation of the component inevitably increases, so that the impact of the elastic deformation of the various robot components must be noted. In addition, the flexible structure has aroused more and more researchers' attention and has been widely applied in robotics [62], industries [115, 347], and aerospace [32]. However, the flexible manipulator must be connected by small cross parts, which will result in flexibility and mechanical vibration [224]. Therefore, the traditional rigid robot research methods are obviously not suitable for flexible robots.

Due to the deformation of the flexible manipulator during the movement, the dynamic analysis method and the control strategy applied to the traditional rigid body cannot satisfy the flexible structure. Generally, the flexible manipulator is a distributed parameter model described by infinite dimension. However, the control design of the distributed parameter system is often based on the finite-dimensional model. Therefore, we have to consider a problem of compromising between the complexity of the model and the control performance of the system. Thus, how to establish a suitable dynamic model and design a high-performance controller to effectively control the movement of the robot manipulator is a problem that must be faced and solved.

There are many methods of dynamic modeling for flexible manipulators [284, 307]. In order to analyze the movement of the flexible structure, Hamilton's principle is introduced to obtain a set

Dynamic Modeling and Neural Network-Based Intelligent Control of Flexible Systems,
First Edition. Hejia Gao, Wei He, and Changyin Sun.

of highly nonlinear partial differential equations, which is difficult to find a solution. The key in the flexible arm modeling is the finite-dimensional approximation of distributed flexibility. Some approximation methods are developed to deal with systems with relatively uniform density and mass, such as the Ritz method, the Galerkin method, the assumed mode method (AMM) [72, 238], and the linearization method. Some methods are employed to handle the centralized parameter system, such as finite element method (FEM) [56, 58], finite segment method (FSM) [100, 305], finite difference method (FDM) [60], and lumped parameter method (LPM) [214, 263]. In addition, the rotation algebra method and singular perturbation method are helpful to establish the governing equations.

Because the vibration will have an impact on the stability of the manipulator, it is essential to search a suitable control strategy to suppress vibration, so as to minimize the harm. There are different control strategies such as PID control [238], sliding model control [149], adaptive impedance control [111, 182], adaptive robust control [231], boundary control [144, 199], fuzzy control [310], and neural dynamical approach [112, 336]. PID control is widely used in control of the flexible manipulator for its simplicity. However, the control gains of the PID controller are not easy to adjust [23]. Subsequently, numerous researchers have studied adaptive method. In [160], the entire state vector is required for the purpose of designing the controller via sliding modes, nevertheless, it may not be able to measure all the states. In [291], the authors proposed the robust stabilization via an approximate optimal control approach. In [215], an air-breathing hypersonic vehicle tracking control based on adaptive dynamic programming was proposed.

Compared with the conventional control methods, neural networks [108, 189, 192, 245] can fully approximate any complex nonlinear system. Neural network (NN) control [59, 187, 283] is proved to be effective in dealing with uncertain nonlinear systems. In [25], by constructing the NN-based state observer for each follower, an adaptive consensus tracking control strategy is developed to solve the unmeasurable state problem of high-order nonlinear multiagent systems. In addition, fuzzy NN control is composed of the NN and fuzzy logic system organically. And it is good at self-learning and self-tuning. In [282], a fuzzy-neural network inherited sliding mode control (FNNISMC) scheme was proposed to relax the requirement of detailed system information and deal with chattering control efforts. In [186], a novel recurrent fuzzy NN was presented for prediction and identification of dynamic systems.

In this chapter, the modeling and control problems are studied for the flexible robotic manipulator. Compared with the previous works, the major contributions of this chapter include:

(1) The n-dimensional model is achieved via the AMM which realizes the conversion from PDE to ODE and facilitates the analysis of dynamic uncertainty and the design of the controller.
(2) The NN controller by full-state feedback for the fuzzy system is designed to achieve the accurate trajectory tracking and vibration suppression. Besides, physical experiments are presented on Quanser platform to validate the effectiveness of the designed NN control and the correction of the theoretical analysis.
(3) The NN controller by output feedback is designed to handle the vibration problem of the flexible manipulator. Besides, a high-gain observer is introduced to estimate the unmeasured signals directly.

The dynamic modeling is developed in Section 3.2. Fuzzy NN controller by full-state and output feedback is designed in Section 3.3. Simulation results are given in Section 3.4. Experiments on the Quanser platform are implemented in Section 3.5. Conclusions are obtained in Section 3.6.

3.2 Problem Formulation

The single-link flexible robotic manipulator, shown in Figure 3.1, is a typical distributed parameter system with complicated dynamics. This type of system is always described by partial differential equations (PDEs). In the control theory of PDEs, the challenge comes from the system dynamics being infinite-dimensional characteristics of the system. In order to solve the problem of control design caused by the infinite dimensional. Motivated by this reason, we established the n-dimensional model of the single-link flexible robotic manipulator and realized the conversion from PDEs to ordinary differential equations (ODEs) using AMM in this chapter.

3.2.1 Dynamic Modeling

Considering the structural characteristics, bending deformation, axial deformation, and shear deformation exist in the Timoshenko beam. Since the length of the single-link flexible manipulator is much larger than the cross-sectional area, the axial deformation, and the shear deformation are quite small compared with the bending deformation. The effect of the axial deformation and the shear deformation can be ignored, the single-link flexible manipulator is thus regarded as a Euler–Bernoulli beam in the dynamic modeling. In the absence of the air resistance and the internal damping of the beam, assumed that the single-link flexible manipulator has the uniform cross-section and the sufficiently small elongation, offset value and slope. Subsequently, the dynamics of the single-link flexible manipulator is studied in the horizontal plane. For convenience, the related modeling parameters and their descriptions are summarized in Table 3.1.

The energy analysis is carried out, and use the Hamilton's principle, the boundary conditions are derived as follows [246]:

$$\rho y'''(L,t) + m_t y''''(L,t) = 0 \tag{3.1}$$

$$I_h\ddot{\theta}(t) - EIy''(0,t) = \tau(t) \tag{3.2}$$

$$y(0,t) = y'(0,t) = y''(L,t) = 0. \tag{3.3}$$

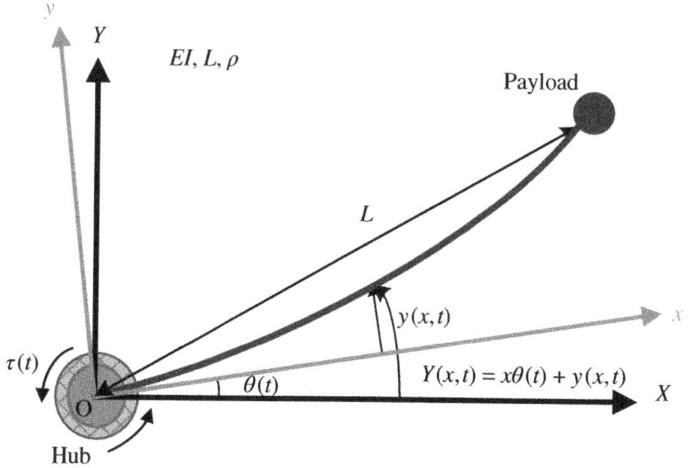

Figure 3.1 System representation of the single-link flexible robotic manipulator.

Table 3.1 Related modeling parameters and their descriptions of the single-link flexible robotic manipulator.

Parameter	Description	Unit
M	Total mass of the single-link flexible manipulator	kg
L	Length of the single-link flexible manipulator	m
m_t	Tip mass of the single-link flexible manipulator	kg
ρ	Density of the single-link flexible manipulator	kg/m
$\theta(t)$	Hub angular position of the single-link flexible manipulator	deg
$y(x,t)$	Elastic deformation of the single-link flexible manipulator	m
I_h	Hub inertia	kg m^2
EI	Bending stiffness	N m^2
$\tau(t)$	Control torque	N m
XOY	Inertial coordinate system	–
xOy	Local coordinate system	–

Also, we get the governing equation [246]:

$$EIy''''(x,t) + \rho\ddot{y}(x,t) + \rho x\ddot{\theta}(t) = 0, \tag{3.4}$$

where $\theta(t)$ denotes the hub angular position which is from 15° up to 75° [369].

3.2.2 Model Discretization

From the modal analysis, we employ the AMM to describe the elastic deformation:

$$y(x,t) = \sum_{i=1}^{\infty} \phi_i(x)p_i(t). \tag{3.5}$$

The AMM is a discrete method for continuous systems, which uses the linear sum of assumed modes to describe the vibration of the elastic body. Where $\phi_i(x)$ represents the model function, $p_i(t)$ is the time dependent function, and $w_i = \frac{A_i^2}{L^2}\sqrt{\frac{EI}{\rho}}$ is the harmonic with frequency. Substituting Eq. (3.5) into Eqs. (3.1)–(3.4):

$$\phi(x) = \alpha\left[\cosh\left(\frac{A}{L}x\right) - \cos\left(\frac{A}{L}x\right) + \beta(\sin\left(\frac{A}{L}x\right) - \sinh\left(\frac{A}{L}x\right)\right]. \tag{3.6}$$

When $m_t = 0$, $\alpha = 1/\sqrt{L}$; when $m_t > 0$,

$$\alpha = \left[L + \frac{\rho L^2}{m_t A^2}\left(\frac{1+\cosh(A)\cos(A)}{\sinh(A)\sin(A)}\right)^2\right]^{-\frac{1}{2}}. \tag{3.7}$$

A satisfies the following function:

$$m_t A\sinh(A)\cos(A) + \rho L\cosh(A)\cos(A) + \rho L = m_t A\cosh(A)\sin(A) \tag{3.8}$$

β can be described as:

$$\beta = \frac{\cosh(A)+\cos(A)}{\sinh(A)+\sin(A)}. \tag{3.9}$$

According the Lagrange equations and define that $Q = [\theta, p_1, ..., p_N]^T$, we obtain

$$M\ddot{Q} + CQ = \tau, \tag{3.10}$$

where the matrix M is given:

$$M = \begin{bmatrix} I_t & b_A^1 & b_A^2 & \cdots & b_A^N \\ b_A^1 & \gamma_A^1 & b_A^{12} & \cdots & b_A^{1N} \\ b_A^2 & b_A^{12} & \gamma_A^2 & \cdots & b_A^{2N} \\ \vdots & \vdots & \vdots & \vdots & \vdots \\ b_A^N & b_A^{1N} & b_A^{2N} & \cdots & \gamma_A^N \end{bmatrix}. \tag{3.11}$$

The matrix C is given:

$$C = \text{diag}\left[0 \;\; w_1^2\rho \;\; w_2^2\rho \;\; \cdots \;\; w_N^2\rho\right], \tag{3.12}$$

where $I_t = I_h + \frac{1}{3}\rho L^3 + m_t L^2$; $b_A^i = \rho \int_0^L x\phi_i(x)dx + m_t L\phi_i(L)$; $b_A^{ij} = 0$; $\gamma_A^i = \rho$.

There are many kinds of function systems which can be used as universal approximation of continuous functions. Fuzzy system is just one of them. However, the special feature of fuzzy system is that it is not based on the traditional accurate mathematical basis, and it can effectively use the language information. Hence, fuzzy system has a strong practical value.

(1) The key point of fuzzy logic is that it changed the 0/1 two value strict distinction and achieved continuous in the range of [0,1]. Based on fuzzy logic, the relationship between an element and a fuzzy set is no longer a simple belonging relation. The degree of membership function is introduced. Fuzzy set is the most basic part of fuzzy system and has a certain continuity from the perspective of mapping. Therefore, the fuzzy system can approximate the continuous function, which is determined by the characteristics of fuzzy logic.
(2) According to universal approximation theory:

$$\|g - f\|_\infty \leq \|\frac{\partial g}{\partial x_1}\|_\infty h_1 + \|\frac{\partial g}{\partial x_2}\|_\infty h_2 \tag{3.13}$$

$$h_i = \max_{1 \leq j \leq N_i - 1} |e_i^{j+1} - e_i^j| \tag{3.14}$$

The approximation accuracy of a fuzzy system is related to the number of rules. The number of rules in the FBF fuzzy system is represented by the number of FBF, and when the number of rules increases, the approximation accuracy will be improved. From [274], the fuzzy logic system can uniformly approximate any continuous function with any degree of accuracy. In this chapter, we use the fuzzy system to approximate the flexible manipulator system equation.

Considering a fuzzy logical system, it consists of the inference rule, singleton fuzzifier, center average defuzzifier, and Gaussian membership function. The N fuzzy IF-THEN rules are described as:

IF x_1 is A_1^j... and x_n is A_n^j, THEN y is $W^j, j = 1..., N$.

F^j represents the jth rule, $1 \leq j \leq N$; $(x_1, x_2...x_n)^T \in U \subset R^n$ represents the input; $y \in R$ represents the output; A_i^j and W^j are the fuzzy sets in U and R. The degree of membership function of the flexible manipulator is chosen as:

$$\mu_{A_i^j}(x_i) = \exp\left[-\frac{(x_i - c_{ij})^2}{\sigma_{ij}^2}\right]. \tag{3.15}$$

The fuzzy logic system is defined as:

$$y(x) = \frac{\sum_{j=1}^{N} y_j(\prod_{i=1}^{n} \mu_{A_i^j}(x_i))}{\sum_{j=1}^{N}(\prod_{i=1}^{n} \mu_{A_i^j}(x_i))}. \tag{3.16}$$

Therefore, the approximation can be represented:

$$f_i(x_i) = W_i^{*T} S(x_i) + \epsilon_i(x_i), \forall x_i \in \Xi_i \subset R^i, \tag{3.17}$$

where W_i^* is unavailable, $S(x_i)$ is the fuzzy basis vector, $\epsilon_i(x_i)$ is the approximation error.

3.3 Fuzzy Neural Network Control

For the fuzzy logic system (3.10), we designed a fuzzy NN control with full-state feedback and output feedback in this section, respectively. In addition, we theoretically prove the stability of the controller by Lyapunov direct method. The control block diagram is revealed in Figure 3.2.

In the process of the control design, we adopt the backstepping method which is to decompose the complex nonlinear system into subsystems that may not exceed the number of system order. Subsequently, by using the iterative design method, through coordinate transformation, introduce a virtual stabilization function, design the control torque starting from the first subsystem without control input signal, and gradually backstepping to the last subsystem with control input, and then the real control input signal is obtained. The detailed design procedure for the proposed fuzzy NN control strategy is revealed in Figure 3.3:

3.3.1 Control Design

Transform the Eq. (3.10), we have the description of the system dynamics:

$$\ddot{Q} = M^{-1}[\tau - CQ]. \tag{3.18}$$

Given the errors of system states z_1 and z_2 as follows

$$z_1 = Q - Q_d, \tag{3.19}$$

$$z_2 = \dot{Q} - \mu_1. \tag{3.20}$$

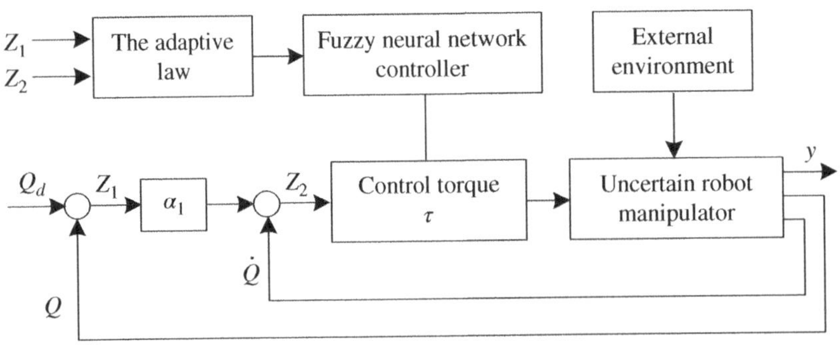

Figure 3.2 Neural network control with fuzzy logic systems.

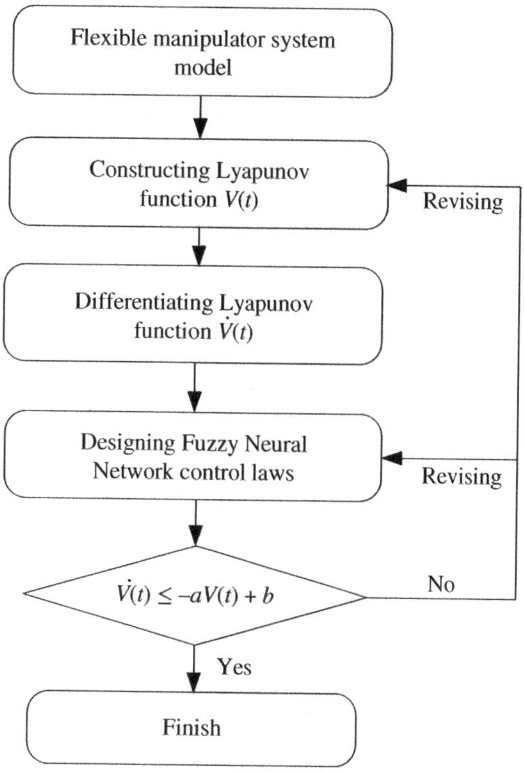

Figure 3.3 Design procedure of the fuzzy neural network controller.

where μ_1 is the virtual control to z_1. We choose $\mu_1 = -K_1 z_1 + \dot{x}_d$, $K_1 = K_1^T > 0$. Then the derivative of the error variables can be described as

$$\dot{z}_1 = z_2 + \alpha_1 - \dot{Q}_d = z_2 - K_1 z_1. \tag{3.21}$$

$$\dot{z}_2 = \ddot{Q} - \dot{\mu}_1 = M^{-1}[\tau - CQ] - \dot{\mu}_1. \tag{3.22}$$

Design V_1 as the initial Lyapunov function:

$$V_1 = \frac{1}{2} z_1^T z_1, \tag{3.23}$$

Further, $\dot{V}_1$ can be obtained:

$$\dot{V}_1 = z_1^T (z_2 - K_1 z_1) = z_1^T z_2 - z_1^T K_1 z_1. \tag{3.24}$$

Aiming at removing the coupling term $z_1^T z_2$ in (3.24), the Lyapunov function as follows can be employed.

$$V_2 = \frac{1}{2} z_1^T z_1 + \frac{1}{2} z_2^T M z_2, \tag{3.25}$$

Further, $\dot{V}_2$ can be obtained:

$$\begin{aligned} \dot{V}_2 &= z_1^T \dot{z}_1 + z_2^T M \dot{z}_2 \\ &= z_1^T z_2 - z_1^T K_1 z_1 + z_2^T (\tau - CQ - M\dot{\mu}_1). \end{aligned} \tag{3.26}$$

Then, the model-based controller τ_0 is proposed

$$\tau_0 = -z_1 - K_2 z_2 + CQ + M\dot{\mu}_1. \tag{3.27}$$

Substituting Eq. (3.27) into Eq. (3.26)

$$\dot{V}_2 = -z_1^T K_1 z_1 - z_2^T K_2 z_2. \tag{3.28}$$

To guarantee the closed-loop system stability, $\dot{V}_2$ requires to be negative definite. Therefore, control parameters $K_1 > 0$, $K_2 > 0$. Since $M\ddot{\mu}_1$ and CQ are uncertain, the control law (3.27) may be infeasible. Therefore, fuzzy NN is introduced to deal with the dynamic uncertainty via the online estimation.

$$\tau = -z_1 - K_2 z_2 + \hat{W}^T S(Z), \tag{3.29}$$

where $\hat{W}_i$ is the estimation of NN weight. $Z = [Q, \dot{Q}, \mu_1, \dot{\mu}_1]^T$ denotes the inputs to the adaptive radial basis function (RBF) NN.

The adaptive law $\dot{\hat{W}}_i$ is proposed

$$\dot{\hat{W}}_i = -\Gamma_i (S_i(Z) z_{2,i} + \sigma_i \hat{W}_i), \tag{3.30}$$

where, because the system needs strong robustness, a small constant $\sigma_i > 0$ is introduced. $\hat{W}_i^T$ is the approximation of fuzzy NN weights W_i^{*T}.

$$W_i^{*T} S_i(Z) = CQ + M\ddot{\mu}_1 - \epsilon(Z), \tag{3.31}$$

Theorem 3.1 For the dynamic model of the flexible manipulator in (3.10), by employing the control input and the adaptive law, because the full state information is known, for each compact set Ω_0 where $(x_1(0), x_2(0), \hat{W}_1(0) \dots \hat{W}_n(0)) \in \Omega_0$, the initial conditions are bounded and the trajectories of closed-loop robotic system are semi-globally uniformly bounded. The error signals z_1, z_2 and $\tilde{W}$ will remain within $\Omega_{z_1}, \Omega_{z_2}, \Omega_W$:

$$\Omega_{z_1} := \left\{ z_1 \in R^n \mid \|z_1\| \le \sqrt{X} \right\}, \tag{3.32}$$

$$\Omega_{z_2} := \left\{ z_2 \in R^n \mid \|z_2\| \le \sqrt{\frac{X}{\lambda_{\min}(M)}} \right\}, \tag{3.33}$$

$$\Omega_{\tilde{W}} := \left\{ \tilde{W} \in R^{l \times n} \mid \|\tilde{W}\| \le \sqrt{\frac{X}{\lambda_{\min}(\Gamma^{-1})}} \right\}, \tag{3.34}$$

where $X = 2(V_3(0) + \frac{b_1}{a_1})$, $b_1 > 0$ and $a_1 > 0$.

Proof: We propose the Lyapunov function candidate as follows:

$$V_3 = V_2 + \frac{1}{2} \sum_{i=1}^{n} \tilde{W}_i^T \Gamma_i^{-1} \tilde{W}_i. \tag{3.35}$$

Then, $\dot{V}_3$ can be derived

$$\begin{aligned} \dot{V}_3 &= z_1^T z_2 - z_1^T K_1 z_1 + z_2^T (\tau - CQ - M\ddot{\mu}_1) \\ &\quad - \sum_{i=1}^{n} \tilde{W}_i^T (S_i(Z) z_{2,i} + \sigma_i \hat{W}_i) \\ &= -z_1^T K_1 z_1 - z_2^T K_2 z_2 + z_2^T \hat{W}^T S(Z) - z_2^T \epsilon(Z) \\ &\quad - z_2^T W^{*T} S(Z) - \sum_{i=1}^{n} \tilde{W}_i^T (S_i(Z) z_{2,i} + \sigma_i \hat{W}_i). \end{aligned} \tag{3.36}$$

Consider the following inequalities:

$$
\begin{aligned}
-\sum_{i=1}^{n}\sigma_i \tilde{W}_i^T \hat{W}_i &= -\sum_{i=1}^{n}\sigma_i \tilde{W}_i^T (W_i^* + \tilde{W}_i) \\
&\leq -\frac{1}{2}\sum_{i=1}^{n}\sigma_i(\|\tilde{W}_i\|^2 + \|W_i^*\|^2) \\
-z_2^T\epsilon(Z) &\leq \frac{1}{2}(\|\bar{\epsilon}(Z)\|^2 + z_2^T z_2)
\end{aligned}
\tag{3.37}
$$

We have:

$$
\begin{aligned}
\dot{V}_3 \leq& -z_1^T K_1 z_1 - z_2^T(K_2 - \frac{1}{2}I)z_2 - \sum_{i=1}^{n}\tilde{W}_i^T S_i(Z)z_{2,i} \\
&- \sum_{i=1}^{n}\frac{\sigma_i}{2}\|\tilde{W}_i\|^2 + \sum_{i=1}^{n}\frac{\sigma_i}{2}\|W_i^*\|^2 + \frac{1}{2}\|\bar{\epsilon}\|^2 \\
\leq& -a_1 V_3 + b_1,
\end{aligned}
\tag{3.38}
$$

where:

$$
a_1 = \min\left(2\lambda_{\min}(K_1), \frac{2\lambda_{\min}(K_2 - \frac{1}{2}I_{nxn})}{\lambda_{\max}(M)}, \min_{i=1,\ldots,n}\left(\frac{\sigma_i}{\lambda_{\max}(\Gamma_i^{-1})}\right)\right), \tag{3.39}
$$

$$
b_1 = \sum_{i=1}^{n}\frac{\sigma_i}{2}\|W_i^*\|^2 + \frac{1}{2}\|\bar{\epsilon}(Z)\|^2, \tag{3.40}
$$

where $\lambda_{\max}$ and $\lambda_{\min}$, respectively are the maximum and minimum eigenvalues. In order to guarantee $a_1 > 0$, the control parameters should be chosen to satisfy $\lambda_{\min}(K_1) > 0$ and $\lambda_{\min}(K_2 - \frac{1}{2}I) > 0$.

Multiplying (3.38) by $e^{a_1 t}$, and integrating, we have

$$
V_3 \leq \left(V_3(0) - \frac{a_1}{b_1}\right)e^{-a_1 t} + \frac{a_1}{b_1} \leq V_3(0) + \frac{a_1}{b_1}. \tag{3.41}
$$

Thus

$$
\frac{1}{2}\|z_1\|^2 \leq V_3(0) + \frac{a_1}{b_1}. \tag{3.42}
$$

From the analysis above, the system of the flexible manipulator is stabilized via Lyapunov function that we designed. ■

3.3.2 Stability Analysis

However, the control law with full-state feedback (3.26) needs to satisfy that both Q and $\dot{Q}$ can be obtained by measurement. In practice, if $\dot{Q}$ is immeasurable, a high-gain observer can be introduced to estimate $\dot{Q}$ by employing separation principle and the certainty equivalence property.

According to [13], $\frac{\pi_{k+1}}{\varepsilon^k}$ converges asymptotically to $Q^{(k)}$. As a result of the high-gain $\frac{1}{\varepsilon}$, ξ_k converges to zero with a small time constant if Q and its k derivatives are bounded. Therefore, the observer $\frac{\pi_{k+1}}{\varepsilon^k}$ is appropriate to estimate the system unmeasurable states up to the nth order. Therefore,

$$
Q = \pi_1, \hat{\dot{Q}} = \frac{\pi_2}{\varepsilon}, \hat{z}_2 = \frac{\pi_2}{\varepsilon} - \alpha_1, \tag{3.43}
$$

where

$$
\varepsilon\dot{\pi}_1 = \pi_2, \varepsilon\dot{\pi}_2 = -\bar{\lambda}_1\pi_2 - \pi_1 + Q. \tag{3.44}
$$

Rewrite the control torque (3.29) and adaptive law (3.30) by backstepping.

$$\tau_2 = -z_1 - K_2\hat{z}_2 + \hat{W}^T S(\hat{Z}), \tag{3.45}$$

$$\dot{\hat{W}}_i = -\Gamma_i(S_i(\hat{Z})\hat{z}_{2,i} + \sigma_i \hat{W}_i), \tag{3.46}$$

Theorem 3.2 For the dynamic model in (3.10), by employing the control input and the adaptive law, because the output information is known, for each compact set Ω_0 where $(x_1(0), x_2(0), \hat{W}_1(0)... \hat{W}_n(0)) \in \Omega_0$, the initial conditions are bounded and the trajectories of closed-loop robotic system are semi-globally uniformly bounded. The error signals z_1, z_2 and $\tilde{W}$ will remain within $\Omega_{z_1}, \Omega_{z_2}, \Omega_W$:

$$\Omega_{z_1} := \{z_1 \in R^n \mid \|z_1\| \leq \sqrt{X}\}, \tag{3.47}$$

$$\Omega_{z_2} := \{z_2 \in R^n \mid \|z_2\| \leq \sqrt{\frac{X}{\lambda_{\min}(M)}}\}, \tag{3.48}$$

$$\Omega_{\tilde{W}} := \{\tilde{W} \in R^{l\times n} \mid \|\tilde{W}\| \leq \sqrt{\frac{X}{\lambda_{\min}(\Gamma^{-1})}}\}, \tag{3.49}$$

where $X = 2(V_3(0) + \frac{b_2}{a_2})$, $b_2 > 0$ and $a_2 > 0$.

Proof: We propose the Lyapunov function candidate as follows:

$$V_4 = \frac{1}{2}z_1^T z_1 + \frac{1}{2}z_2^T M z_2 + \frac{1}{2}\sum_{i=1}^{n} \tilde{W}_i^T \Gamma_i^{-1} \tilde{W}_i. \tag{3.50}$$

Then, $\dot{V}_4$ can be derived:

$$\begin{aligned} \dot{V}_4 \leq & -z_1^T K_1 z_1 - z_2^T(K_2 - \frac{1}{2}I)z_2 - z_2^T K_2 \tilde{z}_2 + \frac{1}{2}\|\bar{\epsilon}\|^2 \\ & + \sum_{i=1}^{n} z_{2,i}[\hat{W}_i^T S_i(\hat{Z}) - W_i^{*T} S_i(Z)] - \sum_{i=1}^{n}[\tilde{W}_i^T S_i(\hat{Z})\hat{z}_{2,i} + \sigma_i \tilde{W}_i^T \hat{W}_i], \end{aligned} \tag{3.51}$$

where

$$\xi_2 = \frac{\pi_2}{\varepsilon} - \dot{Q} = -\varepsilon\psi^{(2)}, \tag{3.52}$$

$$\tilde{z}_2 = \hat{z}_2 - z_2 = \frac{\pi_2}{\varepsilon} - \mu - \dot{Q} + \alpha_1 = \xi_2. \tag{3.53}$$

where ε is an any small constant, $\psi = \pi_2 + \overline{\lambda}_1 \pi_1$. For $\forall t > t^* > 0$ and $h_2 > 0$, then $\|\xi_2\| \leq \varepsilon h_2$. Therefore, $\dot{Q}$ can be estimated by $\frac{\pi_2}{\varepsilon}$.

$$\hat{\dot{Q}} = \frac{\pi_2}{\varepsilon}, \tag{3.54}$$

$$\hat{z}_2 = \frac{\pi_2}{\varepsilon} - \mu_1. \tag{3.55}$$

Using the following inequality:

$$-\sigma_i \tilde{W}_i^T \hat{W}_i \leq -\sigma_i(\frac{\|\tilde{W}_i\|^2}{2} - \frac{\|W_i^*\|^2}{2}), \tag{3.56}$$

$$\|S_i(\hat{Z})\|^2 \leq l_i, \tag{3.57}$$

$$\hat{W}_i^{\ T} S_i(\hat{Z}) = (\tilde{W}_i^{\ T} + W_i^{*T})(S_i(Z) + \epsilon S_{ti}), \tag{3.58}$$

$$= W_i^{*T} \epsilon S_{ti} + W_i^{*T} S_i(Z) + \tilde{W}_i^{\ T} S_i(Z). \tag{3.59}$$

where ϵ and l_i are positive and S_{ti} is bounded.

We have

$$\dot{V}_4 \leq -z_1^T K_1 z_1 - z_2^T (K_2 - \frac{1}{2} I) z_2 - z_2^T K_2 \tilde{z}_2 \tag{3.60}$$

$$+ \sum_{i=1}^{n} z_{2,i} W_i^{*T} \epsilon S_{ti} + \frac{1}{2} \|\bar{\epsilon}\|^2 - \sum_{i=1}^{n} \tilde{W}_i^{\ T} S_i(\hat{Z}) \hat{z}_{2,i}$$

$$+ \frac{1}{2} \sum_{i=1}^{n} \sigma_i \ \|W_i^*\|^2 - \frac{1}{2} \sum_{i=1}^{n} \sigma_i \ \|\tilde{W}_i\|^2 \tag{3.61}$$

Apply the following inequalities:

$$\sum_{i=1}^{n} z_{2,i} W_i^{*T} \epsilon S_{ti} \leq \frac{1}{2} z_2^T z_2 + \sum_{i=1}^{n} \frac{\|W_i^*\|^2 \ \epsilon^2 \|S_{ti}\|^2}{2} \tag{3.62}$$

$$-\sum_{i=1}^{n} \tilde{W}_i^{\ T} S_i(\hat{Z}) \tilde{z}_{2,i} = -\sum_{i=1}^{n} \frac{\sqrt{\sigma_i} \tilde{W}_i^{\ T}}{\sqrt{2}} \frac{\sqrt{2} S_i(\hat{Z}) \tilde{z}_{2,i}}{\sqrt{\sigma_i}}$$

$$\leq \sum_{i=1}^{n} \frac{\sigma_i \ \|\tilde{W}_i\|^2}{4} + \sum_{i=1}^{n} \frac{2 \ \|S_i(\hat{Z})\|^2}{\sigma_i} \frac{1}{2} \tilde{z}_2^{\ T} \tilde{z}_2 \tag{3.63}$$

We can obtain:

$$\dot{V}_4 \leq -z_1^T K_1 z_1 - z_2^T (K_2 - \frac{1}{2} I) z_2 - z_2^T K_2 \tilde{z}_2$$

$$+ \frac{1}{2} z_2^T z_2 + \sum_{i=1}^{n} \frac{\|W_i^*\|^2 \ \epsilon^2 \ \|S_{ti}\|^2}{2} + \frac{1}{2} \|\bar{\epsilon}\|^2$$

$$+ \sum_{i=1}^{n} \frac{\sigma_i \|\tilde{W}_i\|^2}{4} + \sum_{i=1}^{n} \frac{2 \ \|S_i(\hat{Z})\|^2}{\sigma_i} \frac{1}{2} \tilde{z}_2^{\ T} \tilde{z}_2$$

$$+ \frac{1}{2} \sum_{i=1}^{n} \sigma_i \ \|W_i^*\|^2 - \frac{1}{2} \sum_{i=1}^{n} \sigma_i \|\tilde{W}_i\|^2$$

$$\leq -z_1^T K_1 z_1 - z_2^T (K_2 - I) z_2 - z_2^T K_2 \tilde{z}_2 + \frac{1}{2} \|\bar{\epsilon}\|^2$$

$$- \sum_{i=1}^{n} \frac{\sigma_i}{4} \|\tilde{W}_i\|^2 + \sum_{i=1}^{n} \frac{2 \ \|S_i(\hat{Z})\|^2}{\sigma_i} \frac{1}{2} \tilde{z}_2^{\ T} \tilde{z}_2$$

$$+ \sum_{i=1}^{n} (\frac{\sigma_i}{2} + \frac{\epsilon^2 \ \|S_{ti}\|^2}{2}) \ \|W_i^*\|^2 \tag{3.64}$$

Using $z_2^T K_2 \tilde{z}_2 \leq \frac{1}{2} z_2^T z_2 + \frac{1}{2} (K_2 \tilde{z}_2)^T (K_2 \tilde{z}_2)$, we further have:

$$\dot{V}_4 \leq -z_1^T K_1 z_1 - z_2^T (K_2 - \frac{3}{2} I) z_2 - \frac{1}{2} K_2^T K_2 \xi_2^T \xi_2$$

$$+ \frac{1}{2} \|\bar{\epsilon}\|^2 - \sum_{i=1}^{n} \frac{\sigma_i}{4} \|\tilde{W}_i\|^2 + \sum_{i=1}^{n} \frac{2 l_i}{\sigma_i} \frac{1}{2} \xi_2^T \xi_2$$

$$+ \frac{1}{2} \sum_{i=1}^{n} (\sigma_i + \epsilon^2 \|S_{ti}\|^2) \|W_i^*\|^2$$

$$\leq -z_1^T K_1 z_1 - z_2^T (K_2 - \frac{3}{2} I) z_2$$
$$+ \frac{1}{2} \|\bar{\epsilon}\|^2 - \sum_{i=1}^{n} \frac{\sigma_i}{4} \|\tilde{W}_i\|^2$$
$$+ \lambda_{\max} (K_2^T K_2 + \mathrm{diag}[\frac{2l_i}{\sigma_i}]) \frac{1}{2} \xi_2^T \xi_2$$
$$+ \frac{1}{2} \sum_{i=1}^{n} (\sigma_i + \epsilon^2 \|S_{ti}\|^2) \|W_i^*\|^2. \quad (3.65)$$

Applying $\frac{1}{2}\xi_2^T \xi_2 \leq \frac{1}{2}\epsilon^2 h_2^2$, we have:

$$\dot{V}_4 \leq -z_1^T K_1 z_1 - z_2^T (K_2 - \frac{3}{2} I) z_2 - \sum_{i=1}^{n} \frac{\sigma_i}{4} \|\tilde{W}_i\|^2$$
$$+ \frac{1}{2} \|\bar{\epsilon}\|^2 + \lambda_{\max} (K_2^T K_2 + \mathrm{diag}[\frac{2l_i}{\sigma_i}]) \frac{1}{2} \epsilon^2 h_2^2$$
$$+ \frac{1}{2} \sum_{i=1}^{n} (\sigma_i + \epsilon^2 \|S_{ti}\|^2) \|W_i^*\|^2$$
$$\leq -a_2 V_3 + b_2, \quad (3.66)$$

where a_2 and b_2 are two constants defined as:

$$a_2 = \min \left(2\lambda_{\min}(K_1), \frac{2\lambda_{\min}(K_2 - \frac{3}{2} I)}{\lambda_{\max}(M)}, \min_{i=1,2\ldots n} \left(\frac{\sigma_i}{2\lambda_{\max}(\Gamma_i^{-1})} \right) \right), \quad (3.67)$$

$$b_2 = \frac{1}{2} \sum_{i=1}^{n} (\epsilon^2 \|S_{ti}\|^2 + \sigma_i) \|W_i^*\|^2$$
$$+ \lambda_{\max} (K_2^T K_2 + \mathrm{diag}[\frac{2l_i}{\sigma_i}]) \frac{1}{2} \epsilon^2 h_2^2 + \frac{1}{2} \|\bar{\epsilon}\|^2. \quad (3.68)$$

To guarantee that $a_2 > 0$, the control parameters K_1 and K_2 require to meet two conditions as follows, respectively.

$$\lambda_{\min}(K_1) > 0, \lambda_{\min}(K_2 - \frac{3}{2} I) > 0. \quad (3.69)$$ ■

Multiplying (3.66) by $e^{a_2 t}$, and integrating, we have

$$V_4 \leq (V_4(0) - \frac{a_2}{b_2}) e^{-a_2 t} + \frac{a_2}{b_2} \leq V_4(0) + \frac{a_2}{b_2}. \quad (3.70)$$

Thus

$$\frac{1}{2} \|z_1\|^2 \leq V_4(0) + \frac{a_2}{b_2}. \quad (3.71)$$

From the analysis above, the system of the flexible manipulator is stabilized via Lyapunov function that we design.

3.4 Numerical Simulations

In this section, the control results of the fuzzy NN control is shown by MATLAB simulation. The parameters of the flexible manipulator are selected as follows (Table 3.2):

Table 3.2 Parameters of the flexible manipulator.

Symbol	Description	Value
ρ	Flexible link density	0.1 kg/m
L	Flexible link length	1.0 m
I_h	Hub inertia	0.5 kgm^2
EI	Bending rigidity	2.0 Nm^2

Simulation results of open-loop system are shown in the Section 3.4.1, and the influence of the model parameters is verified. In the Section 3.4.2, PD strategy is applied to the flexible manipulator model, and the PD control results are given by changing the PD control parameters K_p and K_d. Adaptive fuzzy NN control with full-state feedback is revealed in the Section 3.4.3 and fuzzy NN control for output feedback is shown in Section 3.4.4.

3.4.1 Without Control

In this section, through the MATLAB simulations, we obtain the time-domain responses of the flexible model. The control torque $u(t)$ is designed

$$u(t) = \begin{cases} 0.5 \text{ N m}, & 0\,\text{s} \le t \le 0.2\,\text{s} \\ -0.5 \text{ N m}, & 0.2\,\text{s} \le t \le 0.4\,\text{s} \\ 0 \text{ N m}, & \text{otherwise.} \end{cases} \tag{3.72}$$

We choose the tip payload $M_t = 0$ kg and then have $\beta = 1.875$ and $\gamma = 0.734$. We select the initial states as zero. Considering $N = 5$ and $N = 10$, the corresponding tip position Y is given in Figures 3.4 and 3.5. Hence, we can get the conclusion that the discrete model is better approximated when N increases.

Considering the effect of m_t, we set $m_t = 0$ kg and $m_t = 0.1$ kg. The results are shown in Figure 3.6. We can get the conclusion that the discrete model is better approximated when m_t decreases.

Considering the effect of L, we set $L = 0.8$ m and $L = 1.0$ m. The results are shown in Figure 3.7. We can get the conclusion that the discrete model is better approximated when L increases.

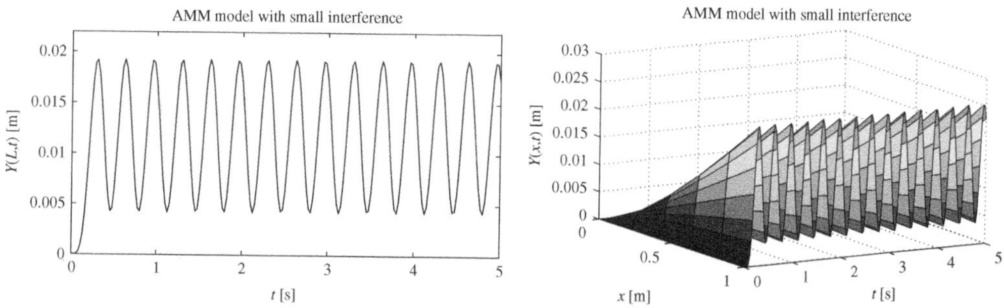

Figure 3.4 The tip position with $m_t = 0$ kg and $N = 5$

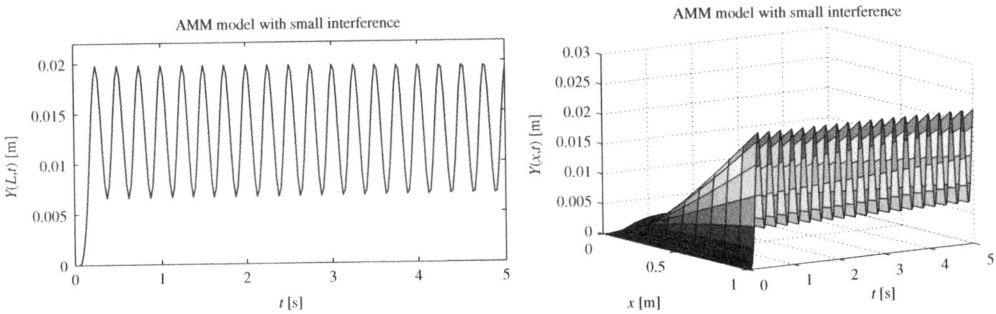

Figure 3.5 The tip position with $m_t = 0$ kg and $N = 10$

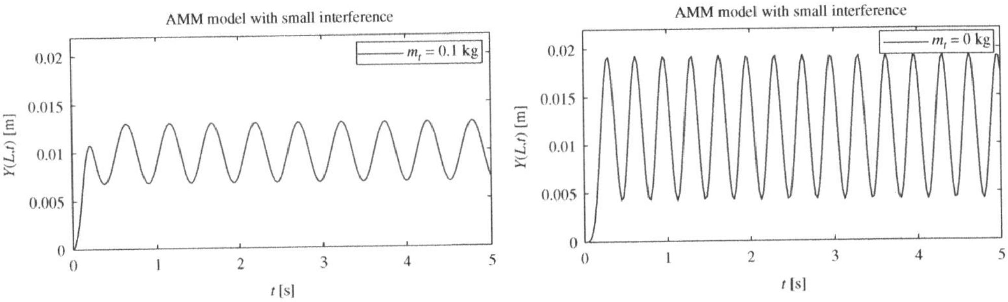

Figure 3.6 Effect of m_t on the dynamic response with $N = 5$ and $L = 1$ m

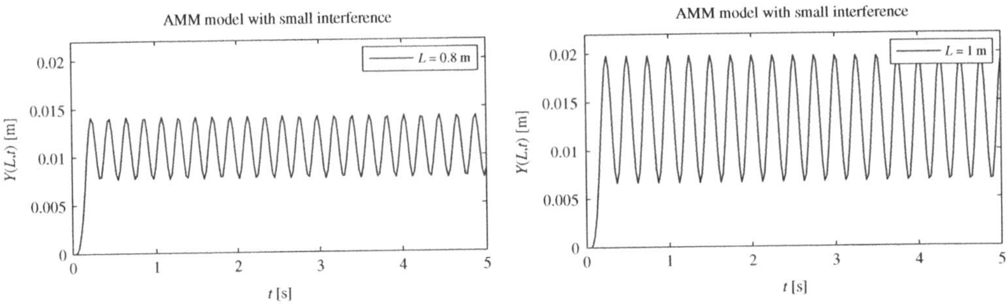

Figure 3.7 Effect of L on the dynamic response with $N = 10$ and $m_t = 0$ kg

3.4.2 PD Control

The PD strategy is applied to the flexible manipulator model with $M_t = 0$ kg, $N = 10$, and then we have $\beta = 1.875$, $\gamma = 0.734$. We select the initial states as zero. Design the PD controller:

$$\tau_0 = -K_p(\theta - \theta_d) - K_d\dot{\theta} \tag{3.73}$$

where K_p and K_d are proportional gain and differential gain, θ and θ_d are the angle position of the hub and the desired trajectory, respectively.

Design $\tau = [\tau_0, 0, \ldots, 0]^T \in R^{(n+1)}$ and consider the following cases:

$$K_p = 3, K_d = 1; K_p = 5, K_d = 2.$$

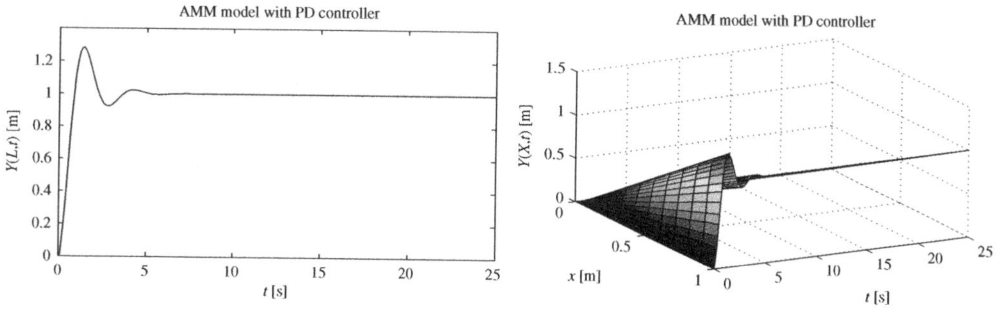

Figure 3.8 Tip position for PD control with $K_p = 3, K_d = 1$

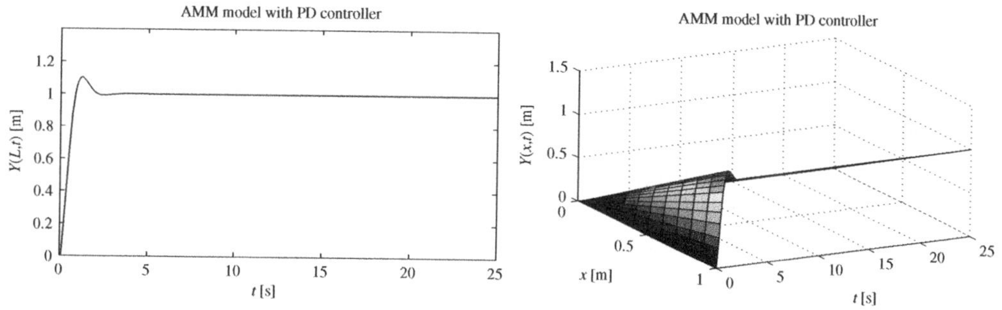

Figure 3.9 Tip position for PD control with $K_p = 5, K_d = 2$

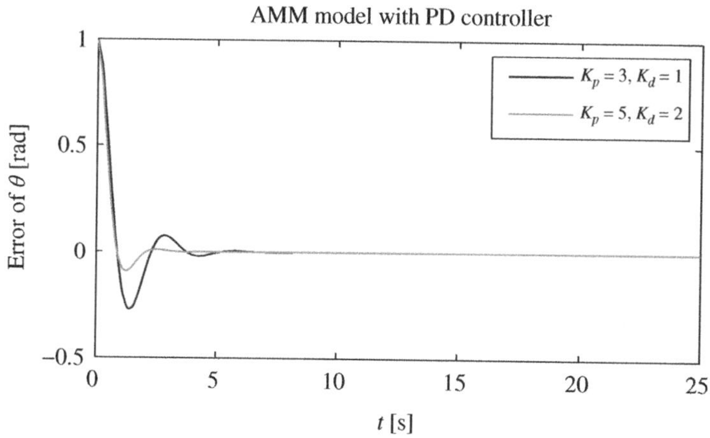

Figure 3.10 The error of θ

The corresponding tip trajectory $Y_N = L + y_N$ using PD strategy are revealed in Figures 3.8 and 3.9. The error of θ and τ_0 are shown in Figures 3.10 and 3.11 respectively. By analyzing the simulation results, we can get the conclusion that in spite of the change of the control parameters K_p and K_d, the flexible system achieves a stable state in 7 s and the real trajectory has a large overshoot which will have an impact on the actual system.

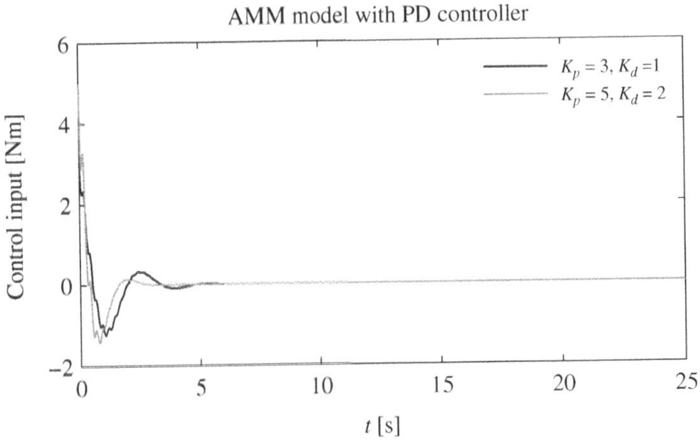

Figure 3.11 PD control inputs

3.4.3 Full-State Feedback

For fuzzy NN controller, the divided elements number $N = 10$. We choose the fuzzy system parameters as: the number of fuzzy rules is 8, c_{ij} is selected as 1 or -1, $\sigma_{ij} = 1$. 2^8 nodes are employed for each $S_i(Z_i)$. Variance of centers $\eta = 2$ as well as $\hat{W}_i = 0$ $(i = 1, 2, \ldots, 256)$. $\theta_d = 1$, $M_t = 0$ kg, $\sigma = 0.0002$, $\Gamma = 100I_{256\times256}$.

θ_d, θ, τ and the error of θ along with that $K_1 = I_{11\times11}$, $K_2 = 10I_{11\times11}$ and $K_1 = 2I_{11\times11}$, $K_2 = 20I_{11\times11}$ are displayed in Figure 3.12. As is shown in Figures 3.13 and 3.14, the flexible system achieves a stable state in 4 s and the transient overshoots are greatly reduced by fuzzy NN controller. Then we set $\theta_d = 0.5\sin(t) + 0.5$. The corresponding control variables are revealed in Figures 3.15–3.17.

By analyzing the simulation results, we can draw a conclusion that the fuzzy NN control for full-state feedback, compared with the PD control, the flexible manipulator can suppress the vibration more quickly and reduce the overshoot maximally.

3.4.4 Output Feedback

The divided elements number $N = 10$. And we choose the fuzzy system parameters as: the number of fuzzy rules is 8, c_{ij} is selected as 1 or -1, $\sigma_{ij} = 1$. 2^8 nodes are employed for each $S_i(Z_i)$. $\theta_d = 1$, $M_t = 0$ kg, $\sigma = 0.0002$, $\Gamma = 100I_{256\times256}$. η is chose as 2 as well as $\hat{W}_i = 0$. The high-gain observer is introduced with $\epsilon = 0.001$ and $\bar{\lambda}_1 = [4, 2]^T$.

θ_d, θ, and τ along with that $K_1 = 1, K_2 = 10$ and $K_1 = 2, K_2 = 20$ are shown in Figure 3.18. As is shown in Figures 3.19 and 3.20, the flexible system achieves a stable state in 4 s and the transient overshoots are greatly reduced by fuzzy NN controller. Then we set $\theta_d = 0.5\sin(t) + 0.5$. The corresponding control variables are revealed in Figures 3.21–3.23.

3.5 Experimental Investigation

3.5.1 Experimental Testbed

In order to make the control strategies more persuasive, we employ the Quanser Flexible Link System in the laboratory experiments, which is the test equipment connected with a FLEXGAGE

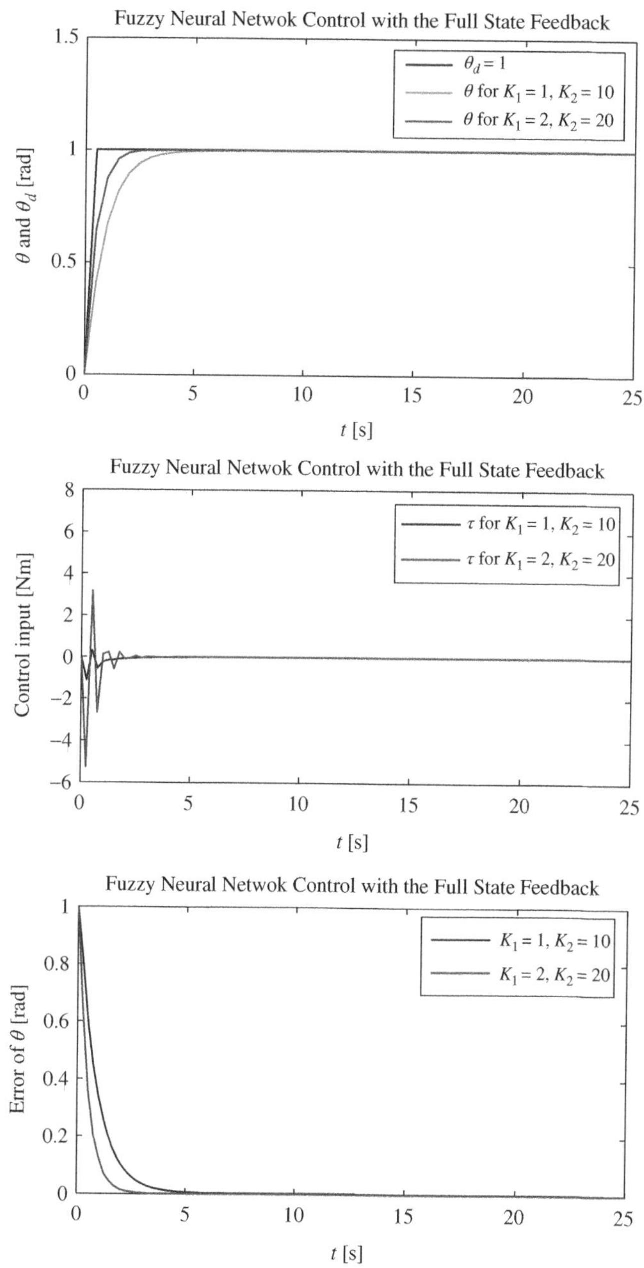

Figure 3.12 Trajectory, control torques, and tracking errors under full-state feedback

module the physical model. The experimental device has the function of real-time monitoring experiment process and it is composed of computer, flexible ruler, SRV02 servo device collected board, power amplifier, and the emergency brake. One end of the flexible ruler with a strain gauge to measure the tip deflection, DC motor servo device is used to drive flexible ruler rotation. The generated angle signal is detected by the optical encoder of the servo device. Control algorithm designed in simulink module connect with SRV02 servo device through the data acquisition board and power

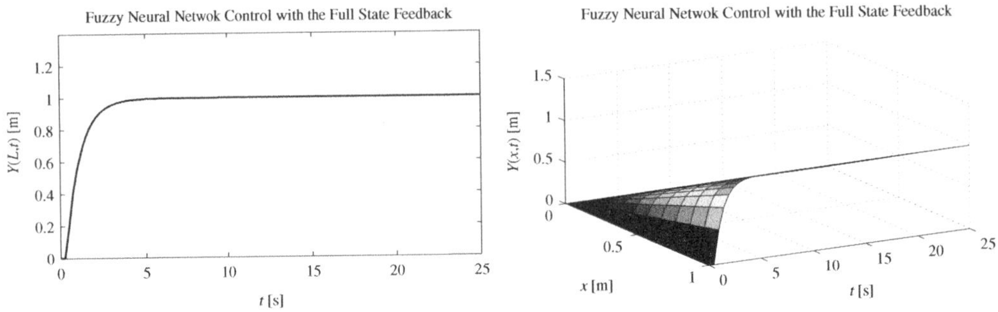

Figure 3.13 Tip position with $K_1 = 1, K_2 = 10$ under full-state feedback

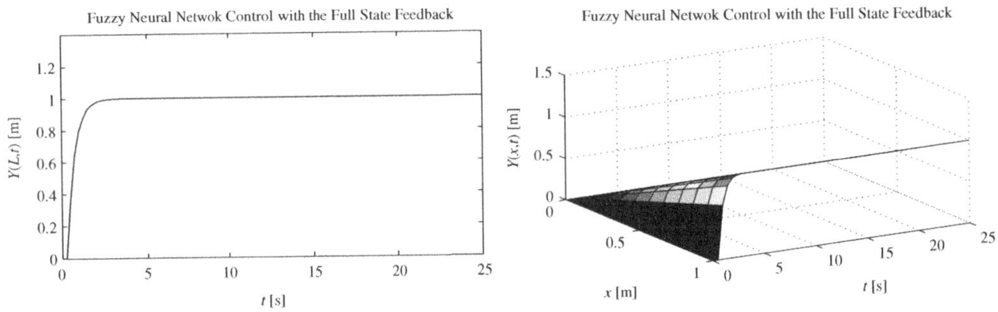

Figure 3.14 Tip position with $K_1 = 2, K_2 = 20$ under full-state feedback

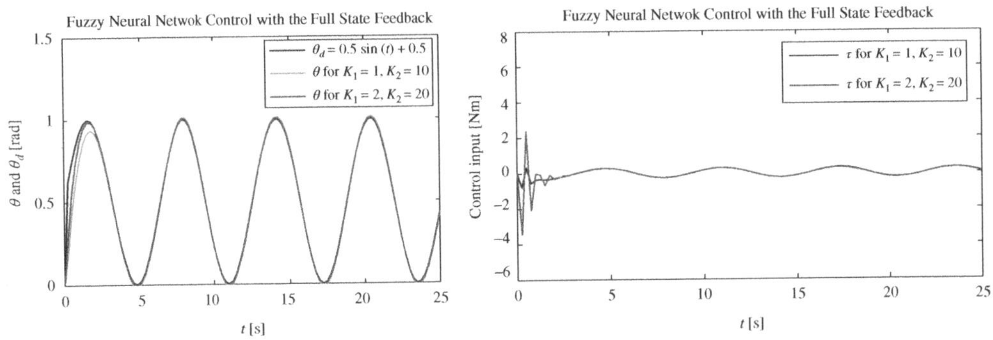

Figure 3.15 Trajectory and control torques under full-state feedback

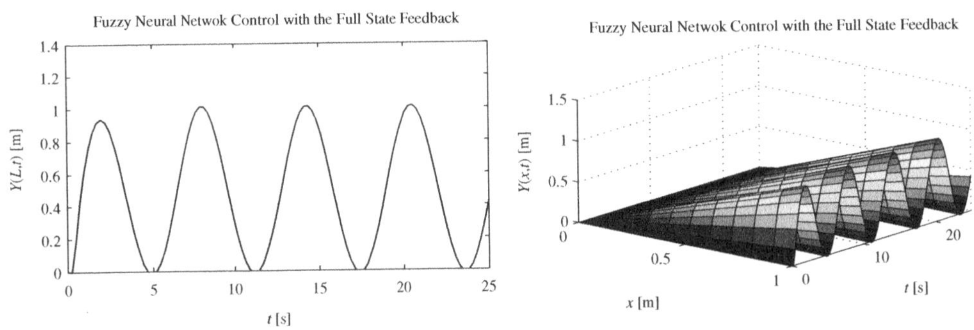

Figure 3.16 Tip position with $K_1 = 1, K_d = 10$ under full-state feedback

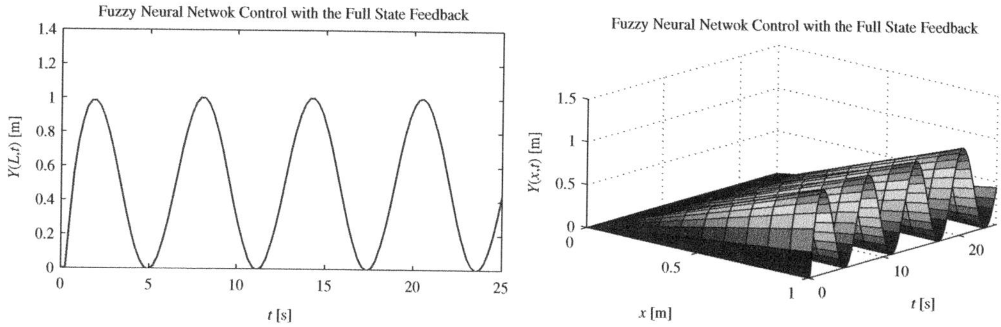

Figure 3.17 Tip position with $K_1 = 2, K_2 = 20$ under full-state feedback

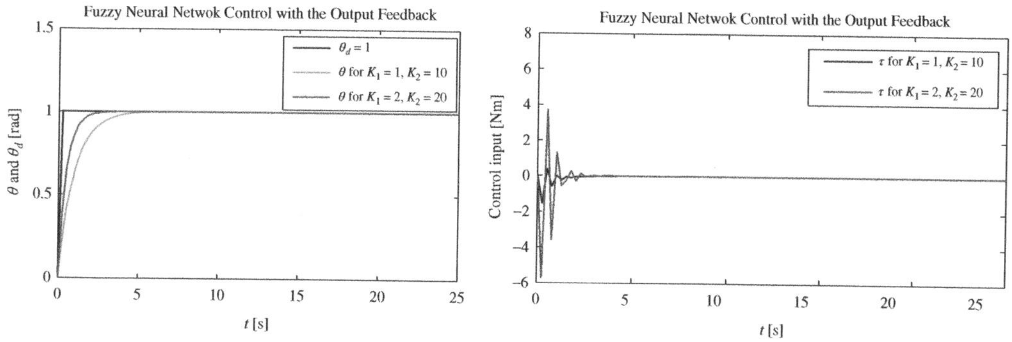

Figure 3.18 Trajectory and control inputs under output feedback

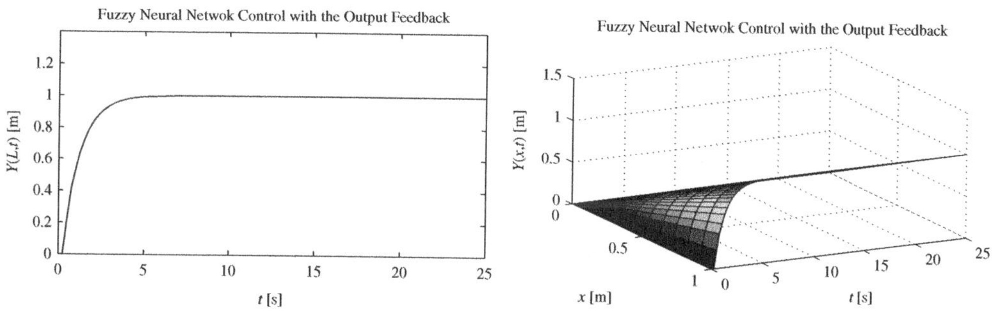

Figure 3.19 Tip position with $K_1 = 1, K_2 = 10$ under output feedback

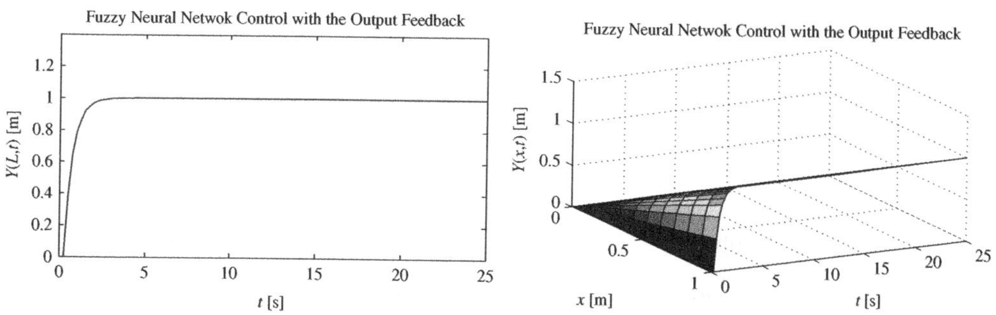

Figure 3.20 Tip position with $K_1 = 2, K_2 = 20$ under output feedback

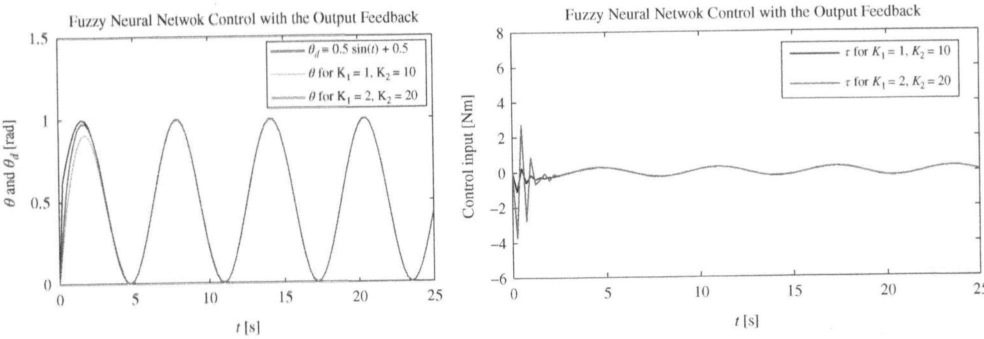

Figure 3.21 Trajectory and control inputs under output feedback

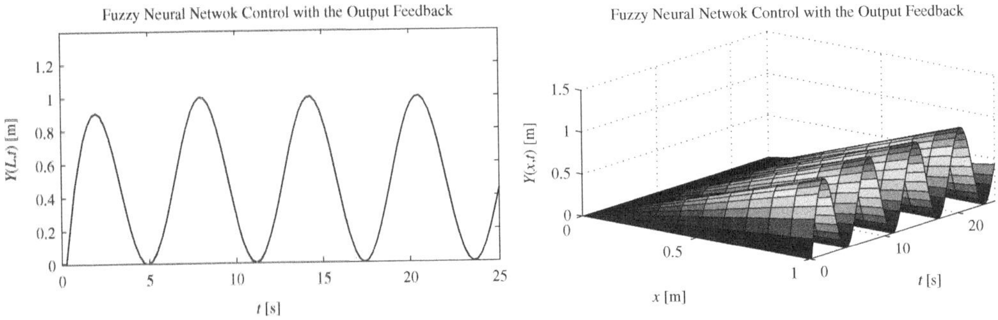

Figure 3.22 Tip position with $K_1 = 1, K_d = 10$ under output feedback

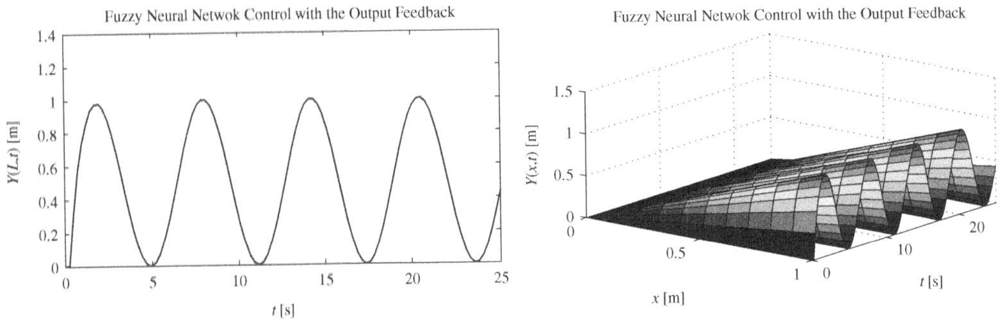

Figure 3.23 Tip position with $K_1 = 2, K_2 = 20$ under output feedback

amplifier. By the above description, the working block diagram of the experiment platform can be obtained as Figure 3.24. The relevant parameters and properties are tabulated in Table 3.3.

3.5.2 Experimental Results

In this section, the open-loop system, the PD strategy and fuzzy NN controller designed by MATLAB Simulink models and implemented in the Quanser Flexible Link Experiment are presented.

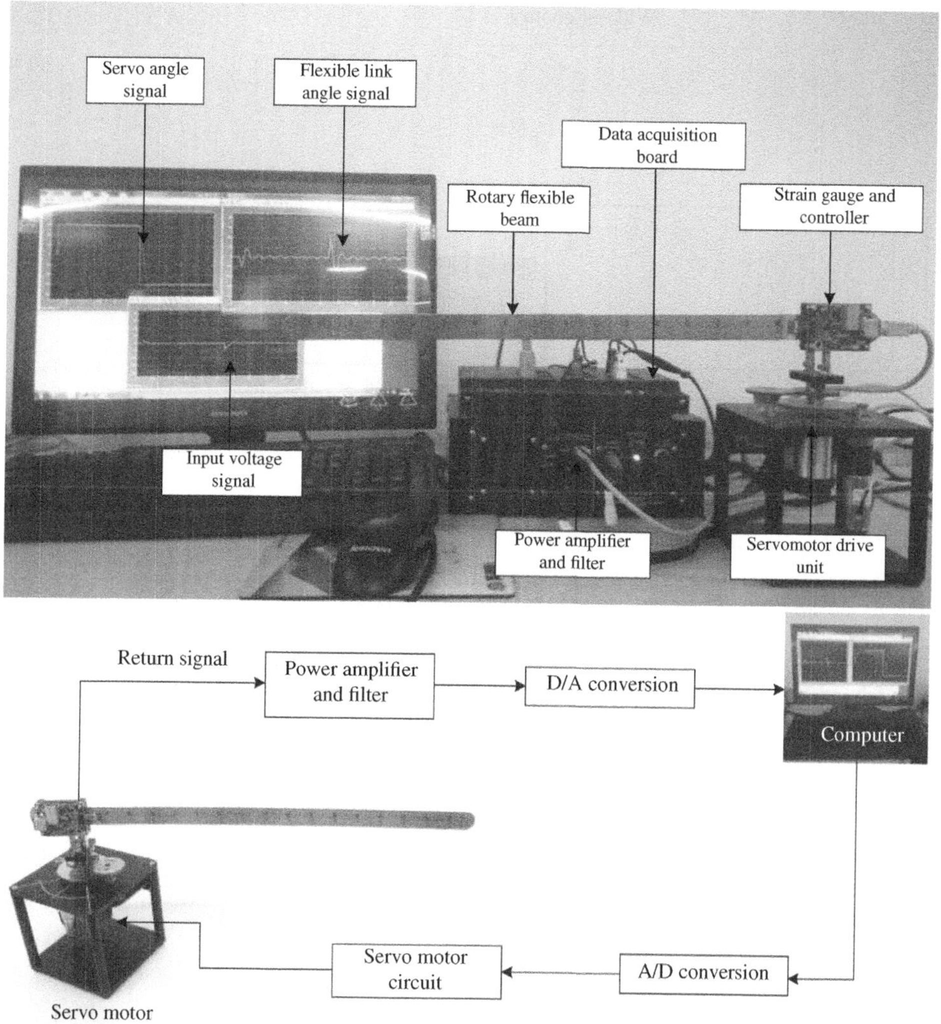

Figure 3.24 Control framework of the experimental flexible link

Table 3.3 Parameters of the Quanser test platform.

Symbol	Description	Value
L	Flexible link length	41.9 cm
m	Flexible link mass	0.065 kg
B_{eq}	Viscous friction coefficient	0.004 N m
J_{eq}	Moment of inertia	0.002 kg m^2

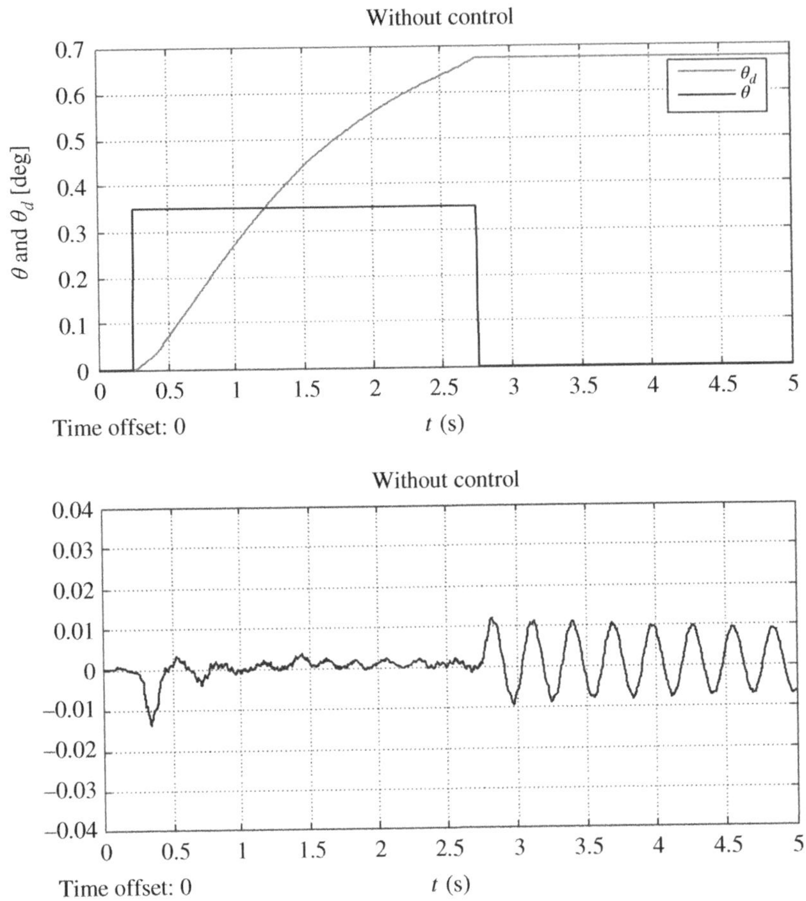

Figure 3.25 Open-loop performance of the experimental flexible link

In experiment, square wave signal through filtering, delay, amplification, angle conversion is sent to the SRV02 servo device. Then the SRV02 servo device drives the flexible manipulator to rotate and obtains the voltage signal. The angle position signal θ and the deflection α measured by the encoder and strain gauge can be observed in real time by the oscilloscope.

In order to observe the effect of excitation signal of flexible manipulator without control, we provide a square wave signal, which has the amplitude of the 1 V and the duty ratio of 55%. From the experimental results shown in Figure 3.25, in the period of 0.25 to 2.75 s, the angular position θ increases at a constant speed; when the square wave signal is 0 V, the angular position θ keeps a constant value. From the deformation α of flexible beam, the flexible beam has a small amplitude vibration irregularly before the flexible ruler stops rotating. After stoping rotating, the flexible ruler keeps the vibration of the same amplitude and has no tendency to decrease. Therefore, in the open-loop system, when the voltage is present, the flexible arm rotates at a constant speed; when there is no voltage, the flexible arm stops rotating and continues the constant amplitude vibration.

With the tip payload $m_t = 0$, in the simulink experiment, the PD control parameters are chosen as $K_p = 14, K_d = 0.4$. The servo angle θ and the deflection α are revealed in Figure 3.26.

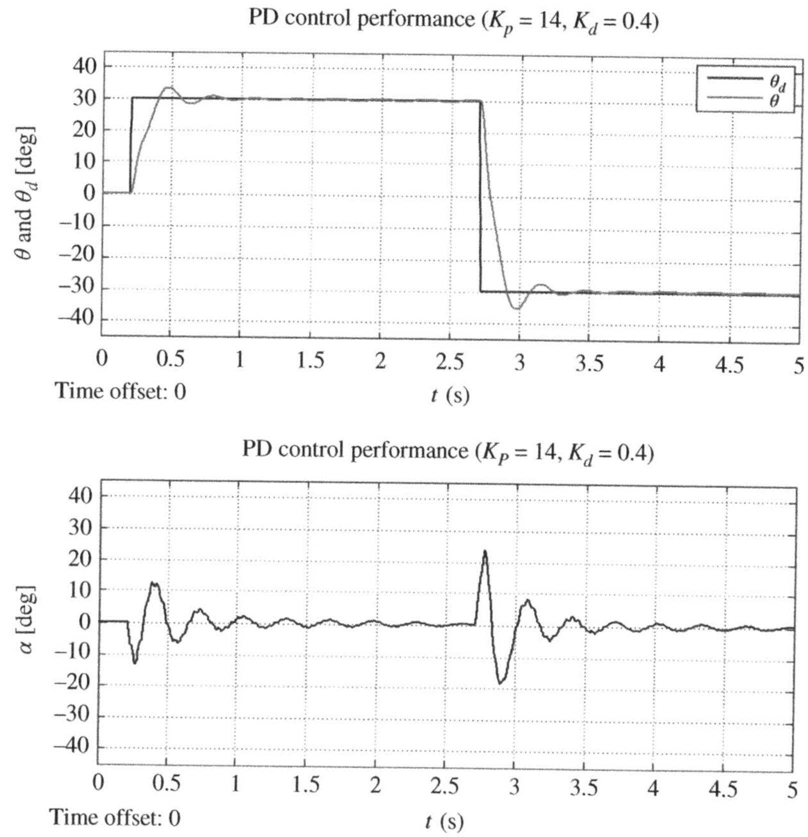

Figure 3.26 PD control of the experimental flexible link

As is shown in Figure 3.26, the flexible system achieves a stable state in 1 s by the PD controller, but there exists overshoot phenomenon. But in some control systems with the risk limit, the overshoot is not allowed. Even if it exceeds the limit in a very short time, it may lead to a disastrous disaster.

Therefore, we add the fuzzy NN controller on the experimental platform with the tip payload $m_t = 0$ and the control parameters $K_p = 14, K_d = 0.4$. The servo angle θ and the deflection α are revealed in Figure 3.27. Compared with the PD controller, the flexible system achieves a stable state in 0.75 s by the fuzzy NN controller. At the same time, the fuzzy NN controller can reduce the overshoot to the maximum.

3.6 Summary

In this chapter, the AMM is employed to develop the dynamic model. Base on the discrete dynamic model, fuzzy NN control is investigated to ensure that the flexible manipulator tracks the desired trajectory accurately and to guarantee that the vibration gets minimum simultaneously. To ensure stability rigorously as the goal, the system is proved to be uniform ultimate boundedness by

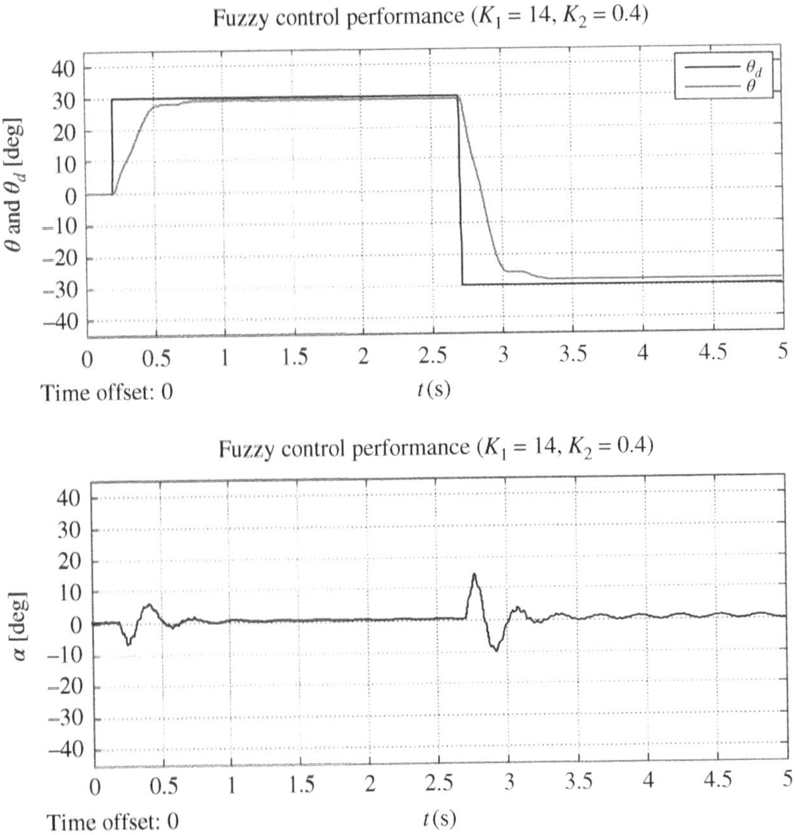

Figure 3.27 Fuzzy neural network control of the experimental flexible link

Lyapunov's stability method. Eventually, simulations verify that the proposed control strategy is effective, and the control performance is compared with PD control. The experiments are implemented on the Quanser platform to further demonstrate the feasibility of the proposed controller.

4

High-Gain Observer-Based Neural Network Control of the Two-Link Flexible Robotic Manipulator

After reading this chapter, you should be able to:

- Establish a finite-dimensional dynamic model of the two-link flexible robotic manipulator
- Master the high-gain observer-based neural network control strategy
- Know the semi-globally uniformly ultimate boundedness

4.1 Introduction

In recent years, the flexible robotic manipulator as the most direct application object of a flexible multibody system dynamics analysis and control theory [280], because of its obvious physical model and the feature of easily being tested, has developed into a key issue for the robot, aviation, and aerospace technology. The research on the flexible manipulator content is divided into two aspects: one is dynamic modeling of flexible manipulators; the other one is the control strategy design.

The flexible manipulator is a system with complicated dynamics. The dynamic model is the foundation of realizing the control of flexible manipulators [284]. Various types of discretized models have been proposed via finite element method (FEM) [345], AMM [81], finite difference method (FDM), and lumped parameter method (LPM) [117]. In [22], the authors proposed a computationally efficient method for modeling flexible robots based on the assumed modes method. In [21], the authors presented a numerically stable and computationally efficient method, which allows the modeling and simulation of a flexible robot with high precision. The appropriate discretized model can ensure the accurate positioning and tracking of the flexible manipulator end during the movement.

In this chapter, the two-link flexible manipulator is modeled by the AMM. We assumed that the elastic deflection is small, then, the two-link flexible manipulator can be expressed as a summation of finite number modes [214]. The proposed discrete model can effectively overcome the hidden defects in the data and make the model result more stable. It is helpful to describe the nonlinear relationship.

According to the proposed dynamic model, the control design can be carried out to achieve the control objectives that the joint motion can converge to the desired position [327] and the elastic vibrations can be effectively suppressed [121]. In literatures, many control strategies have been proposed and applied to various systems [49, 102, 203, 293, 307, 323]. In [144], the boundary control was proposed to solve the boundary external disturbance and input dead-zone nonlinearity.

Dynamic Modeling and Neural Network-Based Intelligent Control of Flexible Systems,
First Edition. Hejia Gao, Wei He, and Changyin Sun.

In [68], robust tracking control was applied to electrically driven flexible-joint robot by voltage control. In [352], an adaptive neural control method was proposed for robotic manipulators with output constraints and uncertainties. In [277] and [276], a novel adaptive fuzzy control approach was developed for nonlinear systems. In [336], experiments on the Baxter Robot have been carried out to demonstrate the performance of the proposed neural control. In [48], the authors proposed a neural network control strategy to solve the learning control problem for ocean surface ship.

However, compared with these conventional control strategies, NN requires relatively less dynamic information. Also, in dealing with the uncertain nonlinear system [47, 59], it has already been proven to be effective [180]. Under certain conditions, artificial neural networks (ANNs) have been widely used in the control system because of their inherent nonlinear simulation ability, high adaptability and fault tolerance [156]. Moreover, NNs have learning capability which makes NNs more widely employed for the controller design of various nonlinear systems, such as [46].

In this chapter, aiming at handling the system uncertainties and improving the robustness, an NN controller is developed for a two-link flexible robotic manipulator based on radial basis function neural networks (RBFNNs). Based on RBFNNs, the flexible robotic manipulator system has strong robustness, memory ability, nonlinear mapping ability, and self-learning ability. Compared with the existing works, the main contributions of this chapter include:

1) Based on the n-dimensional ordinary differential equation (ODE) model, NN control torques by full-state feedback are designed and implemented on double motors mounted at the joint of the flexible link to achieve trajectory tracking and vibration suppression.
2) Output feedback control method and a high-gain observer are adopted to estimate the unmeasured signals directly. Besides, neural network controller is designed to ensure the stability of the flexible system theoretically.
3) Extensive experiments on Quanser Flexible Link System platform for a two-link robotic manipulator are implemented to verify the feasibility of the proposed NN control in engineering practice.

The rest of the chapter is organized as follows. The dynamic modeling is derived in Section 4.2. Adaptive neural network control of a two-link flexible manipulator by full-state and output feedback is proposed in Sections 4.3 and 4.4. Simulations and experiments based on the Quanser platform are implemented to illustration the performance of the proposed controllers in Sections 4.5 and 4.6, respectively. Conclusions are given in Section 4.7.

4.2 Problem Formulation

4.2.1 Dynamic Modeling

In this section, we make some assumptions before the dynamics modeling. In the premise of small deformation, the influence of shear deformation, the moment inertia of the neutral axis and the manipulator's gravity are ignored. Considering the motion of two-degree-of-freedom manipulator in the horizontal plane, we regard it as an Euler–Bernoulli beam in the modeling process [146]. Then the manipulator system will be discretized via AMM in the following part.

Figure 4.1 shows the structure of the two-link manipulator system. The frames XOY and x_iOy_i represent the inertia frame and the local rotating reference frame with the hub, respectively. L_i represents the length of ith link. ρ_i is the mass per unit length. I_{hi} denotes the inertia of the ith hub. I_{ti} is the tip inertia of the ith link. I_{oi} is the inertia of ith link. M_i is the mass of the ith link. m_i

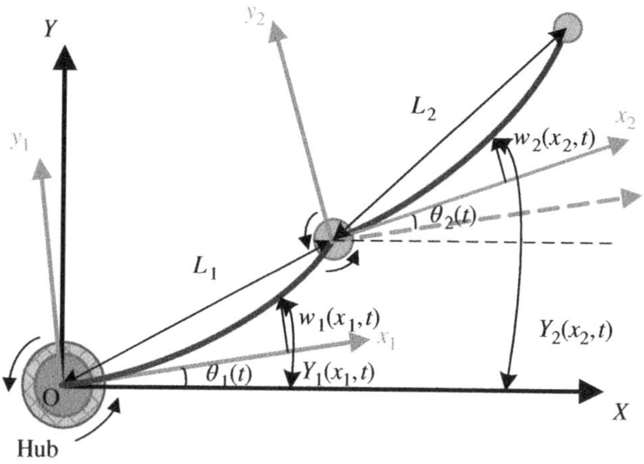

Figure 4.1 The diagram of the two-link flexible manipulator.

represents the tip mass. $\theta_i(t)$ donates the angular position of the ith hub. EI_i is the bending stiffness of the ith flexible link. $w_i(x_i, t)$ represents the transverse movement of point i at the position x_i for time t of the ith link. $Y_i(x_i, t)$ represents the position of the ith flexible link with respect to frame XOY at the position x_i for time t. $\tau_i(t)$ donates the input torque at the ith hub. $x_i \in [0, L_i]$ is the coordinate along the axial center of the ith link. We use derivative notions throughout this chapter as $(\dot{*}) = \frac{\partial *}{\partial t}$, $(\ddot{*}) = \frac{\partial^2(*)}{\partial t^2}$, $(*)' = \frac{\partial *}{\partial x_i}$, $(*)'' = \frac{\partial^2(*)}{\partial x_i^2}$, $(*)''' = \frac{\partial^3(*)}{\partial x_i^3}$, $(*)'''' = \frac{\partial^4(*)}{\partial x_i^4}$. Hamilton's principle which is used to derive the model of the beam system is represented by

$$\int_{t_1}^{t_2} \delta \left[E_k(t) - E_p(t) + W(t)\right] dt = 0, \tag{4.1}$$

where t_1 and t_2 are two time instants, $t_1 < t < t_2$ is the operating interval and δ denotes the variational operator.

Starting from energy analysis, the total kinetic energy of a two-link flexible manipulator system can be obtained:

$$E_k = E_{k1} + E_{k2} + E_{k3} + E_{k4} + E_{k5} + E_{k6}. \tag{4.2}$$

E_{k1}, E_{k2}, E_{k3}, E_{k4}, E_{k5}, E_{k6} are the kinetic energy of the two connecting rods. The kinetic energy of the two joints and the kinetic energy of the tip load of the two connecting rods are respectively determined by the Formula (4.1) and (4.2).

$$E_{k1} = \frac{1}{2}\int_0^{L_1} \rho_1 [\dot{z}_1(x_1)]^2 dx_1, \tag{4.3}$$

$$E_{k2} = \frac{1}{2}\int_0^{L_2} \rho_2 \left[\dot{z}_2\left(x_2\right)\right]^2 dx_2 + \frac{1}{2}\int_0^{L_2} \rho_2 \left[\dot{z}_1\left(L_1\right)\sin\theta_2(t)\right]^2 dx_2, \tag{4.4}$$

$$E_{k3} = \frac{1}{2} I_{h1} [\theta_1(t)]^2, \tag{4.5}$$

$$E_{k4} = \frac{1}{2} I_{h2} [\dot{\theta}_1(t) + \dot{\theta}_2(t)]^2 + \frac{1}{2} m_{h2} [\dot{z}_1(L_1)]^2, \tag{4.6}$$

$$E_{k5} = \frac{1}{2} I_{t2} [\dot{\theta}_1(t) + \dot{y}_1(L_1)]^2 + \frac{1}{2} m_{t1} [\dot{z}_1(L_1)]^2, \tag{4.7}$$

$$E_{k6} = \frac{1}{2}I_{t2}[\dot{\theta}_1(t) + \dot{\theta}_2(t) + \dot{y}_2(L_2)]^2,$$
$$+ \frac{1}{2}m_{t2}[\dot{z}_2(L_2)]^2 + \frac{1}{2}m_{t2}[\dot{z}_1(L_1)\sin\theta_2(t)]^2. \tag{4.8}$$

The total potential energy of the system is as follows:

$$E_p = \frac{1}{2}EI_1\int_0^{L_1}[y_1''(x_1)]^2dx_1 + \frac{1}{2}EI_2\int_0^{L_1}[y_2''(x2)]^2dx_2. \tag{4.9}$$

The dynamic modeling needs to consider the coupling between the first link and the second link. Based on the energy equation and the Hamilton's principle [200], we can get the corresponding governing equations as follows [348]:

$$\rho_1\ddot{Y}_1(x_1) = -EI_1\omega_1''''(x_1), \tag{4.10}$$

$$\rho_2\ddot{Y}_2(x_2) = -EI_2\omega_2''''(x_2). \tag{4.11}$$

The motion of the angular position is as follows:

$$\tau_1 = I_{h1}\ddot{\theta}_1(t) - EI_1\omega_1''(0), \tag{4.12}$$

$$(I_{t1} + I_{h2})\tau_2 = I_{t1}I_{h2}\ddot{\theta}_2(t) - I_{h2}EI_1\omega_1''(L_1)$$
$$- I_{t1}EI_2\omega_2''(L_2). \tag{4.13}$$

The boundary conditions are as follows:

$$g(t)\ddot{Y}_1(L_1) + \dot{g}(t)\dot{Y}_1(L_1) + EI_2\upsilon_1(t) = EI_1\omega_1'''(L_1), \tag{4.14}$$

$$\rho_2L_2[\dot{Y}_1(L_1)]^2\sin\theta_2(t)\cos\theta_2(t) - EI_1\omega_1''(L_1)$$
$$+ EI_2\dot{Y}_1(L_1)\upsilon_2(t)\sin\theta_2(t) + EI_2\omega_2''(0)$$
$$- I_{h2}[\ddot{\theta}_1(t) + \ddot{\theta}_2(t)] - I_{t1}[\ddot{\theta}_1(t) + \ddot{\omega}_1'(L_1)] \tag{4.15}$$

$$- m_2\dot{Y}_1(L_1)\sin\theta_2(t)[2L_2\dot{\theta}_2(t) + \dot{\omega}_2(L_2)] = 0,$$

$$m_2\ddot{Y}_2(L_2) = EI_2\omega_2'''(L_2), \tag{4.16}$$

$$I_{t2}[\ddot{\theta}_1(t) + \ddot{\theta}_2(t) + \ddot{\omega}_2'(L_2)] + EI_2\omega_2''(L_2) = 0, \tag{4.17}$$

$$\omega_i(0) = \omega_i'(0) = 0, i = 1, 2. \tag{4.18}$$

where, in order to facilitate writing, we define the corresponding coefficients in the boundary conditions $g(t)$, $\upsilon_1(t)$, $\upsilon_2(t)$ as follows:

$$Y_1(x_1) = x_1\theta_1(t) + \omega_1(x_1), \tag{4.19}$$

$$Y_2(x_2) = x_2\theta_2(t) + \omega_2(x_2) + \int_0^t[\dot{Y}_1(L_1, \xi)\cos\theta_2(\xi)]d\xi, \tag{4.20}$$

$$g(t) = (M_2 + m_2)\sin^2\theta_2(t) + m_1 + M_2, \tag{4.21}$$

$$\upsilon_1(t) = \cos\theta_2(t)\int_0^t\omega_2'''(0, \xi)d\xi, \tag{4.22}$$

$$\upsilon_2(t) = \int_0^t[\omega_2'''(L_2, \xi) - \omega_2'''(0, \xi)]d\xi. \tag{4.23}$$

4.2.2 Model Discretization

Due to the flexible characteristics of the manipulator, elastic deformation will occur as the movement of the manipulator. And it requires to satisfy the appropriate boundary conditions and governing equations. ω_i is the elastic deformation of link i and can be described via AMM:

$$\omega_i(x_i, t) = \sum_{j=1}^{\mu_i} F_{ij}(x_i)p_{ij}(t), i = 1, 2. \tag{4.24}$$

where $F_{ij}(x_i)$ denotes the modes shape function, $p_{ij}(t)$ is the generalized coordinate. On the premise of small deflection, the dynamic model derived by Lagrange's equation and AMM is as follows:

$$B(q)\ddot{q} + H(q, \dot{q})\dot{q} + K(q) = \tau(t). \tag{4.25}$$

where $q = [\theta, p]^T, \theta = [\theta_1, \theta_2]^T$ is the vector of deflection angle, $p = [p_{11} \cdots p_{1n_1}, p_{21} \cdots p_{2n_2}]^T$ represents the flexible generalized coordinate vector. $\phi = n_1 + n_2$ is the total number of flexible variables. $\tau \in \mathbb{R}^{(\phi+2)} = [\tau_1, \tau_2, 0, \dots, 0]^T$ represents the generalized control torque vector applied at each joint. $B(q) \in \mathbb{R}^{(\phi+2)\times(\phi+2)}$ is the inertia matrix. $H(q, \dot{q}) \in \mathbb{R}^{(\phi+2)\times(\phi+2)}$ represents the matrix of Coriolis and centripetal forces. $K(q) \in \mathbb{R}^{(\phi+2)}$ is the matrix of gravity and elasticity forces.

4.3 High-Gain Observer-Based Neural Network Control

4.3.1 Control Design

In this section, considering full-state information, q and $\dot{q}$ are available, we introduce an ANN controller with full-state feedback to effectively track the state vector q. Lyapunov's direct method and back-stepping method are adopted to propose the desired control and prove the stability of the system under the controller. The diagram of full-state feedback control is displayed in Figure 4.2.

Considering the dynamical Eq. (4.25), we have the description of the system dynamics as:

$$\ddot{q} = B^{-1}(q)[\tau - H(q, \dot{q})\dot{q} - K(q)]. \tag{4.26}$$

We introduce two error variables:

$$z_1(t) = q(t) - q_d(t), \; z_2(t) = \dot{q}(t) - \alpha_1(t). \tag{4.27}$$

where $\alpha_1 = -K_1 z_1 + \dot{q}_d$ is a virtual back-stepping control.

The gain matrix $K_1 = K_1^T > 0 \in \mathbb{R}^{(\phi+2)\times(\phi+2)}$, then,

$$\dot{z}_1 = z_2 - K_1 z_1, \tag{4.28}$$

$$\dot{z}_2 = B^{-1}(q)[\tau - H(q, \dot{q})\dot{q} - K(q)] - \dot{\alpha}_1(t). \tag{4.29}$$

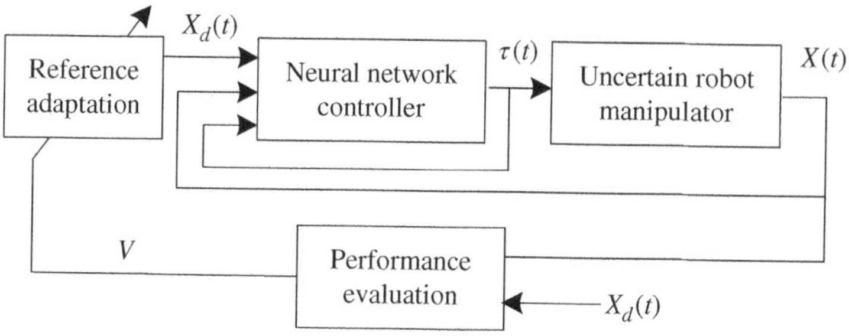

Figure 4.2 The diagram of full-state feedback.

For achieving the stability of the manipulator system, we construct a Lyapunov function. Consider a Lyapunov function candidate as:

$$V_1 = \frac{1}{2} z_1^T z_1. \tag{4.30}$$

Further, the time derivative of Eq. (4.30):

$$\dot{V}_1 = z_1^T z_2 - z_1^T K_1 z_1. \tag{4.31}$$

Because we have got the manipulator system $\dot{z}_1$ and $\dot{z}_2$, we can guarantee the system stability by removing the coupling term $z_1^T z_2$ in Eq. (4.31). Therefore, the Lyapunov function candidate is introduced as:

$$V_2 = \frac{1}{2} z_1^T z_1 + \frac{1}{2} z_2^T B(q) z_2. \tag{4.32}$$

According to the definition of the skew-symmetric matrix, let A be a square matrix, if its transpose A^T plus itself equals exactly zero matrix, then A is called skew-symmetric matrix. Therefore, we obtain:

$$-z_2^T A z_2 = z_2^T A^T z_2 = z_2^T A z_2. \tag{4.33}$$

According to Eq. (4.33), we know that $z_2^T A z_2 = 0$. Because $\frac{1}{2}(\dot{B}(q) - 2H(q,\dot{q}))$ is a skew-symmetric matrix, we can state that:

$$\frac{1}{2} z_2^T (\dot{B}(q) - 2H(q,\dot{q})) z_2 = 0. \tag{4.34}$$

Differentiating with (4.32) respect to time, and substituting (4.34) into this equation, we can obtain:

$$\begin{aligned}\dot{V}_2 = {} & z_1^T z_2 - z_1^T K_1 z_1 \\ & + z_2^T [\tau - K(q) - H(q,\dot{q})\alpha_1(t) - B(q)\dot{\alpha}_1(t)].\end{aligned} \tag{4.35}$$

We need to make $\dot{V}_2 < 0$ by designing control torque τ_0 when $B(q)$, $H(q,\dot{q})$, and $K(q)$ are available. Hence, the control input is as follows:

$$\tau = -z_1 - K_2 z_2 + H(q,\dot{q})\alpha_1(t) + K(q) + B(q)\dot{\alpha}_1(t). \tag{4.36}$$

where the gain matrix $K_2 = K_2^T > 0$.

Substituting (4.36) into (4.35), we have:

$$\dot{V}_2 = -z_1^T K_1 z_1 - z_2^T K_2 z_2. \tag{4.37}$$

where $\dot{V}_2$ is negative definite and τ_0 is feasible.

Due to the uncertainties of $B(q)$, $H(q,\dot{q})$, and $K(q)$, the proposed control input can not be applied to the actual manipulator system. To address this challenge, we propose the following control to approximate the uncertain dynamics and polish up the control performance by the online estimation:

$$\tau = -z_1 - K_2 z_2 + \hat{W}^T R(Z). \tag{4.38}$$

where W_i^*, $\hat{W}_i$, and $\tilde{W}_i$ are the true value, estimation, and error of neural networks weights, respectively. $R(Z) \in \mathbb{R}^l$ is the basis function.

The function approximates $W^{*T} R(Z)$ can be defined as:

$$W^{*T} R(Z) = H(q,\dot{q})\alpha_1(t) + K(q) + B(q)\dot{\alpha}_1(t) - \epsilon(Z). \tag{4.39}$$

where $Z = [q^T, \dot{q}^T, \alpha_1^T, \dot{\alpha}_1^T]$ represents the inputs of the neural networks and $\epsilon(Z) \in \mathbb{R}^{(\phi+2)}$ is the approximation error.

Moreover, the adaptive law is represented as follows:

$$\dot{\hat{W}}_i = -\Gamma_i[S_i(Z)_{z_2,i} + \sigma_i \hat{W}_i]. \tag{4.40}$$

where Γ_i is the constant gain matrix, and σ_i is a small positive constant in order to improve robustness.

Remark 4.1 In this chapter, a more challenging problem that $B(q)$, $H(q, \dot{q})$, and $K(q)$ are totally unavailable is studied. If the individual term can be obtained in practical engineering, it can be removed from the Eq. (4.39) and rewritten explicitly in the Eq. (4.38).

Theorem 4.1 For the dynamic Eq. (4.25), the control law (4.38) with the adaptive law (4.40), because full-state information is known, for each compact set Ω_0 where $(q(0),\ \dot{q}(0),\ \hat{W}_1(0),\ \dots,\ \hat{W}_{\phi+2}(0)) \in \Omega_0$, the initial conditions are bounded, uniformly ultimately bounded (UUB) of the closed-loop system trajectories is achieved. z_1, z_2, and $\tilde{W}$ will remain within the compact sets Ω_{z_1}, Ω_{z_2}, Ω_W:

$$\Omega_{z_1} := \left\{ z_1 \in \mathbb{R}^{\phi+2} \mid \|z_1\| \leq \sqrt{M} \right\}, \tag{4.41}$$

$$\Omega_{z_2} := \left\{ z_2 \in \mathbb{R}^{\phi+2} \mid \|z_2\| \leq \sqrt{\frac{M}{\lambda_{\min}(B)}} \right\}, \tag{4.42}$$

$$\Omega_W := \left\{ \tilde{W} \in \mathbb{R}^{l\times(\phi+2)} \mid \|\tilde{W}\| \leq \sqrt{\frac{M}{\lambda_{\min}(\Gamma^{-1})}} \right\}. \tag{4.43}$$

where $M = 2(V(0) + \frac{\beta}{\gamma})$, $\beta > 0$ and $\gamma > 0$.

4.3.2 Stability Analysis

Construct a Lyapunov function as:

$$V_3 = \frac{1}{2} z_1^T z_1 + \frac{1}{2} z_2^T B z_2 + \frac{1}{2} \sum_{i=1}^{\phi+2} \tilde{W}_i^T \Gamma_i^{-1} \tilde{W}_i. \tag{4.44}$$

Differentiating (4.44) and applying (4.35), we obtain:

$$\begin{aligned} \dot{V}_3 \leq{}& l z_1^T z_2 - z_1^T K_1 z_1 + z_2^T [\tau - H(q, \dot{q})\alpha_1(t) \\ & -K(q) - B(q)\dot{\alpha}_1(t)] + \sum_{i=1}^{\phi+2} \tilde{W}_i^T \Gamma_i^{-1} \dot{\hat{W}}_i. \end{aligned} \tag{4.45}$$

Applying Eq. (4.39) into (4.45):

$$\begin{aligned} \dot{V}_3 \leq{}& z_1^T z_2 - z_1^T K_1 z_1 + z_2^T [\tau - W^{*T} R(Z) - \epsilon(Z)] \\ & + \sum_{i=1}^{\phi+2} \tilde{W}_i^T \Gamma_i^{-1} \dot{\hat{W}}_i. \end{aligned} \tag{4.46}$$

Applying the following inequalities:

$$-\sum_{i=1}^{n}\sigma_i\tilde{W}_i^T\hat{W}_i = -\sum_{i=1}^{n}\sigma_i\tilde{W}_i^T(\tilde{W}_i + W_i^*)$$
$$\leq -\sum_{i=1}^{n}\frac{\sigma_i}{2}(\|\tilde{W}_i\|^2 - \|W_i^*\|^2), \tag{4.47}$$

$$-z_2^T\epsilon(Z) \leq \frac{1}{2}(\|\bar{\epsilon}(Z)\|^2 + z_2^Tz_2). \tag{4.48}$$

Substituting the Eqs. (4.38) and (4.40) into (4.46):

$$\dot{V}_3 \leq -z_1^TK_1z_1 - z_2^T(K_2 - \frac{1}{2}I_{(\phi+2)\times(\phi+2)})z_2$$
$$+\frac{1}{2}\|\bar{\epsilon}(Z)\|^2 - \sum_{i=1}^{\phi+2}\frac{\sigma_i}{2}(\|\tilde{W}_i\|^2 - \|W_i^*\|^2)$$
$$\leq -\gamma_1V_3 + \beta_1. \tag{4.49}$$

where:

$$\gamma_1 = \min\left(2\lambda_{\min}(K_1), \frac{2\lambda_{\min}(K_2 - \frac{1}{2}I_{(\phi+2)\times(\phi+2)})}{\lambda_{\max}(B)}, \min_{i=1,\ldots\phi+2}\left(\frac{\sigma_i}{\Gamma_i^{-1}}\right)\right), \tag{4.50}$$

$$\beta_1 = \sum_{i=1}^{\phi+2}\frac{\sigma_i}{2}\|W_i^*\|^2 + \frac{1}{2}\|\bar{\epsilon}(Z)\|^2. \tag{4.51}$$

where $\lambda_{\max}(A)$ and $\lambda_{\min}(A)$ donate the maximum and minimum eigenvalues of matrix A, respectively. To guarantee γ_1 more than zero, K_1 and K_2 are required to satisfy:

$$\lambda_{\min}(K_1) > 0,\ \lambda_{\min}(K_2 - \frac{1}{2}I) > 0. \tag{4.52}$$

Multiplying (4.49) by $e^{\gamma t}$:

$$\frac{d}{dt}(V_3e^{\gamma_1 t}) \leq \beta_1 e^{\gamma_1 t}. \tag{4.53}$$

Integrating both sides of Eq. (4.53):

$$V_3e^{\rho_1 t}\mid_0^t \leq \frac{\beta_1}{\gamma_1}\mid_0^t, \tag{4.54}$$

$$V_3 \leq (V_3(0) - \frac{\beta_1}{\gamma_1})e^{-\gamma_1 t} + \frac{\beta_1}{\gamma_1} \leq V_3(0) + \frac{\beta_1}{\gamma_1}. \tag{4.55}$$

Then, we have:

$$\frac{1}{2}\|z_1\|^2 \leq V_3(0) + \frac{\beta_1}{\gamma_1}. \tag{4.56}$$

Hence, we can get the conclusion that z_i converges to the compact set Ω_{zi} and $\tilde{W}_i$ converges to Ω_W. The system is stabilized via Lyapunov function (4.44) that we designed.

Remark 4.2 It is apparent that the control performance will be better if the control gains K_1 and K_2 and the adaptive gains Γ_i are increased. Hence, we can obtain the conclusion that the tracking errors z_1 and z_2 will finally converge to a small neighborhood of zero by appropriate design parameters.

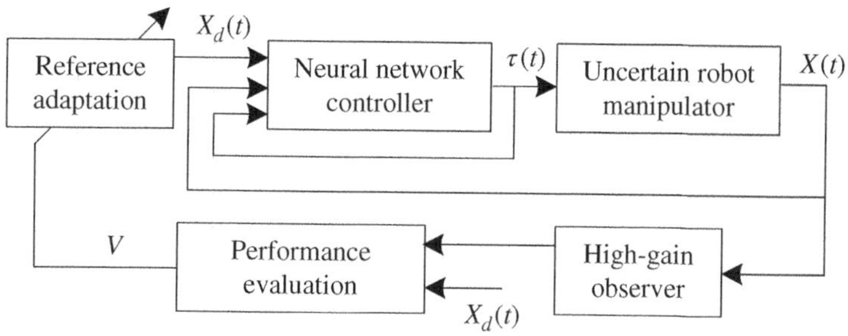

Figure 4.3 The diagram of output feedback.

The proposed full-state feedback control (4.38) requires q and $\dot{q}$ both to be available. On the premise of the speed sensors, a high-gain observer is employed to estimate $\dot{q}$. The diagram of output feedback is shown in Figure 4.3.

According to the Lemma from [13], $\frac{\pi_{k+1}}{\varepsilon^k}$ converges asymptotically to the kth order derivative of q. ξ_k is close to zero in a short time owing to the high gain $\frac{1}{\varepsilon}$ and the boundedness of q and its k derivatives. Thus, $\frac{\pi_{k+1}}{\varepsilon^k}$ is proper as an observer. The estimate of the unmeasurable state vector z_2 is defined as:

$$q = \pi_1,\ \dot{q} = \frac{\pi_2}{\varepsilon},\ \hat{z}_2 = \frac{\pi_2}{\varepsilon} - \alpha_1. \tag{4.57}$$

where $\varepsilon\dot{\pi}_1 = \pi_2$, $\varepsilon\dot{\pi}_2 = -\overline{\lambda}_1\pi_2 - \pi_1 + x_1$.

Update the control torque (4.38) and adaptive law (4.40),

$$\tau = -z_1 - K_2\hat{z}_2 + \hat{W}^T R(\hat{Z}), \tag{4.58}$$

$$\dot{\hat{W}}_i = -\Gamma_i(R_i(\hat{Z})_{\hat{z}_2,i} + \sigma_i \hat{W}_i), \tag{4.59}$$

Theorem 4.2 For the dynamic model (4.25), the controller (4.58) with the adaptive law (4.59), because the output information is available, for each compact set Ω_0 where $(q(0), \dot{q}(0), \hat{W}_1(0), \hat{W}_2(0), \ldots, \hat{W}_{\phi+2}(0)) \in \Omega_0$, the initial conditions are bounded, UUB of the closed-loop system trajectories are achieved. z_1, z_2 and $\tilde{W}$ will remain within Ω_{z_1}, Ω_{z_2} and Ω_W:

$$\Omega_{z_1} := \left\{ z_1 \in \mathbb{R}^{\phi+2} \mid \|z_1\| \le \sqrt{M} \right\}, \tag{4.60}$$

$$\Omega_{z_2} := \left\{ z_2 \in \mathbb{R}^{\phi+2} \mid \|z_2\| \le \sqrt{\frac{M}{\lambda_{\min}(B)}} \right\}, \tag{4.61}$$

$$\Omega_W := \left\{ \tilde{W} \in \mathbb{R}^{l\times(\phi+2)} \mid \|\tilde{W}\| \le \sqrt{\frac{M}{\lambda_{\min}(\Gamma^{-1})}} \right\}, \tag{4.62}$$

where $M = 2(V_4(0) + \frac{\beta_2}{\gamma_2})$, $\beta_2 > 0$ and $\gamma_2 > 0$.

Proof: Construct a Lyapunov function as:

$$V_4 = \frac{1}{2}z_1^T z_1 + \frac{1}{2}z_2^T B z_2 + \frac{1}{2}\sum_{i=1}^{\phi+2} \tilde{W}_i^T \Gamma_i^{-1} \tilde{W}_i. \tag{4.63}$$

The time derivation of the Eq. (4.63) with (4.58) and (4.59):

$$\begin{aligned}\dot{V}_4 \leq & -z_1^T K_1 z_1 - z_2^T(K_2 - \frac{1}{2}I)z_2 - z_2^T K_2 \tilde{z}_2 \\ & + \sum_{i=1}^{\phi+2} z_{2,i}[\hat{W}_i^T R_i(\hat{Z}) - W_i^{*T} R_i(Z)] + \frac{1}{2}\|\overline{\epsilon}\|^2 \\ & - \sum_{i=1}^{\phi+2} [\tilde{W}_i^T R_i(\hat{Z})_{\hat{z}_{2,i}} + \sigma_i \tilde{W}_i^T \hat{W}_i]. \end{aligned} \tag{4.64}$$

Applying the equation $\tilde{z}_2 = \xi_2 = -\epsilon\vartheta^{(2)}$, where ϵ is an any small positive constant, $\vartheta = \pi_2 + \overline{\lambda}_1\pi_1$, and there exist positive constants t^* and h_2 such that $\forall t > t^*$, we have $\|\xi_2\| \leq \epsilon h_2$. Hence, we can use $\frac{\pi_2}{\epsilon}$ to estimate $\dot{q}$, then $\dot{q}$ and z_2 can be estimated:

$$\hat{x}_2 = \frac{\pi_2}{\epsilon},\ \hat{z}_2 = \frac{\pi_2}{\epsilon} - \alpha_1. \tag{4.65}$$

Using the following inequalities and equations:

$$-\sigma_i \tilde{W}_i^T \hat{W}_i \leq -\frac{\sigma_i}{2}(\|\tilde{W}_i\|^2 - \|W_i^*\|^2), \tag{4.66}$$

$$\|R_i(\hat{Z})\|^2 \leq l_i, \tag{4.67}$$

$$\hat{W}_i^T R_i(\hat{Z}) = W_i^{*T}\epsilon R_{ti} + W_i^{*T} R_i(Z) + \tilde{W}_i^T R_i(Z). \tag{4.68}$$

where $l_i > 0$, $\epsilon > 0$, and R_t is a bounded vector function. ■

Substituting the following inequalities:

$$\sum_{i=1}^{n} z_{2,i} W_i^{*T}\epsilon R_{ti} \leq \frac{1}{2}z_2^T z_2 + \sum_{i=1}^{n} \frac{\|W_i^*\|^2\epsilon^2\|R_{ti}\|^2}{2}, \tag{4.69}$$

and

$$\begin{aligned} -\sum_{i=1}^{n} \tilde{W}_i^T R_i(\hat{Z})_{\tilde{z}_2,i} &= -\sum_{i=1}^{n} \frac{\sqrt{\sigma_i}\tilde{W}_i^T}{\sqrt{2}} \frac{\sqrt{2}R_i(\hat{Z}_{\tilde{z}_2,i})}{\sqrt{\sigma_i}} \\ &\leq \sum_{i=1}^{n} \frac{\sigma_i\|\tilde{W}_i\|^2}{4} + \sum_{i=1}^{n} \frac{2\|R_i(\hat{Z})\|^2}{\sigma_i} \frac{1}{2}\tilde{z}_2{}^T\tilde{z}_2. \end{aligned} \tag{4.70}$$

We can obtain:

$$\begin{aligned}\dot{V}_4 \leq & -z_1^T K_1 z_1 - z_2^T(K_2 - I)z_2 - z_2^T K_2 \tilde{z}_2 + \frac{1}{2}\|\overline{\epsilon}\|^2 \\ & - \sum_{i=1}^{\phi+2} \frac{\sigma_i}{4}\|\tilde{W}_i\|^2 + \sum_{i=1}^{\phi+2} \frac{2\|R_i(\hat{Z})\|^2}{\sigma_i} \frac{1}{2}\tilde{z}_2{}^T\tilde{z}_2 \\ & + \sum_{i=1}^{\phi+2} \left(\frac{\sigma_i}{2} + \frac{\epsilon^2\|R_{ti}\|^2}{2}\right)\|W_i^*\|^2. \end{aligned} \tag{4.71}$$

Using (4.66) and $z_2^T K_2 \tilde{z}_2 \leq \frac{1}{2}z_2^T z_2 + \frac{1}{2}(K_2\tilde{z}_2)^T(K_2\tilde{z}_2)$, we further have:

$$\begin{aligned}\dot{V}_4 \leq & -z_1^T K_1 z_1 - z_2^T\left(K_2 - \frac{3}{2}I\right)z_2 + \frac{1}{2}\|\overline{\epsilon}\|^2 \\ & - \sum_{i=1}^{\phi+2} \frac{\sigma_i}{4}\|\tilde{W}_i\|^2 + \lambda_{\max}\left(K_2^T K_2 + diag\left[\frac{2l_i}{\sigma_i}\right]\right)\frac{1}{2}\xi_2^T\xi_2 \\ & + \frac{1}{2}\sum_{i=1}^{\phi+2}(\sigma_i + \epsilon^2\|R_{ti}\|^2)\|W_i^*\|^2. \end{aligned} \tag{4.72}$$

According to the Lemma from [13], applying $\frac{1}{2}\xi_2^T\xi_2 \leq \frac{1}{2}\varepsilon^2 h_2^2$, we have:

$$\begin{aligned}
\dot{V}_4 \leq& -z_1^T K_1 z_1 - z_2^T(K_2 - \frac{3}{2}I)z_2 + \frac{1}{2}\|\bar{\epsilon}\|^2 \\
&-\sum_{i=1}^{\phi+2}\frac{\sigma_i}{4}\|\tilde{W}_i\|^2 + \lambda_{\max}(K_2^T K_2 + \mathrm{diag}[\frac{2l_i}{\sigma_i}])\frac{1}{2}\varepsilon^2 h_2^2 \\
&+\frac{1}{2}\sum_{i=1}^{\phi+2}(\sigma_i + \epsilon^2\|R_{ti}\|^2)\|W_i^*\|^2 \\
\leq& -\gamma_3 V_4 + \beta_3,
\end{aligned} \tag{4.73}$$

where γ_3 and β_3 are two constants defined as:

$$\begin{aligned}
\gamma_3 =& \min\left(2\lambda_{\min}(K_1), \frac{2\lambda_{\min}(K_2 - \frac{3}{2}I)}{\lambda_{\max}(B)},\right. \\
&\left.\min_{i=1,2\ldots\phi+2}\left(\frac{\sigma_i}{2\lambda_{\max}(\Gamma_i^{-1})}\right)\right),
\end{aligned} \tag{4.74}$$

$$\begin{aligned}
\beta_3 =& \frac{1}{2}\sum_{i=1}^{\phi+2}(\epsilon^2\|R_{ti}\|^2 + \sigma_i)\|W_i^*\|^2 + \frac{1}{2}\|\bar{\epsilon}\|^2 \\
&+\lambda_{\max}\left(K_2^T K_2 + \mathrm{diag}\left[\frac{2l_i}{\sigma_i}\right]\right)\frac{1}{2}\varepsilon^2 h_2^2.
\end{aligned} \tag{4.75}$$

To guarantee $\gamma_3 > 0$, K_1 and K_2 are required to satisfy:

$$\lambda_{\min}(K_1) > 0, \lambda_{\min}(K_2 - \frac{3}{2}I) > 0. \tag{4.76}$$

Remark 4.3 Under the hypothesis that the position measurements are perfect, by employing a high-gain observer, we designed a rigorous theoretical treatment of the output feedback problem. The tracking error converges and remains within a small neighborhood of zero. If the residual error is lower than expected, we can reduce the residual error by decreasing $\frac{\beta}{\gamma}$ or increasing control gains K_1, K_2, the high gain $\frac{1}{\varepsilon}$ and the approximation accuracy of the neural networks.

4.4 Numerical Simulations

Simulations of open-loop system will be shown in Section 4.4.1; a PD controller will be introduced in Section 4.4.2; neural network control will be revealed in Section 4.4.3 and the comparison will given in Section 4.4.4. The system parameters are specified as Table 4.1.

4.4.1 Simulation Results for Open-Loop System

In this part, time domain responses of the discrete model are observed through MATLAB simulations. Give the system a small interference:

$$\tau_1(t) = \tau_2(t) = \begin{cases} 0.5\ \mathrm{N\ m}, & 0\,\mathrm{s} \leq t \leq 0.5\,\mathrm{s} \\ -0.5\ \mathrm{N\ m}, & 0.5\,\mathrm{s} \leq t \leq 1\,\mathrm{s} \\ 0\ \ \mathrm{N\ m}, & \text{otherwise} \end{cases} \tag{4.77}$$

Table 4.1 Properties of the two-link flexible robotic manipulator.

Para.	Description	Value	Unit
M_1	Mass of link 1	0.49	kg
M_2	Mass of link 2	0.42	kg
m_{h2}	Mass of drive joint 2	1	kg
m_{t2}	Tip mass of link 2	0.157	kg
L_1	Length of link 1	0.35	m
L_2	Length of link 2	0.30	m
I_{hi}	Inertia of hub 1 and 2	0.1	kg m^2
I_{oi}	Inertia of link 1 and 2	0.0083	kg m^2
I_{t2}	Tip inertia of link 2	0.0005	kg m^2
EI_i	Bending stiffness of link 1 and 2	1	N m^2

The reference inputs of the first and the second link were set to $\theta_{1d} = 0.5$ rad, $\theta_{2d} = 0.25$ rad, respectively. The tracking trajectory θ_i of the first and second link for open-loop system is shown in Figure 4.4. We can see that the outputs of the system are divergent.

4.4.2 Simulation Results for PD Control

We designed two vibration control structures to reduce the vibration of the two-link flexible manipulator model. At first, we applied two separate PD controllers to achieve the tip tracking control performance of the two flexible links. Select the initial states as zero. Design the PD controller:

$$\tau_i = -K_p(\theta_i - \theta_{id}) - K_d\dot{\theta}_i, \ i = 1, 2. \tag{4.78}$$

where K_p and K_d are proportional gain and differential gain.

In the simulations model, the reference inputs of the first and the second link were set to $\theta_{1d} = 0.5$ rad, $\theta_{2d} = 0.25$ rad, respectively. We introduce a PD controller with $K_p = 5$, $K_d = 1.5$. The tracking trajectory θ_i and the control torque τ_i of the first and second link are shown in Figure 4.5. We can conclude that the angular displacement signal θ_i is stable at 3 s, but there is a large overshoot within the range of −0.2 to 0.2 rad in the tracking process. Tip position in 3-D of the two links is shown in Figure 4.6. There is a large vibration of the first link within the range of −0.05 to 0.05 m in the tracking process.

4.4.3 Simulation Results for Neural Network Control

We choose the radial basis function (RBF) neural network system parameters as: the center parameters are selected as 1 or −1, 2^8 nodes are employed for each $S_i(Z_i)$. Variance of centers $\eta = 2$ as well as the initial weights $\hat{W}_i = 0$ $(i = 1, 2, \dots, 256)$, $\sigma = 0.0002$, $\Gamma = 100I_{256\times256}$. We introduce a neural network controller with $K_1 = 2I_{6\times6}, K_2 = 8I_{6\times6}$. The tracking trajectory θ_i and the control torque τ_i of the first and second link are shown in Figure 4.7. Tip position in 3-D of the two links is shown in Figure 4.8. We can conclude that the angular displacement signal θ_i is stable at 1 s, and the overshoot and vibration are removed in the tracking process.

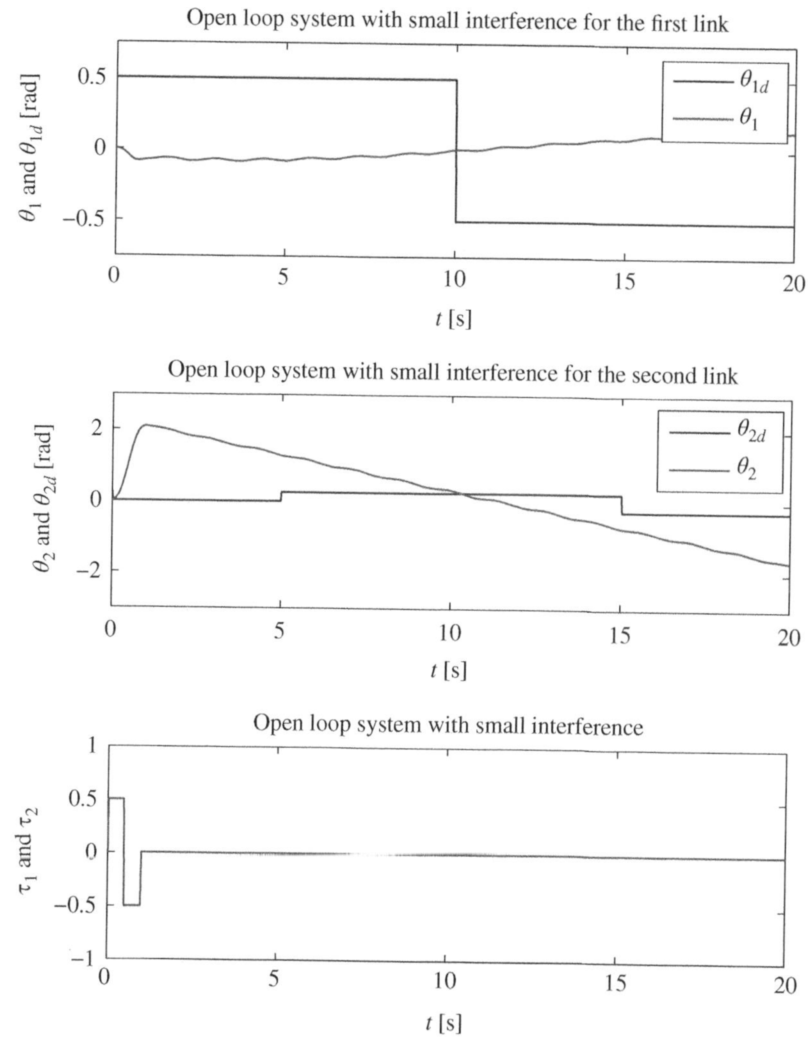

Figure 4.4 Tracking trajectory and control torque for open-loop system.

4.4.4 Comparison Between PD and NN Simulation Results

Notice that between the start of the motion of the second link and the stability of the second link in Figures 4.5 and 4.7, there is a visible vibration in the first link. The vibration is caused by the inertia of the first link when the second link stops its motion. Under the PD controller, this vibration ranges from −0.1 to 0.1 rad, which cannot be neglected in practical engineering. In addition, there exists an overshoot within the range of −0.2 to 0.2 rad for the first link in tracking the desired trajectory process. These vibrations will bring irreparable consequences to practical projects. Therefore, the NN controller is employed in the two-link flexible manipulator. Seen from the simulation results, we can conclude that each of the two flexible links is able to track the desired angular position trajectories in the form of square waves. However, compared with the PD controller, under the NN controller, the flexible manipulator reaches the desired position in a shorter time. Furthermore, the NN controller works better in vibration suppression and the overshoot is removed in the tracking process.

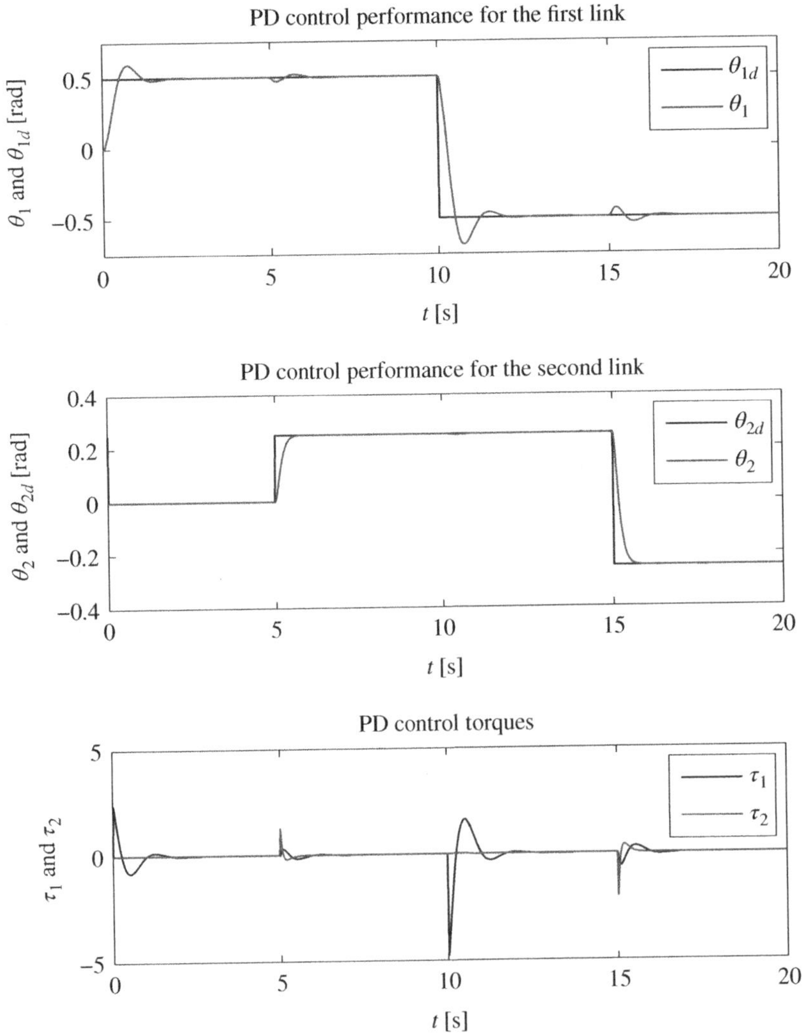

Figure 4.5 Tracking trajectory and control torque for PD control.

Tip position of first link with PD control/tip position os second link with PD control

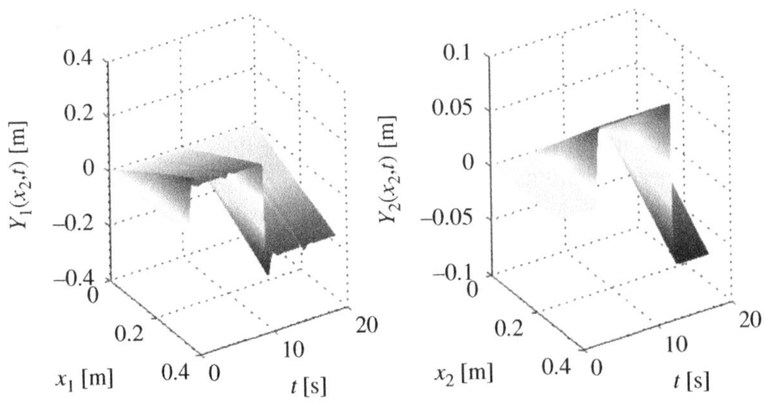

Figure 4.6 Tip position for PD control.

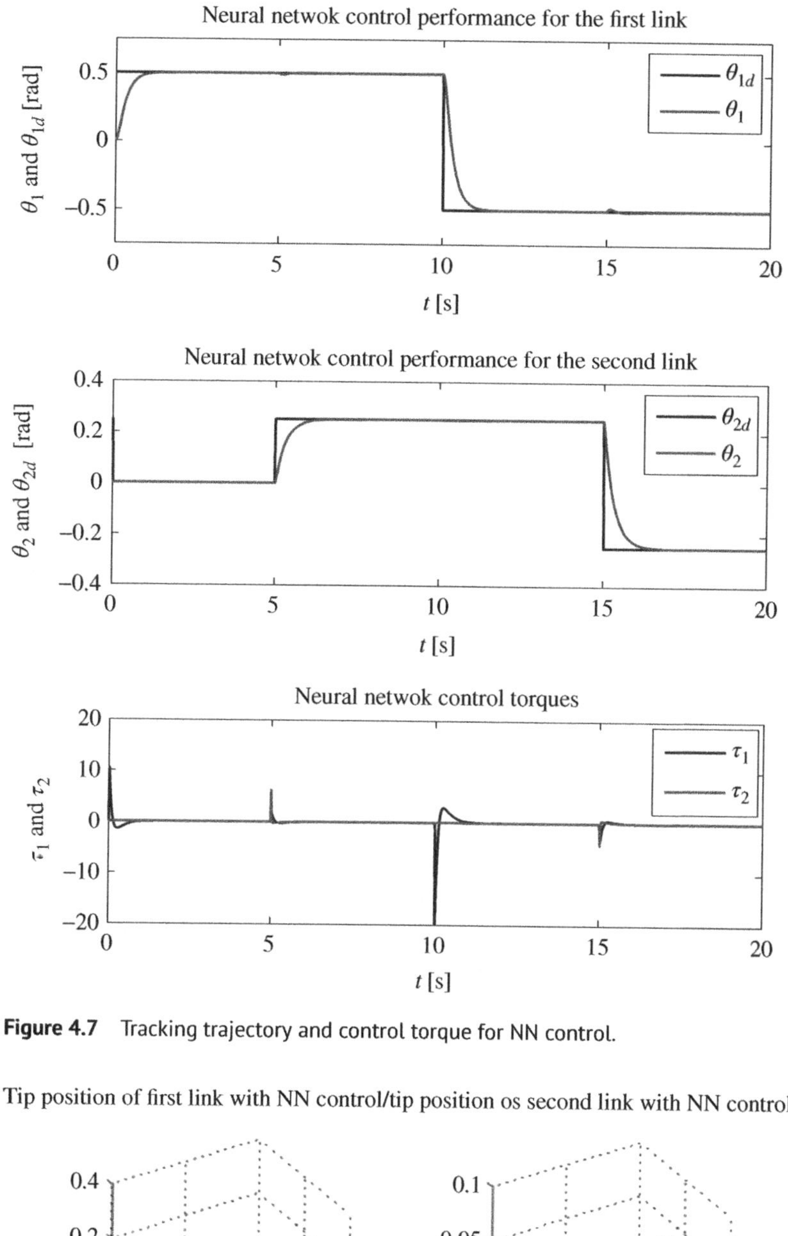

Figure 4.7 Tracking trajectory and control torque for NN control.

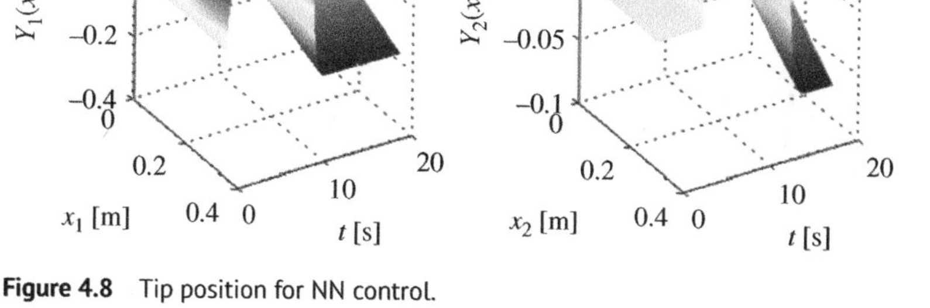

Figure 4.8 Tip position for NN control.

4.5 Experimental Investigation

4.5.1 Introduction of the Experimental Testbed

In order to verify the feasibility of the control strategy in engineering practice, the simulation experiment is implemented in the Quanser Flexible Link System. From Figure 4.9, the flexible rotating manipulator consists of double motors driven gapless harmonic gear reducer and two series connected flexible links. The first stage flexible link is connected to the first motor and equipped with a strain sensor to measure the end displacement. In addition, the end is connected to the second motor. The second flexible link with strain sensor is connected to the second motor. The displacement of the two motors is measured by an incremental photoelectric encoder.

Figure 4.10 shows working principle diagram of flexible experiment platform. The two DOF serial flexible ruler rotates driven by DC motor (servo device). In the process of rotation, the end deflection and the angular displacement signal are measured by strain sensor and optical encoder (4096 times per turn in orthogonal mode), respectively. The derivatives of the angular displacement signal and the end deflection are obtained by a derivative filter. The control algorithm is designed by the Simulink module in PC. Then, the control signal is sent to the SRV02 servo device through Q8-usb data acquisition device and voltPAQ-X1 amplifier. The relevant parameters of this planform are tabulated in Table 4.2.

The purpose of the experiment is to design the controller to track the robot position and minimize the end vibration in the horizontal plane. The planar tracking or position adjusting of the end member is obtained by the motion of a series mechanism with two flexible joints. For the MIMO system, the decoupling controller is used to suppress the vibration due to the flexibility and the coupling effect between the two links.

4.5.2 Experimental Results

In this section, the PD controller and NN controller designed by MATLAB Simulink models and implemented on the Quanser Two-Link Flexible Plant are presented. In order to compare the control performance adequately and intuitively, the superimposed curves of the two controllers are given.

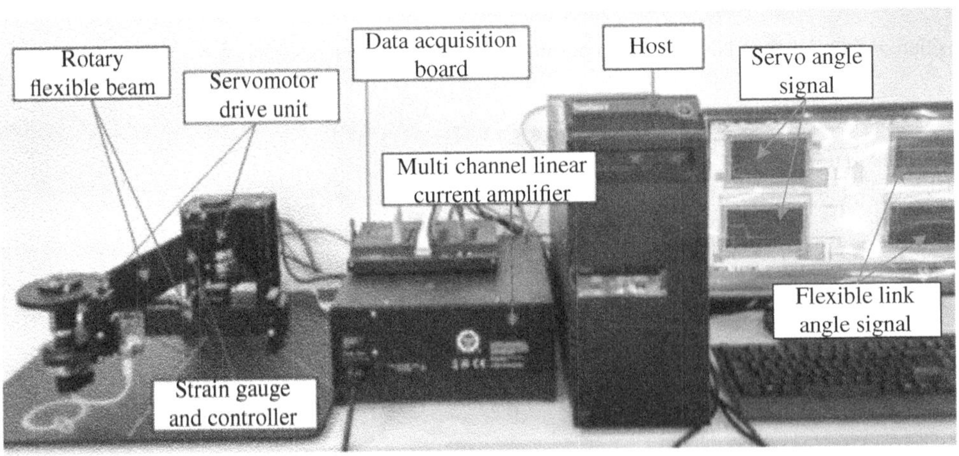

Figure 4.9 Control framework of the experimental flexible link.

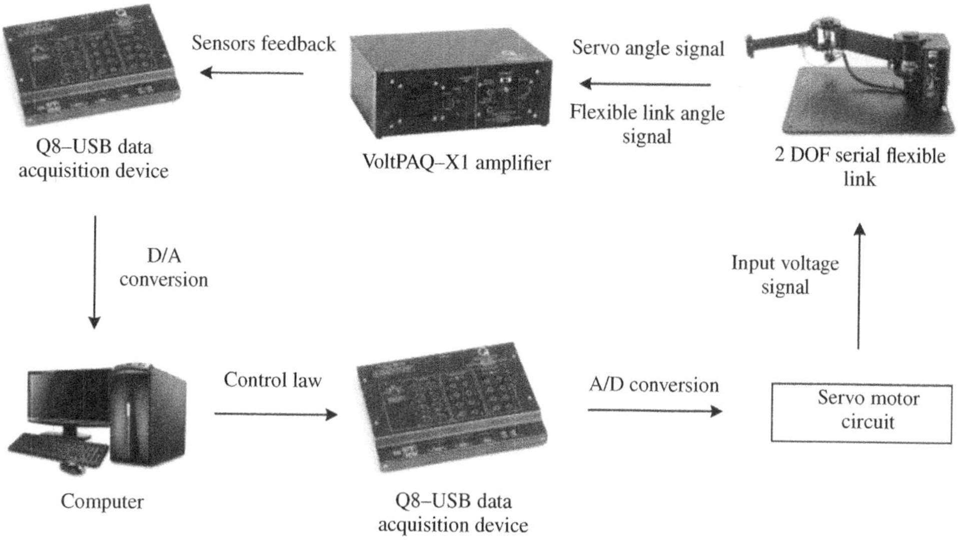

Figure 4.10 Diagram of the experimental flexible link.

Table 4.2 Parameters of the Quanser test platform.

Para.	Description	Value	Unit
M_1	Mass of link 1	0.49	kg
M_2	Mass of link 2	0.42	kg
m_{h2}	Mass of drive joint 2	1.00	kg
m_{t2}	Tip mass of link 2	0.157	kg
L_1	Length of link 1	0.35	m
L_2	Length of link 2	0.30	m
I_{h1}	Inertia of hub 1	0.0063	kg m^2
I_{h2}	Inertia of hub 2	0.0026	kg m^2
I_{o1}	Inertia of link 2	0.02	kg m^2
I_{o2}	Inertia of link 2	0.013	kg m^2
I_{t2}	Tip inertia of link 2	0.0064	kg m^2

A decoupled model is employed in the control system, which is to say that link coupling of the serial mechanism is neglected. In the experiment, we chose square waves as input signals. Through proper amplitude, degrees to radians, and time delay, the square waves are converted to analog outputs. After power amplification, the outputs are applied to SRV02 to drive the flexible manipulator. Both drives are commanded independently of each other, each using a separate control loop. The rotation angle θ and the deflection α measured by the encoder and strain gauge sensor can be observed in real-time plot. The derivatives of the angular displacement signal and the end deflection $\dot{\theta}$, $\dot{\alpha}$, are obtained by a derivative filter.

We designed two vibration control structures to reduce the vibration of the two-link flexible manipulator. At first, we applied two separate PD controllers to achieve the control performance.

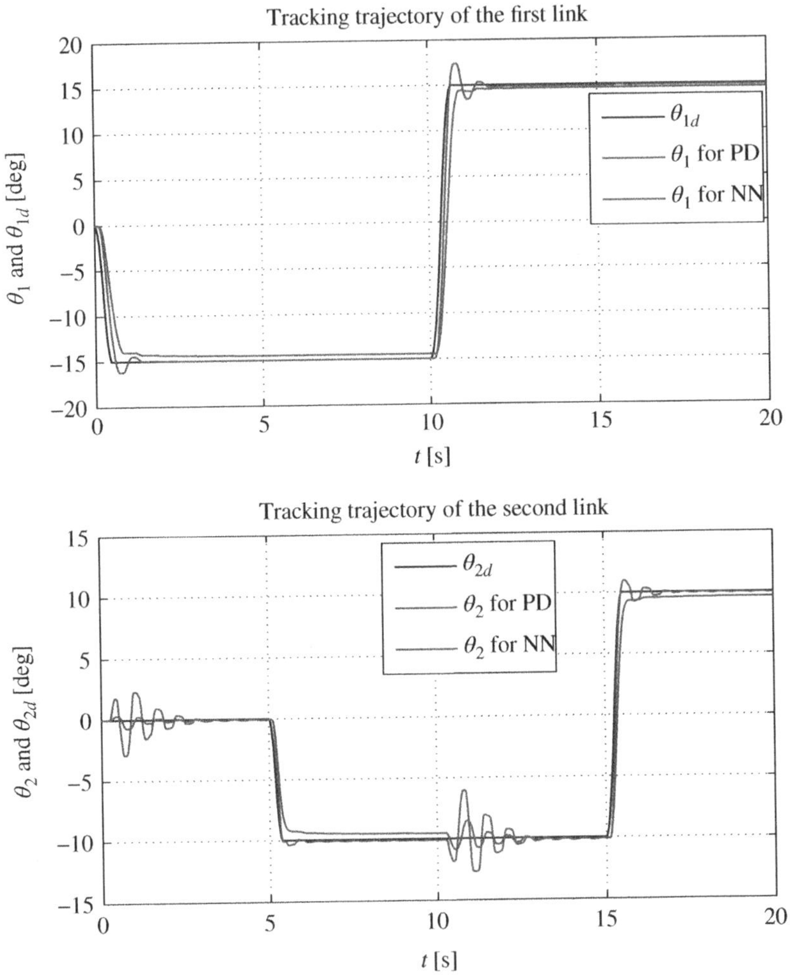

Figure 4.11 Tracking trajectory with PD and NN control.

In the Simulink model, the reference inputs of the first and the second link were set to $\frac{15\pi}{180}$ and $\frac{10\pi}{180}$, respectively. The control parameters are selected as $K_p = 5$, $K_d = 0.01$ in the first link, and $K_p = 8$, $K_d = 0.01$ in the second link. The servo angle θ_i, the end deflection α_i and the control inputs τ_i are revealed in Figures 4.11, 4.12, and 4.13, respectively. Seen from the real-time plot, each of the two flexible links is able to track the desired angular position trajectories in the form of square waves. However, the end-effect (α_1 and α_2) of the flexible manipulator using PD controller remains 0.6°. And the links reach the desired position in 2 s, with a large overshoot whose max amplitude is 4°.

In order to eliminate the flexible vibration, we designed the NN controller in the control system. The inputs of the RBF network contain the tracking errors of two links' angle position, the vibration of the tip payload, and their derivatives. Given the same inputs as the PD models, we choose the RBF neural network parameters as: the center parameters are selected as 1 or −1, 2^8 nodes are employed for each $S_i(Z_i)$. Variance of centers $\eta = 2$ as well as the initial weights $\hat{W}_i = 0$ $(i = 1, 2, \dots, 256)$. $\sigma = 0.0002$, $\Gamma = 100 I_{256\times256}$. We selected the control parameters as $K_1 = 10$, $K_2 = 0.3$ in the first link,

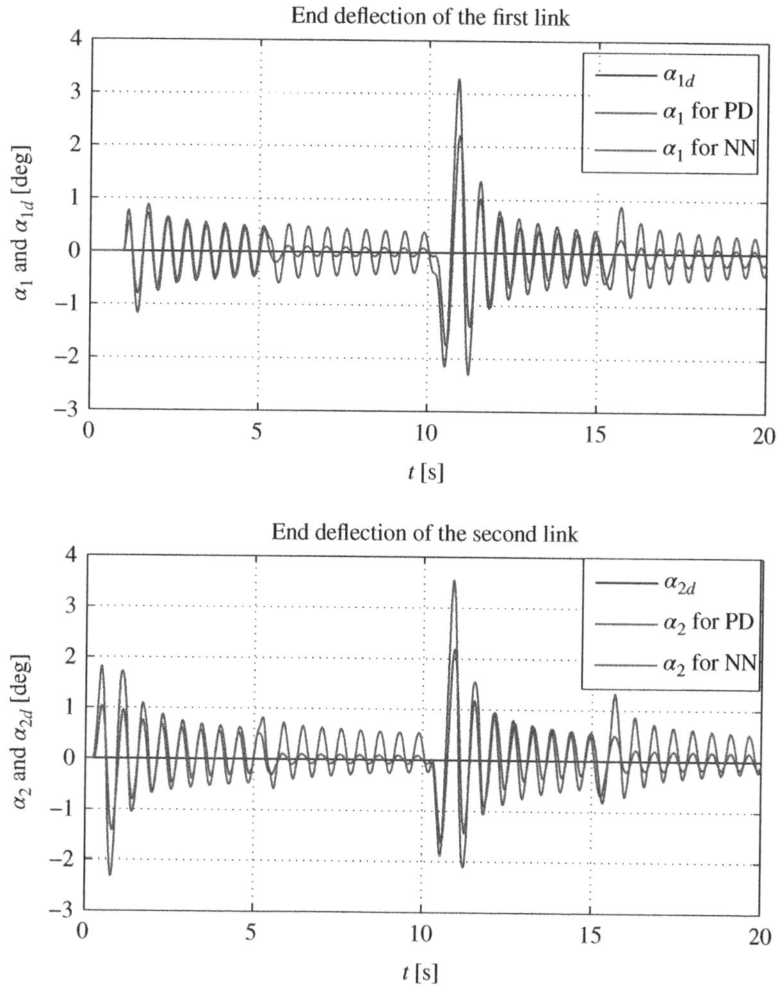

Figure 4.12 Elastic vibration with PD and NN control.

and $K_1 = 12$, $K_2 = 0.7$ in the second link. The servo angle θ_i, the end deflection α_i, and the control inputs τ_i are revealed in Figures 4.11, 4.12, and 4.13, respectively. Seen from the real-time plot, the links reach the desired position in 1 s and the large overshoot is reduced to 1°. More importantly, the elastic vibration (α_1 and α_2) are decreased from 0.6° to 0.05°.

4.5.3 Comparison Between PD and NN Experiment Results

Notice that between the start of the motion of the first link and the stability of the first link in Figure 4.11, there is a vibration ranges from −4° to 4° in the second link and a vibration ranges from −2° to 2° in the first link under PD controller. The vibration is caused by the inertia of the second link when the first link stops its motion, and the magnitude of the vibration is much bigger than that of other vibrations in tracking process. In practice, it will result in serious consequences. Therefore, we would design a more effective control strategy like neural network control to solve the inertial issue in this work. The tracking performance of NN controller is shown in Figure 4.11. Compared with the PD controller, the vibration in the second link is reduced within the range of −1°

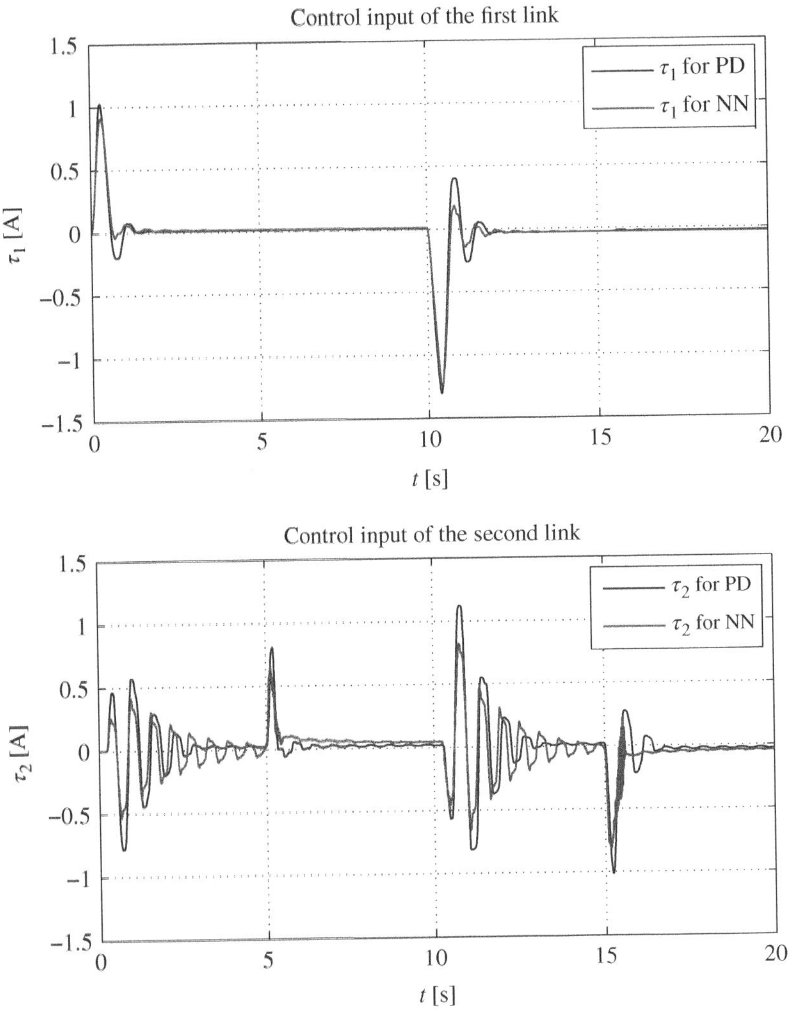

Figure 4.13 Control torques with PD and NN control.

to 1°, and the vibration in the first link is reduced within the range of −0.5° to 0.5° in the tracking process. Furthermore, under the NN controller, the flexible manipulator reaches the stable state in a shorter time. More importantly, as seen form Figure 4.12, the end-effect (α_1 and α_2) of the flexible is decreased from 0.6° to 0.05° which is to say that the neural network controller in the model works better in vibration suppression.

4.6 Summary

In this chapter, the n-dimensional discretized model of the two-link flexible manipulator is achieved via the AMM which realizes the conversion from PDE to ODE and facilitates the analysis of dynamic uncertainty and the design of controller. Subsequently, both full-state and output feedback NN controller are investigated to track the desired trajectory and suppress the flexible vibration. UUB of the closed-loop system is realized by the Lyapunov's stability. Furthermore, extensive simulations and experiments on Quanser platform are implemented to demonstrate the feasibility of the proposed NN controller, and the final performance has been compared with that of PD control strategy. In the future, we will further study more efficient modeling and control techniques to expand the applicability of the proposed method in practice and solve the vibration problems of various flexible structures.

5

Robust Adaptive Vibration Control for a String with Time-Varying Output Constraint

After reading this chapter, you should be able to:

- Understand present the vibration control design for a string with the boundary time-varying output constraint
- Adopt the barrier Lyapunov function with a logarithmic function to prevent the time-varying constraint violations
- Design adaptive control to handle the system parametric uncertainties

5.1 Introduction

The model of the string is widely used to describe the dynamics of flexible systems, such as crane cables [108], marine mooring lines [213], moving strips [42, 332], and some flexible structures used in manipulators and robots [283]. This kind of systems has a number of advantages such as lightweight, efficiency, higher operation speed, and lower cost. However, they may vibrate excessively due to the environmental disturbances. The vibration influences the system performance and even produces premature fatigue problems. Therefore, it is necessary to find an effective control method to suppress the vibration of the flexible systems. What's more, in various practice applications, the tip payload of these flexible structures is restricted and the constraint may vary with the time going on. Thus, avoiding the vibration amplitude exceeding the time-varying constraint should be also taken into account.

The dynamics of the string is essentially a distributed parameter system (DPS), having an infinite-dimensional state space and modeled by the coupled partial differential equations (PDEs) and ordinary differential equations (ODEs). These make it difficult to design control. Approaches including finite-element method, Galerkin's method, and assumed modes method based on truncated finite-dimensional models [14, 86, 285, 289, 308] are provided to solve the control problem of the DPS with finite sensors and actuators. However, the truncated models are approximated by neglecting high-frequency modes, leading to spillover potentially. And the spillover can make the system unstable [208]. In [7], the authors uses a technique – scabbard-like actuator to suppress the vibration of a string to suppress a strings vibration by repeating application and removal of a constraint at one point close to one of its boundaries. It is noted that it is a passive vibration control method. Differing from this method, an active vibration control is developed

Dynamic Modeling and Neural Network-Based Intelligent Control of Flexible Systems,
First Edition. Hejia Gao, Wei He, and Changyin Sun.

to study the vibration problem and constraint problem. Driven by practical needs and with the development of theory, several advanced methods such as sliding mode control [290], feedback control [150], iterative learning control [119], and parallel control [258, 294, 295] are proposed to deal with control problems of distributed parameter systems. The literatures [18, 95, 328] pay great attention to the flexible systems and vibration suppression of the systems. In [69, 98], controllers are designed to deal with finite-time and exponential stabilization for one-dimensional wave equations. Compared with distributed control, boundary control including [96] has many advantages for flexible systems' vibration suppression. It is not practical for the distributed control to be implemented. When the point actuators and sensors are located at nodal points, the distributed system will become uncontrollable and unobservable. However, for boundary control, the dynamic model can be derived based on the dynamics of the system [105, 221] with the sensors and actuators attached at the boundary. In [307], a static output feedback controller via PDE-ODE boundary with ODE measurements utilizing boundary control method is designed, resulting in the closed-loop cascaded system is exponentially stable. For this reason, boundary control has been applied in a number of research fields [44, 125, 222, 311]. In this chapter, we choose boundary control to suppress the vibration of the string due to its advantages.

In practice applications, the tip payload of flexible structures is usually restricted and the output constraint may vary with the time going on. Thus, avoiding the vibration amplitude exceeding the time-varying constraint should be considered. Recently several works have been done to meet such requirements. A high-gain disturbance observer based backstepping control with position tracking error constraint for electro-hydraulic systems is proposed in [303]. A control method for strict feedback nonlinear systems with time-varying output constraints is presented in [272], tracking the desired trajectory. However, these articles do not mention how to control a system described by PDEs, for example, the flexible structures. In [310], a distributed H_∞ fuzzy controller with constraint is designed for a class of nonlinear spatially distributed processes modeled by first-order hyperbolic PDEs. In [109, 123, 239, 271], controllers are designed for flexible structures without violating the symmetrical constant output constraints. And asymmetrical output constraint is taken into account in [209]. However, the output constraints are constant in these works. Adaptive control method [134, 292, 293, 330] is an effective method to cope with parameter uncertainties. Robust control [291] can persist the performance characteristics of the system under certain parametric perturbations. To deal with the asymmetrical time-varying output constraint of flexible system modeled by second-order hyperbolic PDE, in this chapter, we design a controller based on the exact model and a robust adaptive controller coping with the uncertain parameters and analyze the stability via Lyapunov method.

It is well known that Lyapunov function can provide a method for control design and stability analysis, as evidenced by many excellent works [173, 193, 209]. Adaptive controller is constructed for ODE system in [173] utilizing Lyapunov function. In [193], barrier Lyapunov functions are introduced in Nussbaum gain adaptive control for full-state constrained nonlinear systems. In [219], three control inputs are designed to suppresses both longitudinal and transverse vibrations and regulates the transport velocity of the string using Lyapunov method without taking output constraint into consideration. In [209], a Lyapunov function coupled with a piecewise function and a tangent function is introduced in the control design to prevent asymmetrical output constraint violation. However, it does not provide the value range of the controller parameter which can make the parameter selection convenient. In this chapter, we provide the ranges of the controller parameters and the analyze how to select the parameters. The main contributions of this chapter are given as follows:

1) To deal with the parameter uncertainties of the string system, the robust adaptive controller is proposed followed by the exact model-based control. The numerical simulation results verify the effectiveness of both control strategies.
2) An asymmetrical barrier Lyapunov function is employed to prevent the asymmetrical time-varying output constraint violations of the string modeled as a second-order PDE system. The function $q(\cdot)$ to prevent the time-varying constraint violation of different directions. The Lyapunov function $V(t)$ is continuous and differentiable for the displacement of the payload $w(L, t)$ at zero which is necessary to analyze the stability of the closed-loop system.
3) The solvability of the inequality equations is provided, giving the ranges of the parameters where the controller can be found. It can direct us to select parameters of the controller better and avoid a number of useless attempts.

In this chapter, the vibration control design for the string with the asymmetrical time-varying output constraint is developed. In Section 2, dynamics of the string and some lemmas are provided for the subsequent development. In Section 3, both the model-based and adaptive control schemes are proposed based on the barrier Lyapunov function and the uniform boundedness and the uniformly ultimate boundedness of the system are proved with the solvability of the inequality equations provided in Section 4. In Section 5, simulations are provided to illustrate the effectiveness of the proposed control. And the conclusion is provided in Section 6.

5.2 Problem Formulation

5.2.1 Dynamics of the String System

The string (see Figure 5.1) attached with a payload on the bottom and the other end fixed is affected by the unknown time-varying disturbance $d(t)$ on the tip payload. L, M_s, ρ, T, $w(L, t)$, $w(x, t)$, and $u(t)$ represent the length of the string, the mass of the tip payload, uniform mass per unit length of the string, the constant tension of the string, the displacement of the tip payload, the displacement

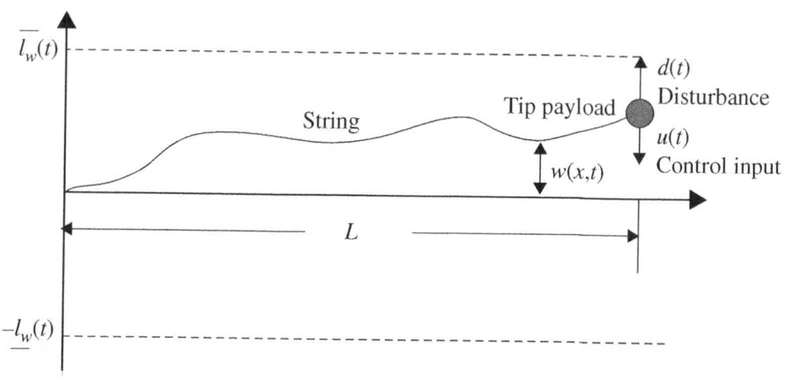

Figure 5.1 The model of the string.

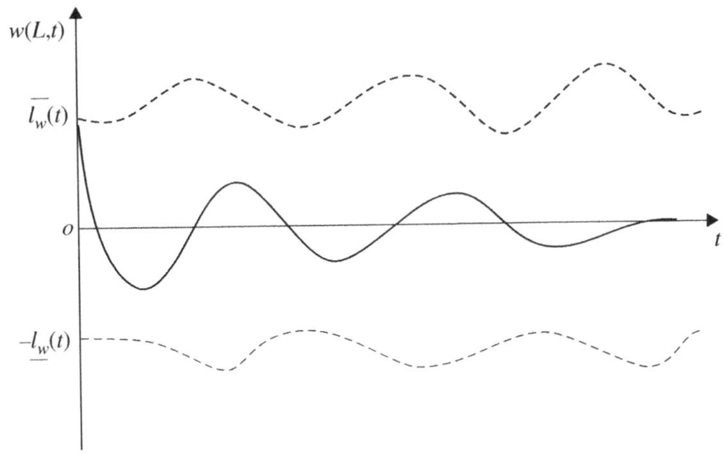

Figure 5.2 Time-varying output constraints on the payload.

of the string at the position x for time t, and the control input, respectively. $\overline{l_w}(t)$ and $\underline{l_w}(t)$ denote the time-varying constraints (see Figure 5.2).

We can obtain the string model from [109]:

$$\rho\ddot{w}(x,t) - Tw''(x,t) = 0. \tag{5.1}$$

$$w(0,t) = 0. \tag{5.2}$$

$$Tw'(L,t) + M_s\ddot{w}(L,t) = l \quad d(t) + u(t). \tag{5.3}$$

Remark 5.1 The time-varying constraint limits the position of the tip payload, i.e., the boundary displacement $w(L,t)$ should be in the scope described by the following equation:

$$\begin{cases} \overline{l_w}(t), & \overline{l_w}(t) > 0 \\ -\underline{l_w}(t), & \underline{l_w}(t) > 0, \end{cases} \tag{5.4}$$

where $\overline{l_w}(t)$ and $\underline{l_w}(t)$ are two different and differentiable functions. It can be seen from the equation that the output constraint is asymmetrical and time-varying.

For clarity, notations $(\cdot)' = \partial(\cdot)/\partial x$ and $(\dot{\cdot}) = \partial(\cdot)/\partial t$ are used throughout this chapter.

Assumption 5.1 For the unknown boundary disturbance $d(t)$, we assume that there exists constant $\bar{d} \in R^+$, such that $|d(t)| \le \bar{d}, \forall t \in [0,\infty]$. This is a reasonable assumption as the time-varying $d(t)$ has finite energy and thus are bounded, i.e., $d(t) \in L_\infty$.

Assumption 5.2 For the time-varying constraints $\overline{l_w}(t)$ and $\underline{l_w}(t)$, we assume that there exist constants l and c, such that $\max(\overline{l_w}(t), \underline{l_w}(t)) \le l$, $\overline{l_w}(t) > 0$, $\max(\dot{\overline{l_w}}(t), \dot{\underline{l_w}}(t)) \le c$ and $\underline{l_w}(t) > 0$. This is a reasonable assumption since the constraints have the finite energies. i.e., $\overline{l_w}(t) \in L_\infty$, $\underline{l_w}(t) \in L_\infty$, $\dot{\overline{l_w}}(t) \in L_\infty$, and $\dot{\underline{l_w}}(t) \in L_\infty$.

5.2.2 Preliminaries

Lemma 5.1 [52, 101] Let $\Phi(x,t) \in R$ be a function defined on $x \in [0, L]$ and $t \in [0, \infty)$ that satisfies the boundary condition $\Phi(0,t) = 0, \forall t \in [0, \infty)$ then the following inequalities hold:

$$\Phi^2 \leq L \int_0^L [\Phi']^2 dx, \ \forall (x,t) \in [0, L] \times [0, \infty). \tag{5.5}$$

Property 1: [52] If the kinetic energy of the system (5.7)–(5.9) given by (5.1) is bounded $\forall t \in [0, \infty)$, then $\dot{w}(x,t)$, $\dot{w}'(x,t)$ and $\dot{w}''(x,t)$ are bounded $\forall (x,t) \in [0, L] \times [0, \infty)$.

Property 2: [52] If the potential energy of the system (5.7)–(5.9) given by (5.2) is bounded $\forall t \in [0, \infty)$, then $w'(x,t)$ and $w''(x,t)$ are bounded $\forall (x,t) \in [0, L] \times [0, \infty)$.

5.3 Control Design

The control objectives are: (1) to guarantee the stability of the string system without violating of the time-varying boundary constraint, i.e., to guarantee the boundary output signal $-\underline{l_w}(t) \leq w(L,t) \leq \overline{l_w}(t)$ and (2) designing adaptive law to achieve the same objective as 1) when the payload M_s and the tension T are uncertain.

5.3.1 Exact Model-Based Boundary Control

Choose the Lyapunov function as

$$V(t) = V_1(t) + V_2(t) + V_3(t). \tag{5.6}$$

where

$$V_1(t) = \frac{\rho}{2} \int_0^L [\dot{w}(x,t)]^2 dx + \frac{T}{2} \int_0^L [w'(x,t)]^2 dx, \tag{5.7}$$

$$\begin{aligned} V_2(t)l = {} & \frac{q[w(L,t)]}{2} M_s [\dot{w}(L,t) + w'(L,t)]^2 \ln \frac{2[\overline{l_w}(t)]^2}{[\overline{l_w}(t)]^2 - w^2(L,t)}, \\ & + \frac{1 - q[w(L,t)]}{2} M_s [\dot{w}(L,t) + w'(L,t)]^2 \ln \frac{2[\underline{l_w}(t)]^2}{[\underline{l_w}(t)]^2 - w^2(L,t)}, \end{aligned} \tag{5.8}$$

$$V_3(t) = \beta \rho \int_0^L x \dot{w}(x,t) w'(x,t) dx, \tag{5.9}$$

the function $q(\bullet)$ is defined as [272]:

$$q(\bullet) = \begin{cases} 1, & \bullet > 0 \\ 0, & \bullet \leq 0, \end{cases} \tag{5.10}$$

and the constant β is positive.

Lemma 5.2 The Lyapunov function $V(t)$ is continuous and differentiable for $w(L,t) \in [-\underline{l_w}(t), \overline{l_w}(t)]$.

Proof: $V_1(t)$ and $V_3(t)$ is continuous and differentiable for $w(L,t) \in [-\underline{l_w}(t), \overline{l_w}(t)]$. And $V_2(t)$ is continuous and differentiable for $w(L,t) \in [-\underline{l_w}(t), 0) \bigcap (0, \overline{l_w}(t)]$. When $w(L,t)$ tends to zero, the following equations can be derived:

$$\lim_{w(L,t)\to 0^+} V_2(t) = \frac{\ln 2}{2} M_s[\dot{w}(L,t) + w'(L,t)]^2,$$

$$\lim_{w(L,t)\to 0^-} V_2(t) = \frac{\ln 2}{2} M_s[\dot{w}(L,t) + w'(L,t)]^2,$$

$$\lim_{w(L,t)\to 0^+} \dot{V}_2(t) = M_s[\dot{w}(L,t) + w'(L,t)] \ln 2[\ddot{w}(L,t) + \dot{w}'(L,t)],$$

$$\lim_{w(L,t)\to 0^-} \dot{V}_2(t) = M_s[\dot{w}(L,t) + w'(L,t)] \ln 2[\ddot{w}(L,t) + \dot{w}'(L,t)],$$

We can obtain that $\lim_{w(L,t)\to 0^+} V_2(t) = \lim_{w(L,t)\to 0^-} V_2(t)$, $\lim_{w(L,t)\to 0^+} \dot{V}_2(t) = \lim_{w(L,t)\to 0^-} \dot{V}_2(t)$. Thus $V_2(t)$ is continuous and differentiable for $w(L,t) \in [-\underline{l_w}(t), \overline{l_w}(t)]$. And the Lyapunov function $V(t)$ is continuous and differentiable for $w(L,t) \in [-\underline{l_w}(t), \overline{l_w}(t)]$. ■

The controller is designed as follows:

$$\begin{aligned}
u(t) = {} & -q[w(L,t)] \Bigg\{ k_1[\dot{w}(L,t) + w'(L,t)] + M_s[\dot{w}(L,t) + w'(L,t)] \\
& \times \frac{\overline{l_w}^2(t) w(L,t)\dot{w}(L,t) - \overline{l_w}(t)\dot{\overline{l_w}}(t) w^2(L,t)}{\overline{l_w}^2(t)[\overline{l_w}^2(t) - w^2(L,t)]} \Bigg\} \Bigg/ \ln \frac{2\overline{l_w}^2(t)}{\overline{l_w}^2(t) - w^2(L,t)} \\
& - \Bigg\{ 1 - q[w(L,t)]\} \{ k_1[\dot{w}(L,t) + w'(L,t)] + M_s[\dot{w}(L,t) + w'(L,t)] \\
& \times \frac{\underline{l_w}^2(t) w(L,t)\dot{w}(L,t) - \underline{l_w}(t)\dot{\underline{l_w}}(t) w^2(L,t)}{\underline{l_w}^2(t)[\underline{l_w}^2(t) - w^2(L,t)]} \Bigg\} \Bigg/ \ln \frac{2\underline{l_w}^2(t)}{\underline{l_w}^2(t) - w^2(L,t)} \\
& - k_2[\dot{w}(L,t) + w'(L,t)] - \mathrm{sgn}[\dot{w}(L,t) + w'(L,t)]\overline{d} \\
& + Tw'(L,t) - M_s\dot{w}'(L,t),
\end{aligned} \tag{5.11}$$

where $k_1 > 0$ and $k_2 > 0$ are control gains.

Theorem 5.1 For the system described by governing equation (5.1) and boundary conditions (5.2) and (5.3), under Assumption 5.1 and Assumption 5.2, and the boundary control (5.11), given that the initial conditions are bounded, we can conclude that the state of the closed-loop system $w(x,t)$ will converge to zero without violating the boundary constraint with $(x,t) \in [0,L] \times [0,\infty)$.

Proof: From (5.9), we can obtain

$$V_3(t) \le \beta\rho L \int_0^L [\dot{w}(x,t)]^2 dx + \beta\rho L \int_0^L [w'(x,t)]^2 dx \le \beta_1 V_1(t), \tag{5.12}$$

where

$$\beta_1 = \frac{2\beta L}{T} \max(T, \rho). \tag{5.13}$$

■

Let $\beta_2 = 1 - \beta_1 > 0$ and $\beta_3 = 1 + \beta_1 > 1$, we have

$$\beta_2 V_1(t) \leq V_1(t) + V_3(t) \leq \beta_3 V_1(t). \tag{5.14}$$

We further have

$$0 \leq \lambda_1[V_1(t) + V_2(t)] \leq V(t) \leq \lambda_2[V_1(t) + V_2(t)], \tag{5.15}$$

where $\lambda_1 = \min(\beta_2, 1)$ and $\lambda_2 = \max(\beta_3, 1)$.

From (5.6), we can obtain

$$\dot{V}(t) = \dot{V}_1(t) + \dot{V}_2(t) + \dot{V}_3(t). \tag{5.16}$$

Substituting the governing equation and the boundary conditions (5.1)–(5.3), we can obtain $\dot{V}_1(t)$ as

$$\dot{V}_1(t) = \frac{T}{2}[w'(L,t) + \dot{w}(L,t)]^2 - \frac{T}{2}[w'(L,t)]^2 - \frac{T}{2}[\dot{w}(L,t)]^2. \tag{5.17}$$

Differentiating $V_2(t)$ and substituting the control force (5.11), we can obtain

$$\begin{aligned}
\dot{V}_2(t) \leq & -q[w(L,t)]k_1[w'(L,t) + \dot{w}(L,t)]^2 \\
& -q[w(L,t)]k_2[w'(L,t) + \dot{w}(L,t)]^2 \ln \frac{2[\overline{l}_w(t)]^2}{[\overline{l}_w(t)]^2 - w^2(L,t)} \\
& -\{1 - q[w(L,t)]\}k_1[w'(L,t) + \dot{w}(L,t)]^2 \\
& -\{1 - q[w(L,t)]\}k_2[w'(L,t) + \dot{w}(L,t)]^2 \ln \frac{2[\underline{l}_w(t)]^2}{[\underline{l}_w(t)]^2 - w^2(L,t)}.
\end{aligned} \tag{5.18}$$

Differentiating $V_3(t)$ and substituting the governing equation and the boundary conditions (5.1)–(5.3), we have

$$\begin{aligned}
\dot{V}_3(t) = & \frac{\beta TL}{2}[w'(L,t)]^2 + \frac{\beta \rho L}{2}[\dot{w}(L,t)]^2 - \frac{\beta T}{2}\int_0^L [w'(x,t)]^2 dx \\
& -\frac{\beta\rho}{2}\int_0^L [\dot{w}(x,t)]^2 dx.
\end{aligned} \tag{5.19}$$

Substituting (5.17)–(5.19) into (5.16), considering (5.15), we have

$$\begin{aligned}
\dot{V}(t) \leq & -\frac{\beta T}{2}\int_0^L [w'(x,t)]^2 dx - \frac{\beta\rho}{2}\int_0^L [\dot{w}(x,t)]^2 dx \\
& -q[w(L,t)]k_2[w'(L,t) + \dot{w}(L,t)]^2 \ln \frac{2[\overline{l}_w(t)]^2}{[\overline{l}_w(t)]^2 - w^2(L,t)} \\
& -\{1 - q[w(L,t)]\}k_2[w'(L,t) + \dot{w}(L,t)]^2 \ln \frac{2[\underline{l}_w(t)]^2}{[\underline{l}_w(t)]^2 - w^2(L,t)} \\
& -\left(k_1 - \frac{T}{2}\right)[w'(L,t) + \dot{w}(L,t)]^2 \\
& -\left(\frac{T}{2} - \frac{\beta TL}{2}\right)[w'(L,t)]^2 - \left(\frac{T}{2} - \frac{\beta\rho L}{2}\right)[\dot{w}(L,t)]^2 \\
\leq & -\lambda_3[V_1(t) + V_2(t)] \\
\leq & -\lambda V(t),
\end{aligned} \tag{5.20}$$

where $\lambda_3 = \min(\beta, 2k_2/M_s)$, $\lambda = \lambda_3/\lambda_2$. And the following inequations should be satisfied.

$$k_1 - \frac{T}{2} \geq 0$$

$$T - \beta TL \geq 0$$

$$T - \beta \rho L \geq 0.$$

5.3.2 Robust Adaptive Boundary Control for System Parametric Uncertainty

In this section, the adaptive controller is designed to guarantee the constraint satisfaction and the uniform boundedness of the closed-loop system despite the uncertainty of the parameters M_s and T. Let $\widehat{M}_s(t)$ and $\hat{T}(t)$ be the estimates of M_s and T respectively and define $\widetilde{M}_s(t) := M_s - \widehat{M}_s(t)$ and $\widetilde{T}(t) := T - \hat{T}(t)$ as the estimation errors.

We rewrite the Lyapunov candidate as

$$V_a(t) = V(t) + \frac{\gamma_1}{2}\widetilde{M}_s^{\,2}(t) + \frac{\gamma_2}{2}\widetilde{T}^2(t), \tag{5.21}$$

where $\gamma_1 > 0$ and $\gamma_2 > 0$ are constants and $V(t)$ is defined as same as (5.6)–(5.10).

The controller is designed similar to (5.11):

$$\begin{aligned}
u_a(t) = &-q[w(L,t)]\Bigg\{k_1[\dot{w}(L,t) + w'(L,t)] + \widehat{M}_s(t)[\dot{w}(L,t) + w'(L,t)] \\
&\times \frac{\overline{l}_w^{\,2}(t)w(L,t)\dot{w}(L,t) - \overline{l}_w(t)\dot{\overline{l}}_w(t)w^2(L,t)}{\overline{l}_w^{\,2}(t)[\overline{l}_w^{\,2}(t) - w^2(L,t)]}\Bigg\} \Bigg/ \ln\frac{2\overline{l}_w^{\,2}(t)}{\overline{l}_w^{\,2}(t) - w^2(L,t)} \\
&-\Bigg\{1 - q[w(L,t)]\}\{k_3[\dot{w}(L,t) + w'(L,t)] + \widehat{M}_s(t)[\dot{w}(L,t) + w'(L,t)] \\
&\times \frac{\underline{l}_w^{\,2}(t)w(L,t)\dot{w}(L,t) - \underline{l}_w(t)\dot{\underline{l}}_w(t)w^2(L,t)}{\underline{l}_w^{\,2}(t)[\underline{l}_w^{\,2}(t) - w^2(L,t)]}\Bigg\} \Bigg/ \ln\frac{2\underline{l}_w^{\,2}(t)}{\underline{l}_w^{\,2}(t) - w^2(L,t)} \\
&- k_4[\dot{w}(L,t) + w'(L,t)] - \mathrm{sgn}[\dot{w}(L,t) + w'(L,t)]\overline{d} \\
&+ \hat{T}(t)w'(L,t) - \widehat{M}_s(t)\dot{w}'(L,t),
\end{aligned} \tag{5.22}$$

where $k_3 > 0$ and $k_4 > 0$ are control gains, playing similar roles to k_1 and k_2. And the adaptive laws are designed as follows:

$$\begin{aligned}
\dot{\widehat{M}}_s(t) = &\gamma_1^{-1}\dot{w}'(L,t)q[w(L,t)][\dot{w}(L,t) + w'(L,t)]\ln\frac{2\overline{l}_w^{\,2}(t)}{\overline{l}_w^{\,2}(t) - w^2(L,t)} \\
&+ \gamma_1^{-1}q[w(L,t)][\dot{w}(L,t) + w'(L,t)]^2 \\
&\times \frac{\overline{l}_w^{\,2}(t)w(L,t)\dot{w}(L,t) - \overline{l}_w(t)\dot{\overline{l}}_w(t)w^2(L,t)}{\overline{l}_w^{\,2}(t)[\overline{l}_w^{\,2}(t) - w^2(L,t)]} \\
&+ \gamma_1^{-1}\dot{w}'(L,t)\{1 - q[w(L,t)]\}[\dot{w}(L,t) + w'(L,t)]\ln\frac{2\underline{l}_w^{\,2}(t)}{\underline{l}_w^{\,2}(t) - w^2(L,t)}
\end{aligned}$$

$$
\begin{aligned}
&+\gamma_1^{-1}\{1-q[w(L,t)]\}[\dot{w}(L,t)+w'(L,t)]^2\\
&\times\frac{\underline{l}_w^{2}(t)w(L,t)\dot{w}(L,t)-\underline{l}_w(t)\dot{\underline{l}}_w(t)w^2(L,t)}{\underline{l}_w^{2}(t)[\underline{l}_w^{2}(t)-w^2(L,t)]}\\
&-\sigma_1\gamma_1^{-1}\widehat{M}_s(t),
\end{aligned}
\tag{5.23}
$$

$$
\begin{aligned}
\dot{\hat{T}}(t)={}&-\gamma_2^{-1}w'(L,t)q[w(L,t)][\dot{w}(L,t)+w'(L,t)]\ln\frac{2\overline{l}_w^{2}(t)}{\overline{l}_w^{2}(t)-w^2(L,t)}\\
&-\gamma_2^{-1}w'(L,t)\{1-q[w(L,t)]\}[\dot{w}(L,t)+w'(L,t)]\ln\frac{2\underline{l}_w^{2}(t)}{\underline{l}_w^{2}(t)-w^2(L,t)}\\
&-\sigma_2\gamma_2^{-1}\hat{T}(t),
\end{aligned}
\tag{5.24}
$$

where σ_1 and σ_2 are positive constants.

Theorem 5.2 For the system dynamics described by the governing equation (5.1) and boundary conditions (5.2)–(5.3), under Assumption 5.1 and Assumption 5.2, and the boundary control (5.22) and adaptive laws (5.23)–(5.24), given that the initial conditions are bounded, we can conclude the following.

1) The closed-loop system is uniformly ultimate bounded;
2) all closed-loop signals are bounded; and
3) if the initial boundary condition $w(L,0)$ is within the scope of the constraint, then $w(L,t)$ will remain in the scope of the constraint.

Proof: Substituting (5.21) in (5.15), we obtain

$$
\begin{aligned}
0&\le\lambda_{1a}[V_1(t)+V_2(t)+\widetilde{M}_s^{2}(t)+\tilde{T}^2(t)]\le V_a(t)\\
&\le\lambda_{2a}[V_1(t)+V_2(t)+\widetilde{M}_s^{2}(t)+\tilde{T}^2(t)],
\end{aligned}
\tag{5.25}
$$

in which $\lambda_{1a}=\min(\lambda_1,\frac{\gamma_1}{2},\frac{\gamma_2}{2})$ and $\lambda_{2a}=\max(\lambda_2,\frac{\gamma_1}{2},\frac{\gamma_2}{2})$. ■

Differentiating (5.21) and substituting (5.22)–(5.24), we have

$$
\begin{aligned}
\dot{V}_a(t)={}&\dot{V}(t)+\gamma_1\widetilde{M}_s(t)\dot{\widetilde{M}}_s(t)+\gamma_2\tilde{T}(t)\dot{\tilde{T}}(t)+q[w(L,t)]\widetilde{M}_s(t)[\dot{w}(L,t)+w'(L,t)]^2\\
&\times\frac{\overline{l}_w^{2}(t)w(L,t)\dot{w}(L,t)-\overline{l}_w(t)\dot{\overline{l}}_w(t)w^2(L,t)}{\overline{l}_w^{2}(t)[\overline{l}_w^{2}(t)-w^2(L,t)]}\\
&+q[w(L,t)]\widetilde{M}_s(t)\dot{w}'(L,t)[\dot{w}(L,t)+w'(L,t)]\ln\frac{2\overline{l}_w^{2}(t)}{\overline{l}_w^{2}(t)-w^2(L,t)}\\
&+\{1-q[w(L,t)]\}\widetilde{M}_s(t)[\dot{w}(L,t)+w'(L,t)]^2\\
&\times\frac{\underline{l}_w^{2}(t)w(L,t)\dot{w}(L,t)-\underline{l}_w(t)\dot{\underline{l}}_w(t)w^2(L,t)}{\underline{l}_w^{2}(t)[\underline{l}_w^{2}(t)-w^2(L,t)]}
\end{aligned}
$$

$$
\begin{aligned}
&+ \{1 - q[w(L,t)]\}\widetilde{M}_s(t)\dot{w}'(L,t)[\dot{w}(L,t) + w'(L,t)] \ln \frac{2\underline{l}_w^{2}(t)}{\underline{l}_w^{2}(t) - w^2(L,t)} \\
&- q[w(L,t)]\tilde{T}(t)w'(L,t)[\dot{w}(L,t) + w'(L,t)] \ln \frac{2\overline{l}_w^{2}(t)}{\overline{l}_w^{2}(t) - w^2(L,t)} \\
&- \{1 - q[w(L,t)]\}\tilde{T}(t)w'(L,t)[\dot{w}(L,t) + w'(L,t)] \ln \frac{2\underline{l}_w^{2}(t)}{\underline{l}_w^{2}(t) - w^2(L,t)} \\
= {} & \dot{V}(t) - \gamma_1 \widetilde{M}_s(t)\dot{\widehat{M}}_s(t) - \gamma_2 \tilde{T}(t)\dot{\hat{T}}(t) + q[w(L,t)]\widetilde{M}_s(t)[\dot{w}(L,t) + w'(L,t)]^2 \\
&\times \frac{\overline{l}_w^{2}(t)w(L,t)\dot{w}(L,t) - \overline{l}_w(t)\dot{\overline{l}}_w(t)w^2(L,t)}{\overline{l}_w^{2}(t)[\overline{l}_w^{2}(t) - w^2(L,t)]} \\
&+ q[w(L,t)]\widetilde{M}_s(t)\dot{w}'(L,t)[\dot{w}(L,t) + w'(L,t)] \ln \frac{2\overline{l}_w^{2}(t)}{\overline{l}_w^{2}(t) - w^2(L,t)} \\
&+ \{1 - q[w(L,t)]\}\widetilde{M}_s(t)[\dot{w}(L,t) + w'(L,t)]^2 \\
&\times \frac{\underline{l}_w^{2}(t)w(L,t)\dot{w}(L,t) - \underline{l}_w(t)\dot{\underline{l}}_w(t)w^2(L,t)}{\underline{l}_w^{2}(t)[\underline{l}_w^{2}(t) - w^2(L,t)]} \\
&+ \{1 - q[w(L,t)]\}\widetilde{M}_s(t)\dot{w}'(L,t)[\dot{w}(L,t) + w'(L,t)] \ln \frac{2\underline{l}_w^{2}(t)}{\underline{l}_w^{2}(t) - w^2(L,t)} \\
&- q[w(L,t)]\tilde{T}(t)w'(L,t)[\dot{w}(L,t) + w'(L,t)] \ln \frac{2\overline{l}_w^{2}(t)}{\overline{l}_w^{2}(t) - w^2(L,t)} \\
&- \{1 - q[w(L,t)]\}\tilde{T}(t)w'(L,t)[\dot{w}(L,t) + w'(L,t)] \ln \frac{2\underline{l}_w^{2}(t)}{\underline{l}_w^{2}(t) - w^2(L,t)} \\
\leq {} & -\lambda_4[V_1(t) + V_2(t)] - \frac{\sigma_1}{2}\widetilde{M}_s^{2}(t) - \frac{\sigma_2}{2}\tilde{T}^2(t) + \frac{\sigma_1}{2}M_s^2 + \frac{\sigma_2}{2}T^2 \\
\leq {} & -\lambda_{3a}[V_1(t) + V_2(t) + \widetilde{M}_s^{2}(t) + \tilde{T}^2(t)] + \varepsilon \\
\leq {} & -\lambda_a V_a(t) + \varepsilon, \qquad (5.26)
\end{aligned}
$$

where

$$
\begin{aligned}
\lambda_4 &= \min(\beta, 2k_4/M_s) \\
\lambda_{3a} &= \min\left(\lambda_4, \frac{\sigma_1}{2}, \frac{\sigma_2}{2}\right) \\
\varepsilon &= \frac{\sigma_1}{2}M_s^2 + \frac{\sigma_2}{2}T^2
\end{aligned}
$$

$$\lambda_a = \frac{\lambda_{3a}}{\lambda_{2a}}$$
$$k_1 - \frac{T}{2} \geq 0.$$

Multiplying (5.26) by $e^{\lambda_a t}$ yields

$$\frac{\partial[V_a(t)e^{\lambda_a t}]}{\partial t} \leq \varepsilon e^{\lambda_a t}. \tag{5.27}$$

Integrating the inequality (5.27), we have

$$V_a(t) \leq V_a(0)e^{-\lambda_a t} - \frac{\varepsilon}{\lambda_a}e^{-\lambda_a t} + \frac{\varepsilon}{\lambda_a}. \tag{5.28}$$

Combining (5.5), (5.7), (5.8) and (5.28), we can obtain

$$\begin{aligned}\frac{T}{2L}[w(x,t)]^2 &\leq \frac{T}{2}\int_0^L [w'(x,t)]^2 dx \leq V_1(t)\\ &\leq V_1(t) + V_2(t) \leq \frac{V_a(t)}{\lambda_{1a}}\\ &\leq \frac{V_a(0)e^{-\lambda_a t}}{\lambda_{1a}} + \frac{\varepsilon}{\lambda_{1a}\lambda_a}.\end{aligned} \tag{5.29}$$

From (5.29), we further have

$$|w(x,t)| \leq \sqrt{\frac{2LV_a(0)e^{-\lambda_a t}}{\lambda_{1a}T} + \frac{2L\varepsilon}{\lambda_{1a}\lambda_a T}}, \tag{5.30}$$

and

$$\lim_{t\to\infty}|w(x,t)| \leq \sqrt{\frac{2L\varepsilon}{\lambda_{1a}\lambda_a T}}. \tag{5.31}$$

And we can obtain the following equations utilizing the similar approach

$$|\widetilde{M}_s(t)| \leq \sqrt{\frac{V_a(0)e^{-\lambda_a t}}{\lambda_{1a}} + \frac{\varepsilon}{\lambda_a\lambda_{1a}}}. \tag{5.32}$$

$$\lim_{t\to\infty}|\widetilde{M}_s(t)| \leq \sqrt{\frac{\varepsilon}{\lambda_a\lambda_{1a}}}. \tag{5.33}$$

$$|\tilde{T}(t)| \leq \sqrt{\frac{V_a(0)e^{-\lambda_a t}}{\lambda_{1a}} + \frac{\varepsilon}{\lambda_a\lambda_{1a}}}. \tag{5.34}$$

$$\lim_{t\to\infty}|\tilde{T}(t)| \leq \sqrt{\frac{\varepsilon}{\lambda_a\lambda_{1a}}}. \tag{5.35}$$

The inequation (5.28) reveals that $V_a(t)$ is bounded $\forall t \in [0, \infty)$. Concerning the inequation (5.22), we can know that $V_a(t)$ is also bounded $\forall t \in [0, \infty)$. From the definition of $V_2(t)$, we know that when $w(L,t) \to \overline{l_w}(t)$ or $w(L,t) \to -\underline{l_w}(t)$, $V_2(t)$ tend to infinity, where conflict arises. Supposing that $-\underline{l_w}(t) < w(L,0) < \overline{l_w}(t)$, we infer that $w(L,t)$ remains in the scope of the constraint, $\forall t \in [0, \infty)$.

5.4 The Solvability of the Inequality Equations

From the Section 5.3, we note that the following equations hold:

$$\begin{cases} k_1 - \dfrac{T}{2} \geq 0 \\ k_3 - \dfrac{T}{2} \geq 0 \\ T - \beta TL \geq 0 \\ T - \beta\rho L \geq 0 \\ \beta_1 = \dfrac{2\beta L}{T}\max(T,\rho) \\ \beta_2 = 1 - \beta_1 > 0 \\ \beta_3 = 1 + \beta_1 > 1 \\ \lambda_1 = \min\left(\beta_2, 1\right) = \beta_2 \\ \lambda_2 = \max\left(\beta_3, 1\right) = \beta_3 \\ \lambda_3 = \min\left(\beta, \dfrac{2k_2}{M_s}\right) \\ \lambda_4 = \min\left(\beta, \dfrac{2k_4}{M_s}\right) \\ \lambda = \dfrac{\lambda_3}{\lambda_2} \\ \lambda_{1a} = \min\left(\lambda_1, \dfrac{\gamma_1}{2}, \dfrac{\gamma_2}{2}\right) \\ \lambda_{2a} = \max\left(\lambda_2, \dfrac{\gamma_1}{2}, \dfrac{\gamma_2}{2}\right) \\ \lambda_{3a} = \min\left(\lambda_4, \dfrac{\sigma_1}{2}, \dfrac{\sigma_2}{2}\right) \\ \lambda_a = \dfrac{\lambda_{3a}}{\lambda_{2a}}, \end{cases} \tag{5.36}$$

in which we can infer that $2\beta L \max(T, \rho) < T$. We divide them in two cases:

Case 1: $T \geq \rho$. In this case, the inequations and equations (5.36) are equivalent to the following:

$$\begin{cases} \beta_1 = 2\beta L < 1 \\ \lambda_1 = \beta_2 = 1 - 2\beta L \\ \lambda_2 = \beta_3 = 1 + 2\beta L \\ \lambda = \min\left(\dfrac{\beta}{1+2\beta L}, \dfrac{2k_2}{M_s(1+2\beta L)}\right) \\ \lambda_{1a} = \min\left(1-2\beta L, \dfrac{\gamma_1}{2}, \dfrac{\gamma_2}{2}\right) \\ \lambda_{2a} = \max\left(1+2\beta L, \dfrac{\gamma_1}{2}, \dfrac{\gamma_2}{2}\right) \\ \lambda_{3a} = \min\left(\beta, \dfrac{2k_4}{M_s}, \dfrac{\sigma_1}{2}, \dfrac{\sigma_2}{2}\right) \\ \lambda_a = \min\left(\dfrac{\beta}{1+2\beta L}, \dfrac{2k_4}{M_s(1+2\beta L)}, \dfrac{2\beta}{\gamma_1}, \dfrac{2\beta}{\gamma_2}, \dfrac{4k_4}{\gamma_1 M_s}, \dfrac{4k_4}{\gamma_2 M_s}, \dfrac{\sigma_1}{2(1+2\beta L)}, \right. \\ \qquad \left. \dfrac{\sigma_1}{\gamma_1}, \dfrac{\sigma_1}{\gamma_2}, \dfrac{\sigma_2}{2(1+2\beta L)}, \dfrac{\sigma_2}{\gamma_1}, \dfrac{\sigma_2}{\gamma_2}\right). \end{cases} \tag{5.37}$$

From the equations (5.37), we can infer that the value of k_2 does not effect λ when $k_2 > \beta M_s/2$. Similarly, the value of k_4 does not effect λ_a when $k_4 > \beta M_s/2$. λ_a and σ_1, σ_2 are positive correlation, a negative correlation with the γ_1, γ_2 since σ_1 and σ_2 are denominators, γ_1 and γ_2 are numerators. We illustrate numerical simulation to show the solvability of the equations. Here we choose the system parameters as $L = 1.00\,\mathrm{m}$, $M_s = 1\,\mathrm{kg}$, $T = 10\,\mathrm{N}$, and $\rho = 0.1\,\mathrm{kg/m}$. In the simulation, we use a_1, a_2, b_1, and b_2 replacing σ_1, σ_2, γ_1, and γ_2, respectively for convenience. From the analysis above, we can get $2(\beta M_s/2) < 0.5$. For $k_2 \leq 0.5$, $k_2 \leq 0.5$, we draw the solvability region in Figures 5.3 and 5.4. And for $k_2 > 0.5$, $k_4 > 0.5$, we draw the solvability region in Figures 5.5 and 5.6.

Case 2: $T < \rho$. In this case, the inequations and equations (5.35) are equivalent to the following:

$$
\begin{cases}
\beta_1 = \dfrac{2\beta L\rho}{T} < 1 \\
\lambda_1 = \beta_2 = 1 - \dfrac{2\beta L\rho}{T} \\
\lambda_2 = \beta_3 = 1 + \dfrac{2\beta L\rho}{T} = \dfrac{2\beta L\rho + T}{T} \\
\lambda = \min\left(\dfrac{T\beta}{T + 2\beta L\rho}, \dfrac{2Tk_2}{M_s(T + 2\beta L\rho)}\right) \\
\lambda_{1a} = \min\left(1 - \dfrac{2\beta L\rho}{T}, \dfrac{\gamma_1}{2}, \dfrac{\gamma_2}{2}\right) \\
\lambda_{2a} = \max\left(\dfrac{2\beta L\rho + T}{T}, \dfrac{\gamma_1}{2}, \dfrac{\gamma_2}{2}\right) \\
\lambda_{3a} = \min\left(\beta, \dfrac{2k_4}{M_s}, \dfrac{\sigma_1}{2}, \dfrac{\sigma_2}{2}\right) \\
\lambda_a = \min\left(\dfrac{\beta T}{T + 2\beta L\rho}, \dfrac{2k_4 T}{M_s(T + 2\beta L\rho)}, \dfrac{2\beta}{\gamma_1}, \dfrac{2\beta}{\gamma_2}, \dfrac{4k_4}{\gamma_1 M_s}, \dfrac{4k_4}{\gamma_2 M_s}, \dfrac{\sigma_1 T}{2(T + 2\beta L\rho)}, \right. \\
\qquad \left. \dfrac{\sigma_1}{\gamma_1}, \dfrac{\sigma_1}{\gamma_2}, \dfrac{\sigma_2 T}{2(T + 2\beta L\rho)}, \dfrac{\sigma_2}{\gamma_1}, \dfrac{\sigma_2}{\gamma_2}\right).
\end{cases}
\tag{5.38}
$$

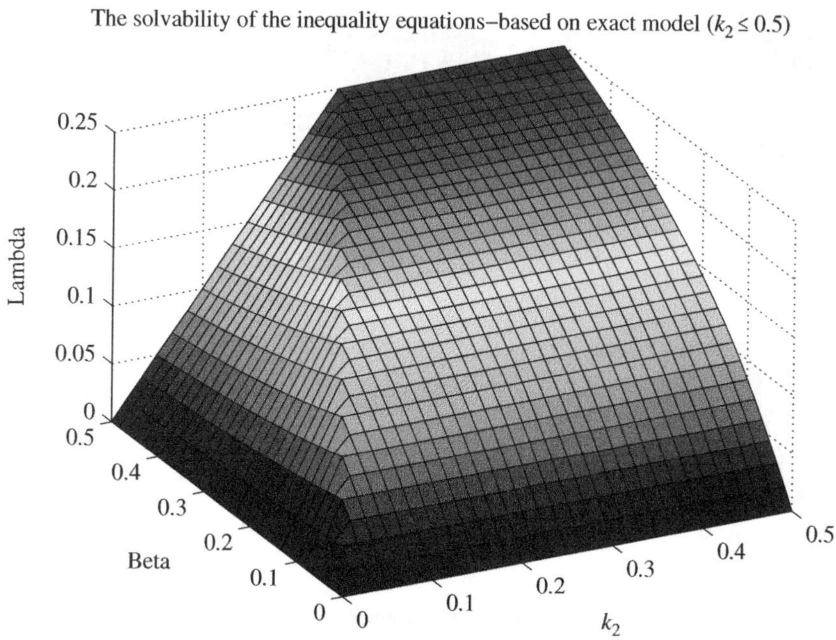

Figure 5.3 The solvability of the equations (based on exact model control with $k_2 \leq 0.5$).

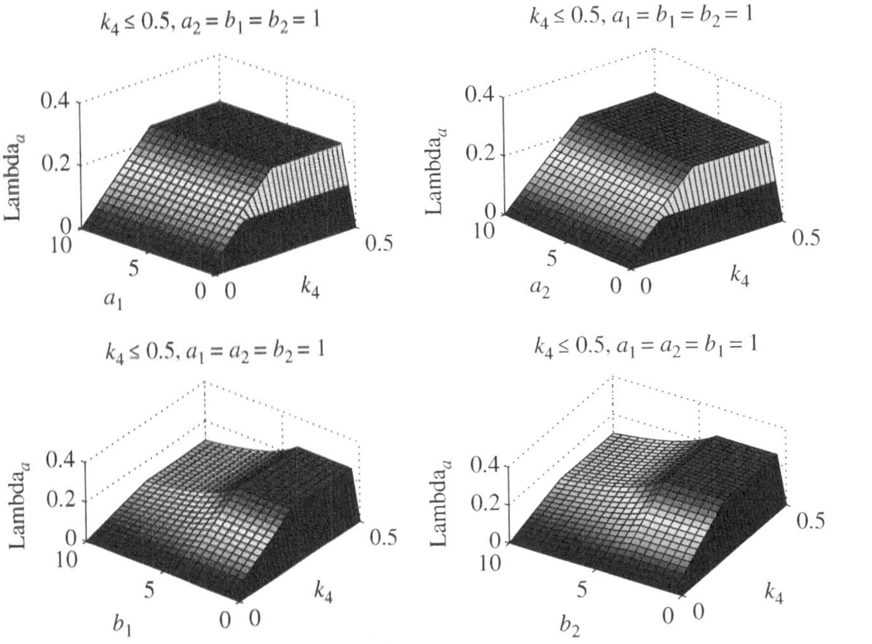

Figure 5.4 The solvability of the equations (based on adaptive control with $k_4 \leq 0.5$).

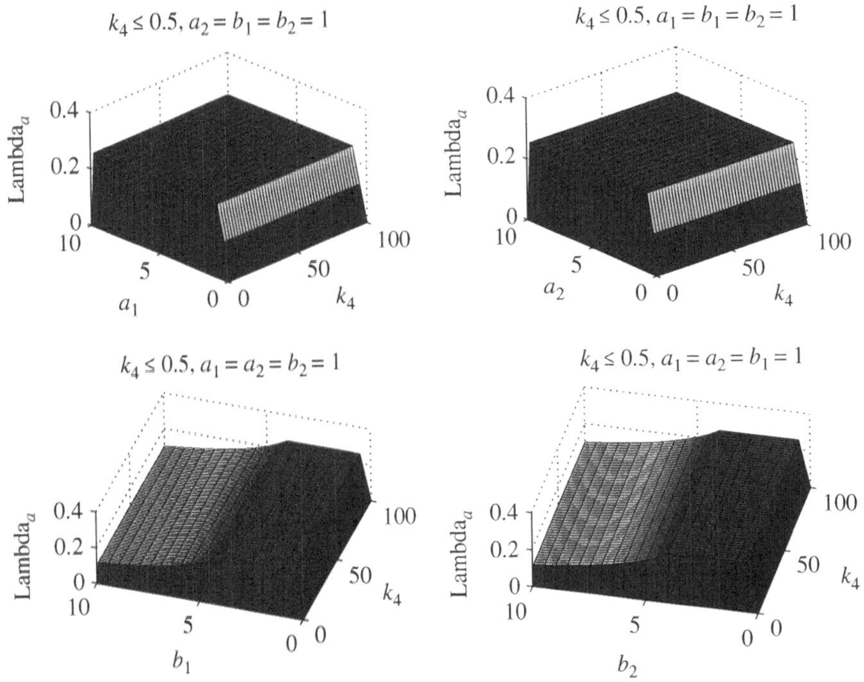

Figure 5.5 The solvability of the equations (based on adaptive control with $k_4 > 0.5$).

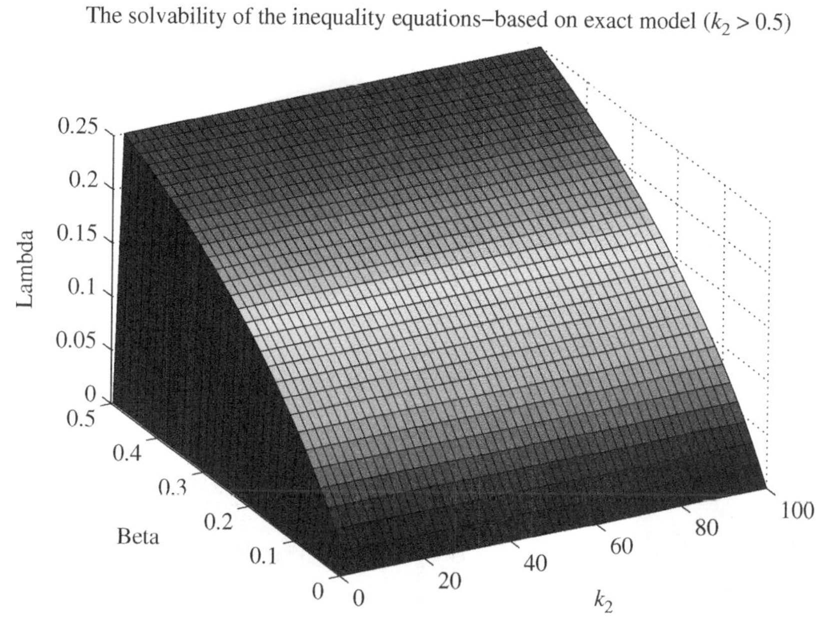

Figure 5.6 The solvability of the equations (based on exact model control with $k_2 > 0.5$).

From the equations (5.38), we can infer that the value of k_2 does not effect λ. Similarly, the value of k_4 does not effect λ_a when $k_4 > \beta M_s/2$. λ_a and σ_1, σ_2 are positive correlation, a negative correlation with the γ_1, γ_2 since σ_1 and σ_2 are denominators, γ_1 and γ_2 are numerators. We illustrate numerical simulation to show the solvability of the equations. Here we choose the system parameters as $L = 1.00\,\text{m}, M_s = 1\,\text{kg}, T = 0.1\,\text{N}$, and $\rho = 0.2\,\text{kg/m}$. In the simulation, we use a_1, a_2, b_1, and b_2 replacing

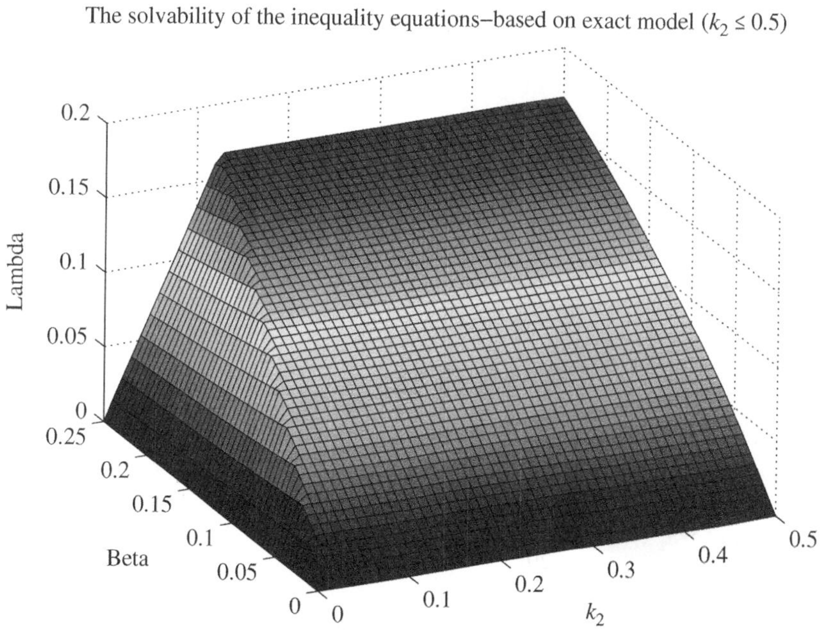

Figure 5.7 The solvability of the equations (based on exact model control with $k_2 \leq 0.5$).

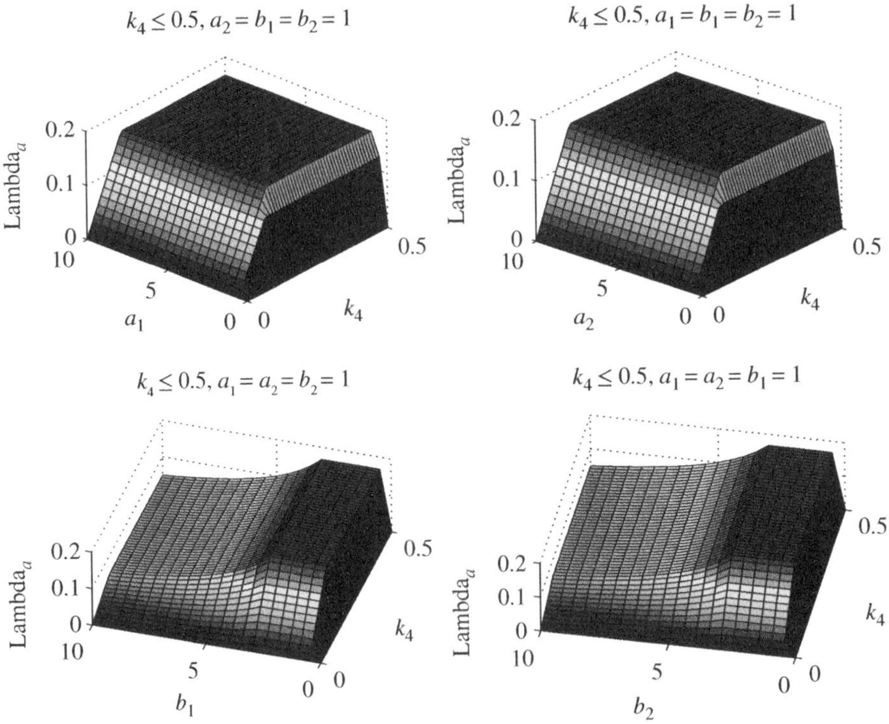

Figure 5.8 The solvability of the equations (based on adaptive control with $k_4 \leq 0.5$).

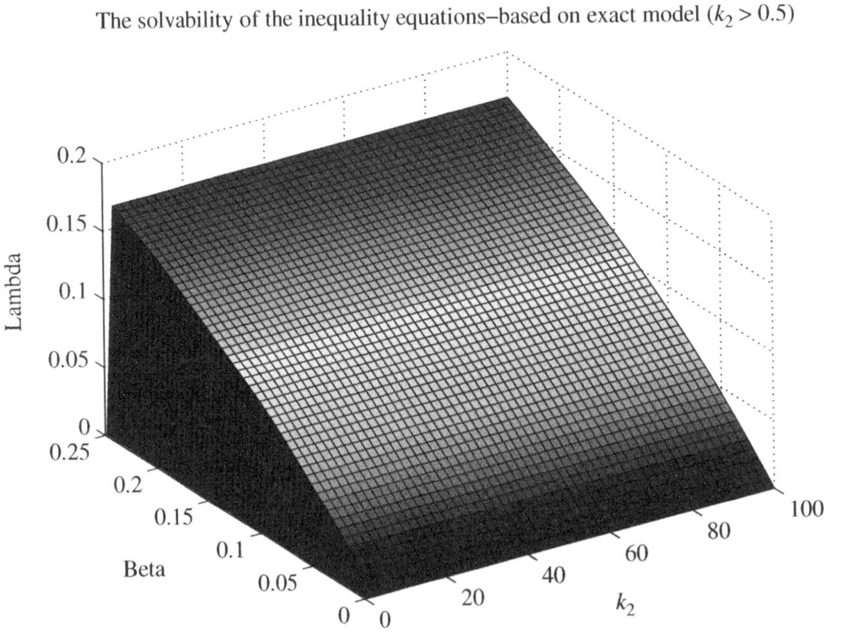

Figure 5.9 The solvability of the equations (based on exact model control with $k_2 > 0.5$).

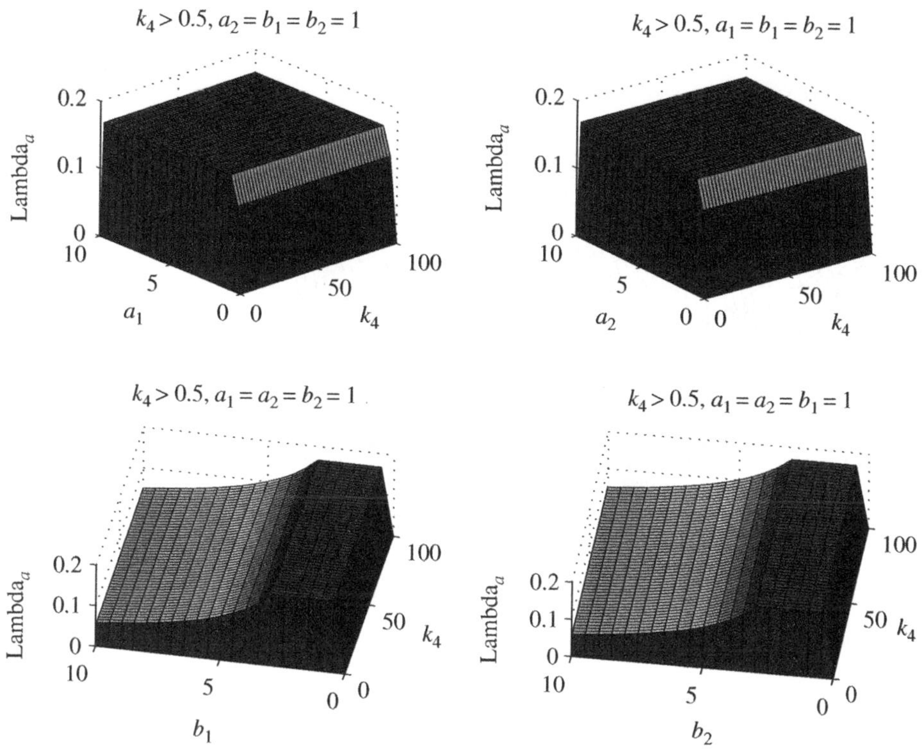

Figure 5.10 The solvability of the equations (based on adaptive control with $k_4 > 0.5$).

σ_1, σ_2, γ_1, and γ_2, respectively for convenience. For $k_2 \leq 2(M_sT/4L\rho)$, $k_4 \leq 2(M_sT/4L\rho)$, we draw the solvability region by Figures 5.7 and 5.8. And for $k_2 > 2(M_sT/4L\rho)$, $k_4 > 2(M_sT/4L\rho)$, we draw the solvability region by Figures 5.9 and 5.10.

5.5 Numerical Simulations

The effectiveness of the proposed control method is illustrated by numerical simulations utilizing finite-difference method. The initial conditions of the string are given as $w(x,0) = x$, $\dot{w}(x,0) = 0, x \in [0, L]$. The boundary disturbance is given as $d(t) = 1 + 0.1\sin(0.1t) + 0.3\sin(0.3t) + 0.5\sin(0.5t)$. Thus we choose $\overline{d} = 2$. And detailed parameters of the string are given in Table 5.1.

Table 5.1 Parameters of the string.

Parameter	Description	Value
L	Length of the string	1.00 m
M_s	Mass of the tip payload	1 kg
T	Tension of the string	10 N
ρ	Mass per unit length of the flexible link	0.1 kg/m

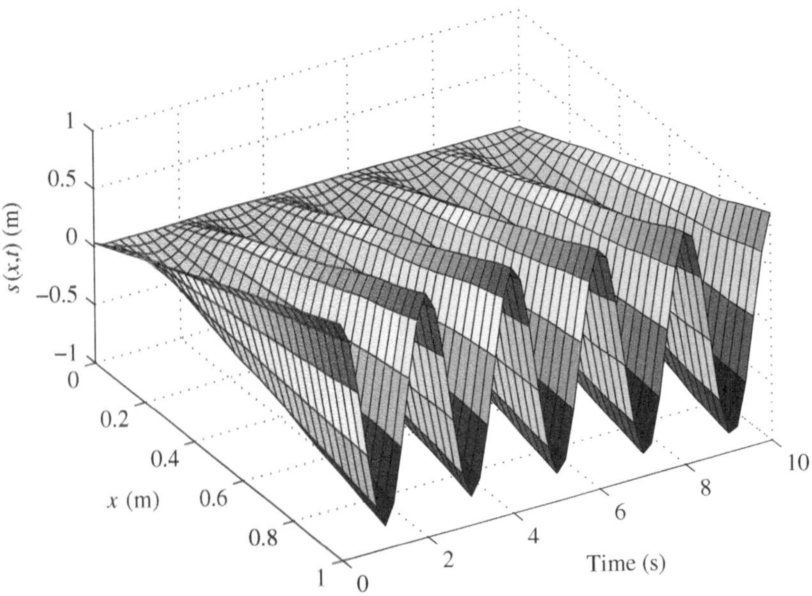

Figure 5.11 Displacement of string without control.

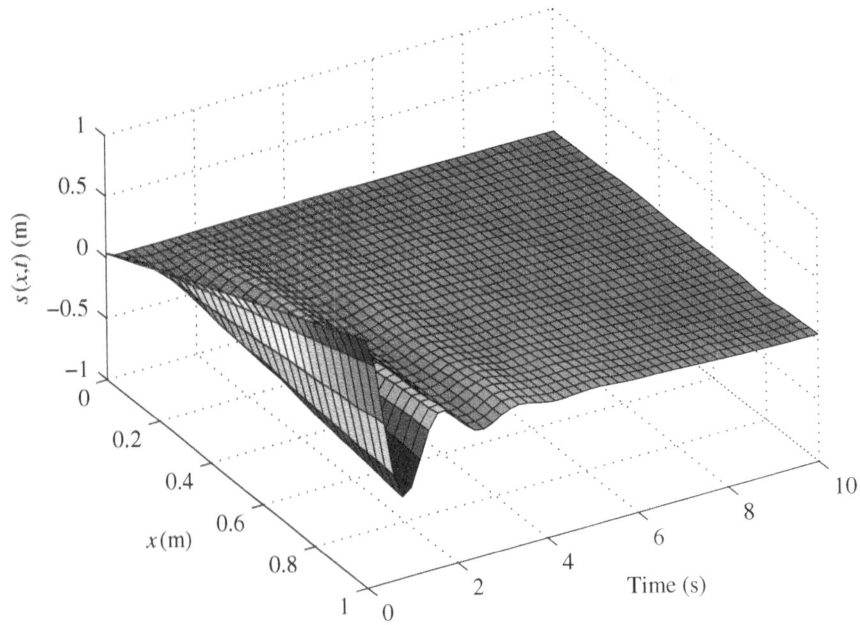

Figure 5.12 Displacement of string with PD control.

Figure 5.11 shows the simulation of the string without control, i.e., $u(t) = 0$. In Figure 5.12, we choose $k_p = 5$, $k_d = 2$ as the parameters of the PD control. While in exact model-based control, we choose $k_1 = 100$, $k_2 = 50$, and the spatial time representation is shown in Figure 5.13. And we simulate the string under adaptive control choosing $k_3 = 5$, $k_4 = 100$, and $\gamma_1 = \gamma_2 = \sigma_1 = \sigma_2 = 1$ with the result shown in Figure 5.14. In Figure 5.11, we can observe that there are significant vibrations

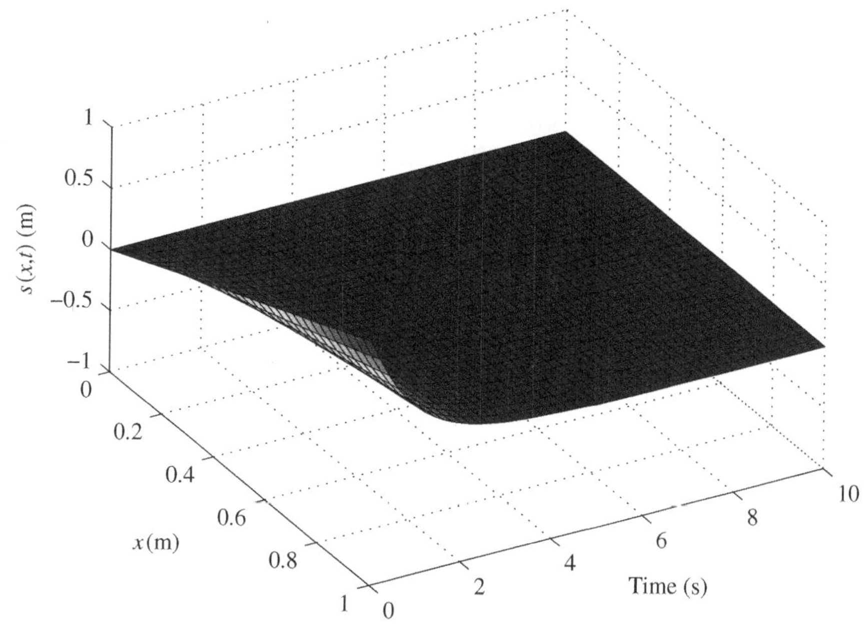

Figure 5.13 Displacement of string with exact model control.

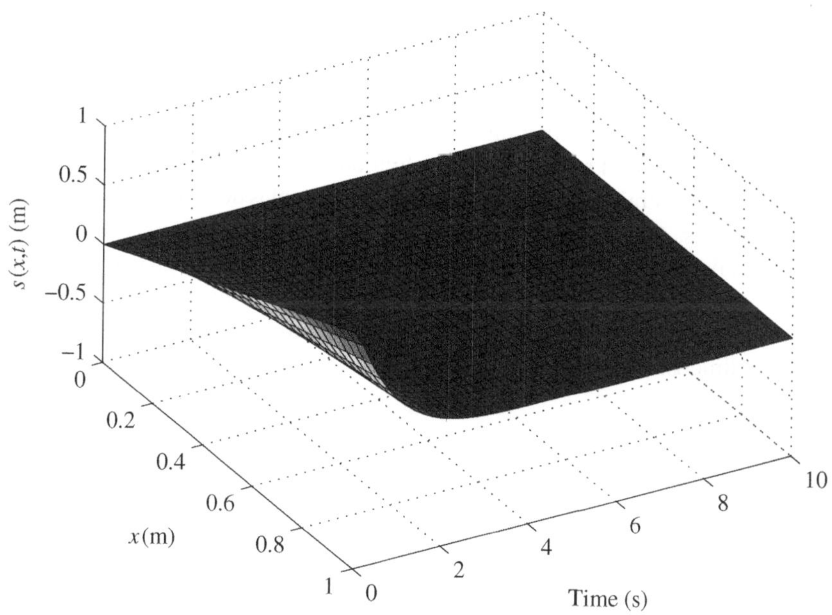

Figure 5.14 Displacement of string with adaptive control.

along the string due to the boundary disturbance $d(t)$ acting on the payload. Figures 5.12–5.14 illustrate that all the three approaches can suppress the vibrations of the string. Figure 5.15 shows the displacement of payload attached on the string under the three control methods. Comparing the effectiveness of PD control, exact model-based control and adaptive control converge quite quickly, having better performances and not violating the time-varying constraint.

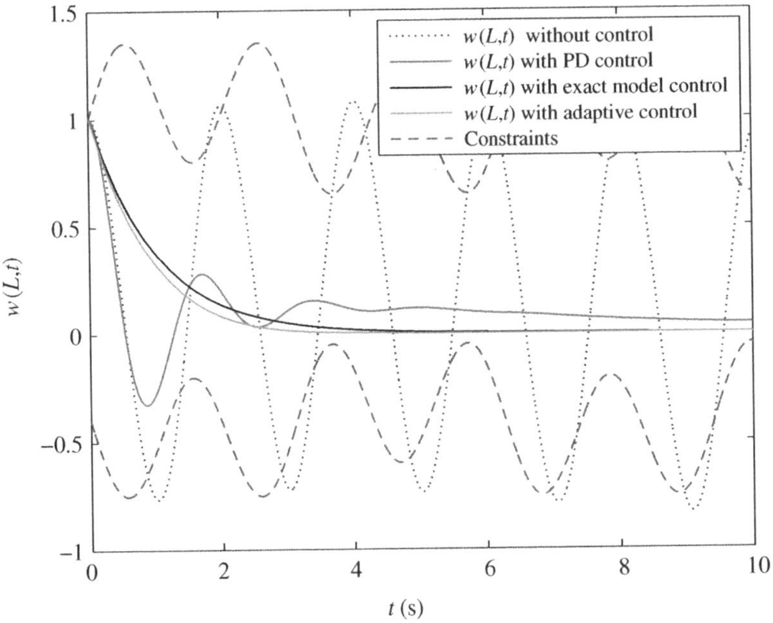

Figure 5.15 Displacement of payload.

5.6 Summary

In this chapter, taking the boundary time-varying output constraint into account, we design boundary control with barrier Lyapunov function to prevent the constraint violations. Firstly, model-based barrier control is employed to suppress the vibration of a nonlinear string under the boundary time-varying output constraint. Subsequently, adaptive control is designed to handle the system parametric uncertainties. And the stability analysis of both methods is proposed. Finally, the provided numerical simulations demonstrate that the control strategy we design can suppress the vibrations along the string and the displacement of the tip payload remains in the scope of time-varying constraint. In future work, we will consider how to control a flexible structure with output constraint in three-dimensional space.

6

Neural Network Vibration Control of a Stand-Alone Tall Building-Like Structure with an Eccentric Load

After reading this chapter, you should be able to:

- Propose a neural network control approach to suppress vibrations
- Consider the output constraint on the angle of the eccentric load
- Focus on a stand-alone tall building-like structure with an eccentric load

6.1 Introduction

High-rise building is the symbol of modern economic development and scientific progress. With the increase of the building height, the wind speed in the upper part of the building increases continuously, and the upper wind speed still has high turbulence [202]. In addition, the high-rise building structure will produce the separation whirlpool under the wind. These cause high wind loads on high-rise buildings, thus affects the safety and comfort of buildings [364]. Even when the energy that earthquake ground motion passes to the building exceeds the maximum energy dissipated by the structure, the building will be destroyed [329]. With the increase in height, the influence of earthquake load on high-rise building structures is also increasing [128]. Because of the large oscillation period and small damping, the high-rise building structure is subjected to the external load under the action of the earthquake and the wind will produce vibrations, which will damage the structure, shorten the life of the building, or make the occupant feel uncomfortable. In order to effectively suppress these vibrations caused by earthquakes and winds, and to ensure the safety, usability and comfortableness of the high-rise building structure, it is necessary to study the effective vibration control of the high-rise building [106].

At present, a great deal of research has been conducted on the vibration control of earthquakes and wind-driven high-rise buildings, and many control strategies have been proposed [283]. Due to the huge size of the high-rise and super high-rise buildings, mathematical modeling and control design are of great importance [343]. A stand-alone tall building-like structure is generally regarded as a flexible structure [165, 323]. There are many modeling methods for flexible structures [354], such as finite element method (FEM) [92], finite difference method (FDM) [157], finite segment method (FSM) [370], modes synthesis method (MSM) [142], and lumped parameter method (LPM) [263]. In [168], a discrete dynamic model of a flexible arm is proposed by space-time FEM. In [157], the authors focus on modeling for a rectangular flexible thin plate through FDM. In [370], the authors focus on the beam-typed flexible element characteristics of SMA flexible torsion actuator and apply FSM to establish its discrete analysis model.

Dynamic Modeling and Neural Network-Based Intelligent Control of Flexible Systems,
First Edition. Hejia Gao, Wei He, and Changyin Sun.

Various control strategies are also applied to flexible structures [19, 325], which are mainly divided into passive control [259], active control, semi-active control [319], and hybrid control [223]. Passive control is a method to reduce the flexible vibration by installing a mechanical device, which has been tested in some earthquakes [259]. Active control needs external energy to realize vibration control [57]. The current active control studies mainly focus on intelligent structures based on piezoelectric sensors and actuators [266], and most of these control strategies come from adaptive control [17, 247, 338, 359], NN control [47, 337], genetic algorithm control, etc. [312, 355, 360]. Semi-active control systems have been developed to exploit the best features of both passive and active control systems with a prominent property of being adjusted in real time without injecting energy into the controlled system [236]. It mainly includes semi-active variable stiffness device and semi-active variable damping device, variable friction control device, semi-active tuned mass (liquid) damper control device, electromagnetic fluid damper control device, and semi-active isolation device. Hybrid control combining active and passive control is proposed to utilize advantages of both controls and avoid their shortcomings. Hence, a higher level of performance is possibly achievable, although it is often also more complicated. Recently, many scholars begin to study active control strategies. In [138], a fuzzy control which consists of an active tuned mass damper is presented for tall buildings under cross-wind excitation. In [231], a robust active control is investigated to suppress vibration for base-isolated structures in the event of an earthquake. In [273], the problem of proportional-integral-derivative (PID) control for a two-story building prototype is solved.

Some tall buildings implement active-mass dampers to suppress the vibration under conditions such as strong winds or earthquakes. In some cases, a large pendulum with a huge inertial load is used to compensate for the swaying motions of the building [234]. In [4], a new tuned mass damper with lever pendulum control system is put forward for controlling the translation-torsion vibration of tall building under earthquakes. In [253], the effectiveness of multi-stage pendulum mass damper is examined by an experimental investigation using a 1:100 scale model of the Stonecutters Bridge Tower with a specially designed two-stage pendulum mass damper.

In practical engineering, the working environment of flexible systems is very complex [31, 39, 181, 183, 314]. For example, the system performance is affected by the mechanical structure or by the technical constraints resulting in the nonlinear characteristics of the system (saturation, dead zone [36, 180], backlash [241], etc. [169, 309]). These nonlinear characteristics [346, 352] will make it difficult to analyze and control the flexible system. According to safety requirements of high-rise buildings, the output constraint problem must be solved. In addition, considering NN's ability to deal with nonlinearities and the fact that its realization does not require an accurate mathematical model [24, 191, 318, 339, 356], NN control of high-rise buildings with output constraints has great advantages and research value.

This chapter's research object is based on a smart structure (see Figure 6.1). This device consists of a flexible beam mounted on a base support plate. The beam is fitted with a strain gage sensor to measure the deflection of the beam. A motorized inertial load is placed at the other end of the beam. The inertial load angle is measured using an encoder and is actuated using a DC motor via a gearbox. As known some tall buildings implement active-mass dampers to reduce the vibration under conditions such as strong winds or earthquakes. In some cases, a large pendulum with a huge inertial load is used to compensate for the swaying motions of the building. The smart structure shown in Figure 6.1 can capture this active-mass damping methodology. It is thus scaled as a stand-alone tall building-like structure equipped with an eccentric load.

This study aims to bring forward the mathematical model of the flexible building structure and design a controller to dampen the swaying of the flexible beam by adjusting the angle of the

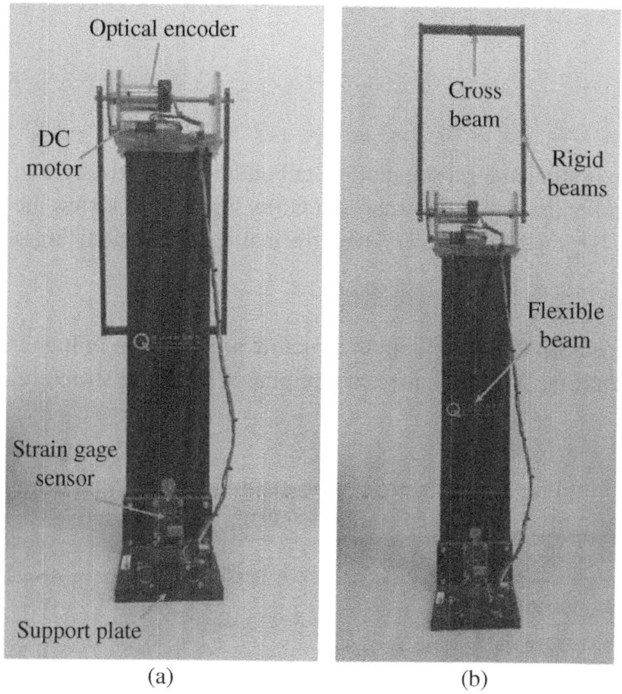

Figure 6.1 Stand-alone tall building-like structure in the laboratory. (a) Smart structure. (b) Position of equilibrium.

pendulum. By considering the output constraint on the angle of the pendulum, smaller deflections in flexible beam can be realized through a smaller angle of the pendulum. In this chapter, the modeling and vibration control problems are studied for a stand-alone tall building-like structure with unknown varying disturbances and output constraints. Compared with the previous works, the novelties include:

1) A finite-dimensional dynamic model of a stand-alone tall building-like structure with an eccentric load is proposed by using AMM.
2) A new NN control strategy in the field of suppressing vibrations is designed to compensate for the dynamic uncertainties of the tall buildings.
3) The output constraint on the angle of the pendulum is considered and such angle can be ensured within the safety limit by incorporating a barrier Lyapunov function.
4) Experiments on Quanser Smart Structure are carried out and the results indicate that the new NN approach is very promising for suppressing vibrations in the flexible system.

In view of the current research status and existing problems of vibration control of high-rise buildings under earthquake and wind excitation, the following work has been done in this chapter: Section 6.2 shows the dynamic modeling of the stand-alone tall building-like structure with unknown varying disturbance. Section 6.3 presents a NN control to solve the vibration problem of the stand-alone tall building-like structure with output constraints under conditions such as strong winds or earthquakes. Sections 6.4 and 6.5 analyze and discuss the simulation results and the experimental results, respectively. Section 6.6 is the summary and prospect of the chapter.

6.2 Dynamic Modeling

6.2.1 Dynamic Modeling

These notations are used: $(\dot{*}) = \frac{\partial *}{\partial t}$, $(\ddot{*}) = \frac{\partial^2(*)}{\partial t^2}$, $(*)' = \frac{\partial *}{\partial x_i}$, $(*)'' = \frac{\partial^2(*)}{\partial x_i^2}$, $(*)''' = \frac{\partial^3(*)}{\partial x_i^3}$, $(*)'''' = \frac{\partial^4(*)}{\partial x_i^4}$.

In this section, the modeling problem is studied for a stand-alone tall building-like structure in the laboratory shown in Figure 6.1. Figure 6.1a shows the smart structure. Figure 6.1b shows the position of equilibrium of the structure. For such a practical system, the mathematical modeling is presented as Figure 6.2.

Assumption 6.1 [234] The external loads on the flexible smart structure are the main influencing factors that determine structural vibration. The low inherent damping of the main structure, therefore, is ignored in this chapter.

Consider the flexible beam of the tall building-like structure system as an Euler-Bernoulli beam [148]. Based on Hamilton's principle, the partial differential equations (PDEs) model is derived by energy analysis.

$$\rho\ddot{w}(x,t) + EIw''''(x,t) + f(x,t) = 0, \tag{6.1}$$

$$w'(0,t) = w(0,t) = 0, \tag{6.2}$$

$$\begin{aligned} &EIw'''(L_b,t) - \left(\frac{1}{2}m_{rb} + m_p\right) l_p\ddot{w}'(L_b,t) \\ &-(m_{rb} + m_p)\ddot{w}(L_b,t) - \left(\frac{1}{2}m_{rb} + m_p\right) l_p\ddot{\theta}(t) = 0, \end{aligned} \tag{6.3}$$

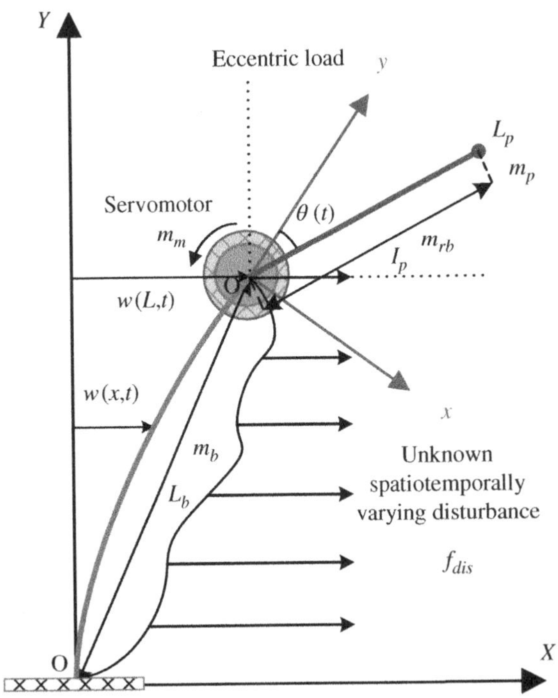

Figure 6.2 Coordinate axes and symbols in dynamic model.

$$EIw''(L_b,t)+\left(\frac{2}{3}m_{rb}l_p^2+2m_pl_p^2+I_b\right)\ddot{w}'(L_b,t)$$
$$+\left(\frac{1}{2}m_{rb}+m_p\right)l_p\ddot{w}(L_b,t)+\left(\frac{1}{3}m_{rb}+m_p\right)l_p^2\ddot{\theta}(t)=0, \tag{6.4}$$

$$u(t)=\left(\frac{1}{2}m_{rb}+m_p\right)l_p\ddot{w}(L_b,t)$$
$$+\left(\frac{1}{3}m_{rb}+2m_p\right)l_p^2\ddot{w}'(L_b,t)+\left(\frac{2}{3}m_{rb}+2m_p\right)l_p^2\ddot{\theta}(t). \tag{6.5}$$

6.2.2 Model Discretization

In the AMM modeling, the elastic vibration of the flexible beam is generally assumed to be of the form:

$$w(x,t)=\Sigma_{i=1}^{\infty}\varphi_i(x)p_i(t). \tag{6.6}$$

According to the space eigenfunction of a cantilever beam, the assumed mode shape function is introduced as $\varphi_i(x)=C_1\sin(\beta_ix)+C_2\cos(\beta_ix)+C_3\sinh(\beta_ix)+C_4\cosh(\beta_ix)$, $p_i(t)=\exp(jw_it)$ is the generalized coordinate, where, w_i is the ith natural angular frequency of the eigenvalue problem, and $\beta_i^4=w_i^2\rho/EI$. In the constrained modes method, $\ddot{\theta}=0$ is assumed. Based on the PDEs model,

$$C_1=-C_3,\ C_2=-C_4, \tag{6.7}$$

$$EIw'''(L_b,t)-(m_{rb}+m_p)\ddot{w}(L_b,t)$$
$$-\left(\frac{1}{2}m_{rb}+m_p\right)l_p\ddot{w}'(L_b,t)=0, \tag{6.8}$$

$$EIw''(L_b,t)+\left(\frac{1}{2}m_{rb}+m_p\right)l_p\ddot{w}(L_b,t)$$
$$+\left(\frac{2}{3}m_{rb}l_p^2+2m_pl_p^2+I_b\right)\ddot{w}'(L_b,t)=0, \tag{6.9}$$

$$EI\beta^4=-\rho j^2w^2. \tag{6.10}$$

The derivatives for $\varphi(x)$ and $p(t)$,

$$\varphi'(x)=\beta[C_1\cos(\beta x)-C_2\sin(\beta x)-C_1\cosh(\beta x)$$
$$-C_2\sinh(\beta x)] \tag{6.11}$$

$$\varphi''(x)=\beta^2[-C_1\sin(\beta x)-C_2\cos(\beta x)-C_1\sinh(\beta x)$$
$$-C_2\cosh(\beta x)], \tag{6.12}$$

$$\varphi'''(x)=\beta^3[-C_1\cos(\beta x)+C_2\sin(\beta x)-C_1\cosh(\beta x)$$
$$-C_2\sinh(\beta x)], \tag{6.13}$$

$$\varphi''''(x)=\beta^4[C_1\sin(\beta x)+C_2\cos(\beta x)-C_1\sinh(\beta x)$$
$$-C_2\cosh(\beta x)], \tag{6.14}$$

$$\ddot{p}(t)=j^2w^2\exp(jwt). \tag{6.15}$$

Substitute $p_i(t)$ and (6.10) into (6.8),

$$\begin{aligned}&\rho[C_1\cos(\beta L_b)-C_2\sin(\beta L_b)+C_1\cosh(\beta L_b)+C_2\sinh(\beta L_b)]\\&\quad=(m_{rb}+m_p)\beta[C_1\sin(\beta L_b)+C_2\cos(\beta L)-C_1\sinh(\beta L_b)\\&\quad-C_2\cosh(\beta L_b)]+\left(\frac{1}{2}m_{rb}+m_p\right)l_p\beta^2[C_1\cos(\beta L_b)\\&\quad-C_2\sin(\beta L_b)-C_1\cosh(\beta L_b)-C_2\sinh(\beta L_b)],\end{aligned}\tag{6.16}$$

and define $s_1=\sin(\beta L_b)+\sinh(\beta L_b)$, $s_2=\sin(\beta L_b)-\sinh(\beta L_b)$, $c_1=\cos(\beta L_b)+\cosh(\beta L_b)$, $c_2=\cos(\beta L_b)-\cosh(\beta L_b)$, that is

$$\begin{aligned}&C_1\left[\rho c_1-\beta(m_{rb}+m_p)s_2-\beta^2\left(\frac{1}{2}m_{rb}+m_p\right)l_pc_2\right]=\\&C_2\left[\beta(m_{rb}+m_p)c_2-\beta^2\left(\frac{1}{2}m_{rb}+m_p\right)l_ps_1+\rho s_2\right].\end{aligned}\tag{6.17}$$

Substituting $p_i(t)$ and (6.10) into (6.9),

$$\begin{aligned}&\rho[C_1\sin(\beta L_b)+C_2\cos(\beta L_b)+C_1\sinh(\beta L_b)\\&\quad+C_2\cosh(\beta L_b)]=\beta^2l_p\left(\frac{1}{2}m_{rb}+m_p\right)[C_1\sin(\beta L_b)\\&\quad+C_2\cos(\beta L_b)-C_1\sinh(\beta L_b)-C_2\cosh(\beta L_b)]\\&\quad+\beta^3\left(\frac{2}{3}m_{rb}l_p^2+2m_pl_p^2+I_b\right)[C_1\cos(\beta L_b)\\&\quad-C_2\sin(\beta L_b)-C_1\cosh(\beta L_b)-C_2\sinh(\beta L_b)],\end{aligned}\tag{6.18}$$

that is

$$\begin{aligned}&C_1[\rho s_1-\beta^2l_p(1/2m_{rb}+m_p)s_2-\beta^3(2/3m_{rb}l_p^2+2m_pl_p^2+I_b)c_2]=\\&C_2[\beta^2l_p(1/2m_{rb}+m_p)c_2+\beta^3(2/3m_{rb}l_p^2+2m_pl_p^2+I_b)s_1-\rho c_1].\end{aligned}\tag{6.19}$$

Therefore, $\varphi(x)$ can further written as:

$$\varphi(x)=C_1[\sin(\beta x)-\sinh(\beta x)+\gamma(\cos(\beta x)-\cosh(\beta x))],\tag{6.20}$$

where

$$\gamma=\frac{\rho s_1-\beta^2l_p(\frac{1}{2}m_{rb}+m_p)s_2-\beta^3(\frac{2}{3}\rho l_p^3+2m_pl_p^2+I_b)c_2}{\rho c_1-\beta^2l_p(\frac{1}{2}m_{rb}+m_p)c_2-\beta^3(\frac{2}{3}\rho l_p^3+2m_pl_p^2+I_b)s_1}.\tag{6.21}$$

Design $[F(\beta)]=\begin{bmatrix}A & B\\ C & D\end{bmatrix}$. Considering (6.17) and (6.19), thus, $[F(\beta)]\begin{bmatrix}C_1\\ C_2\end{bmatrix}=0$, that is $|F(\beta)|=0$.

$$\begin{aligned}&[\rho c_1-\beta(m_{rb}+m_p)s_2-\beta^2(1/2m_{rb}+m_p)l_pc_2]*\\&[\beta^2l_p(\frac{1}{2}m_{rb}+m_p)c_2+\beta^3(\frac{2}{3}m_{rb}l_p^2+2m_pl_p^2+I_b)s_1\\&-\rho c_1]-[\rho s_1-\beta^2l_p(1/2m_{rb}+m_p)s_2\\&-\beta^3(\frac{2}{3}m_{rb}l_p^2+2m_pl_p^2+I_b)c_2]*[\beta(m_{rb}+m_p)c_2\\&-\beta^2(1/2m_{rb}+m_p)l_ps_1+\rho s_2]=0,\end{aligned}\tag{6.22}$$

Note that only positive values of β are used. Applying the following orthogonal conditions:

$$\int_0^{L_b}\varphi_i(x)\varphi_j(x)dx+\frac{m_m}{\rho}\varphi_i(L_b)\varphi_j(L_b)=\begin{cases}0, & i\neq j\\ 1, & i=j\end{cases}$$

$$EI\int_0^{L_b}\varphi_i''(x)\varphi_j''(x)dx=\begin{cases}0, & i\neq j\\ \omega_i^2\rho, & i=j\end{cases}.$$

We have:

$$C_1=1/[L_b+\frac{\rho L_b^2}{m_m\beta^2}(\frac{1+\cosh\beta\cos\beta}{\sinh\beta+\sin\beta})^2]^{\frac{1}{2}}. \tag{6.23}$$

The dynamic model is obtained by satisfying the Lagrangian–Euler equation:

$$\frac{d}{dt}\frac{\partial(E_k-E_p)}{\partial\dot{q}_i}-\frac{\partial(E_k-E_p)}{\partial q_i}=f_i,\ i-1,\ldots,N. \tag{6.24}$$

where f_i are the generalized forces performing work on q_i. The total kinetic energy is given by:

$$E_k=E_F+E_R+E_I+E_M, \tag{6.25}$$

where

$$E_F=\frac{1}{2}\rho\sum_{i=1}^{N}\dot{p}_i^2(t)\left[1-\frac{m_m}{\rho}\varphi_i^2(L_b)\right]-m_m\sum_{i,j=1\ i\neq j}^{N}\dot{p}_i(t)\dot{p}_j(t)\varphi_i(L_b)\varphi_j(L_b), \tag{6.26}$$

$$\begin{aligned}E_R+E_M={}&\frac{1}{2}(m_{rb}+m_p)\sum_{i=1}^{N}\varphi_i^2(L_b)\dot{p}_i^2(t)+(m_{rb}+m_p)\sum_{i,j=1\ i\neq j}^{N}\varphi_i(L_b)\varphi_j(L_b)\dot{p}_i(t)\dot{p}_j(t)\\&+\left(\frac{1}{2}m_{rb}+m_p\right)l_p\sum_{i=1}^{N}\varphi_i(L_b)\varphi_i'(L_b)\dot{p}_i^2(t)+(m_{rb}+2m_p)l_p\sum_{i,j=1\ i\neq j}^{N}\varphi_i(L_b)\varphi_j'(L_b)\dot{p}_i(t)\dot{p}_j(t)\\&+\left(\frac{1}{2}m_{rb}+m_p\right)l_p\sum_{i=1}^{N}\varphi_i(L_b)\dot{p}_i(t)\dot{\theta}(t)+\left(\frac{1}{3}m_{rb}+\frac{1}{2}m_p\right)l_p^2\left[\sum_{i=1}^{N}\varphi_i'^2(L_b)\dot{p}_i^2(t)+\dot{\theta}^2(t)\right]\\&+\left(\frac{2}{3}m_{rb}+m_p\right)l_p^2\sum_{i,j=1\ i\neq j}^{N}\varphi_i'(L_b)\varphi_j'(L_b)\dot{p}_i(t)\dot{p}_j(t)\\&+\left(\frac{1}{3}m_{rb}+\frac{1}{2}m_p\right)l_p^2\sum_{i=1}^{N}\varphi_i'(L_b)\dot{p}_i(t)\dot{\theta}(t),\end{aligned} \tag{6.27}$$

$$E_I=\frac{I_b}{2}\sum_{i=1}^{N}\varphi_i'^2(L_b)\dot{p}_i^2(t)+I_b\sum_{i,j=1\ i\neq j}^{N}\varphi_i'(L_b)\varphi_j'(L_b)\dot{p}_i(t)\dot{p}_j(t). \tag{6.28}$$

The potential energy is given by

$$E_p=\frac{1}{2}\Sigma_{i=1}^{N}p_i^2(t)\omega_i^2\rho. \tag{6.29}$$

As a result of this procedure, the equation of motion for the smart structure can be expressed:

$$B\ddot{q}+Kq+f_{dis}=U. \tag{6.30}$$

where $q=(\theta,p_1,\ldots p_N)\in\mathbb{R}^{N+1}$ is the generalized coordinates, and $U=(u,0,\ldots 0)\in\mathbb{R}^{N+1}$ is the joint torques. $B\in\mathbb{R}^{(N+1)\times(N+1)}$ is the inertia matrix and $K\in\mathbb{R}^{(N+1)\times(N+1)}$ is the stiffness matrix. $f_{dis}\in\mathbb{R}^{N+1}$ represents an external disturbance. The elements of B and K are defined as follows.

$$B[1,1]=\left(\frac{2}{3}m_{rb}+m_p\right)l_p^2, \tag{6.31}$$

$$B[1,j]=B[j,1]=\left(\frac{1}{2}m_{rb}+m_p\right)l_p\sum_{i=j-1}^{N}\varphi_i(L_b)$$

$$+\left(\frac{1}{3}m_{rb}+\frac{1}{2}m_p\right)l_p^2\sum_{i=j-1}^{N}\varphi_i'(L_b), j=2,\ldots,N+1, \tag{6.32}$$

$$B[j,j]=\rho+(m_{rb}+m_p-m_m)\sum_{i=j-1}^{N}\varphi_i^2(L_b)+(m_{rb}+2m_p)l_p\sum_{i=j-1}^{N}\varphi_i(L_b)\varphi_i'(L_b)$$

$$+\left(\frac{2}{3}m_{rb}l_p^2+m_pl_p^2+I_b\right)\sum_{i=j-1}^{N}\varphi_i'^2(L_b), j=2,\ldots,N+1, \tag{6.33}$$

$$B[j,n]=B[n,j]=(m_{rb}+m_p-m_m)\sum_{\substack{i=j-1\\ j=n}}^{N}\varphi_i(L_b)\varphi_j(L_b)$$

$$+(m_{rb}+2m_p)l_p\sum_{i=j-1,j=n}^{N}\varphi_i(L_b)\varphi_j'(L_b)$$

$$+\left(\frac{2}{3}m_{rb}l_p^2+m_pl_p^2+I_b\right)\sum_{i=j-1,j=n}^{N}\varphi_i'(L_b)\varphi_j'(L_b)$$

$$j=2,\ldots,N+1;\ n=j+1,\ldots,N+1, \tag{6.34}$$

$$K=\text{diag}[0\ \omega_1^2\rho\ \omega_2^2\rho\ \ldots\ \omega_N^2\rho]. \tag{6.35}$$

Lemma 6.1 [179] If a continuously differentiable function $f(t)$ satisfies $|f(t)|\le\zeta$, $\forall t\in[t_1,\ t_2]$, where ζ is a positive constant, then $\dot f(t)$ is bounded, $\forall t\in[t_1,\ t_2]$.

6.3 Neural Network Vibration Control

6.3.1 Control Design

In practice, unknown distributed spatiotemporally varying disturbance exists in the tall building-like structure (6.30). Thus, compounded disturbance observer-based adaptive NN control is introduced to achieve vibration suppression, which uses the disturbance observer to approximate the unknown disturbance. Because of output constraints brought about by safety requirements of high-rise buildings, a barrier Lyapunov function is introduced to handle output constraints. In addition, we theoretically prove the stability of the controller by Lyapunov direct method. The compounded disturbance observer-based adaptive NN control structure is shown in Figure 6.3.

Transform (6.30), the stand-alone tall building-like structure system dynamics can be expressed as:

$$\ddot q=B^{-1}[U-f_{dis}-Kq]. \tag{6.36}$$

Define two variables $e_1=q-q_d$ and $e_2=\dot q-\mu$ as the state errors, where $q_d\in\mathbb{R}^N$ is the equilibrium state of the system. Subsequently,

$$\dot e_1=e_2+\mu-\dot q_d, \tag{6.37}$$

$$\dot e_2=B^{-1}[U-f_{dis}-Kq]-\dot\mu. \tag{6.38}$$

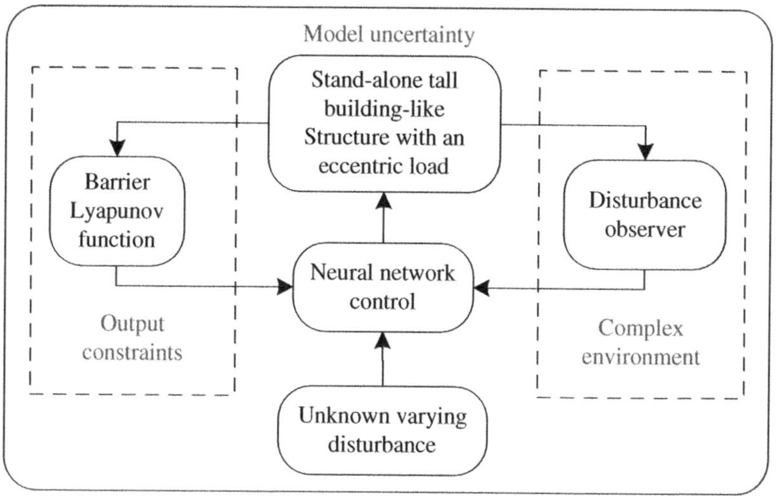

Figure 6.3 Compounded disturbance observer-based adaptive neural network control structure.

6.3.2 Stability Analysis

To analyze the stability of the system, design V_1 as the initial barrier Lyapunov function:

$$V_1 = \frac{1}{2}\sum_{i=1}^{N+1} \ln \frac{b_i^2}{b_i^2 - e_{1i}^2}, \tag{6.39}$$

where $| e_{1i} |<| b_i |$, b_i is the constraint of e_1. Further, $\dot{V}_1$ can be obtained:

$$\dot{V}_1 = \sum_{i=1}^{N+1} \frac{e_{1i}(e_{2i} + \mu_i - \dot{q}_d)}{b_i^2 - e_{1i}^2}. \tag{6.40}$$

We can choose the virtual control μ as

$$\mu = -\begin{bmatrix} k_1 e_{11} - \dot{q}_{d1} \\ k_2 e_{12} - \dot{q}_{d2} \\ \vdots \\ k_{N+1} e_{1N+1} - \dot{q}_{dN+1} \end{bmatrix}. \tag{6.41}$$

Thus,

$$\dot{V}_1 = -\sum_{i=1}^{N+1} \frac{k_i e_{1i}^2}{b_i^2 - e_{1i}^2} + \sum_{i=1}^{N+1} \frac{e_{1i} e_{2i}}{b_i^2 - e_{1i}^2}. \tag{6.42}$$

Considering the unknown disturbance f_{dis} caused by earthquake or wind load, a disturbance observer (DO) is introduced. An auxiliary variable is introduced:

$$e_3 = f_{dis} + \Theta(e_2), \tag{6.43}$$

where $\Theta(e_2) \in \mathbb{R}^N$. Then,

$$\dot{e}_3 = \dot{f}_{dis} + D(e_2)B^{-1}[U - f_{dis} - Kq] - D(e_2)\dot{\mu},$$

where $D(e_2) = \partial\Theta(e_2)/\partial e_2^T \in \mathbb{R}^{(N+1)\times(N+1)}$. In order to approximate f_{dis}, it is essential to estimate e_3. Therefore,

$$\dot{\hat{e}}_3 = D(e_2)B^{-1}[U - \hat{f}_{dis} - Kq] - D(e_2)\dot{\mu}, \tag{6.44}$$

where $\hat{e}$ is the estimation of e_3. Thus,

$$\hat{f}_{dis} = \hat{e}_3 - \Theta(e_2). \tag{6.45}$$

Apparently,

$$\tilde{e}_3 = \hat{e}_3 - e_3 = \hat{f}_{dis} - f_{dis} = \tilde{f}_{dis}. \tag{6.46}$$

Hence,

$$\dot{\tilde{f}}_{dis} = -D(e_2)B^{-1}\tilde{f}_{dis} - \dot{f}_{dis}. \tag{6.47}$$

Considering a slow disturbance f_{dis},

$$\|\dot{f}_{dis}\| \leq \varsigma_0, \tag{6.48}$$

where the constant ς_0 is positive. Rewrite the initial Lyapunov function V_1,

$$V_2 = V_1 + \frac{1}{2}e_2^T B e_2 + \frac{1}{2}\tilde{f}_{dis}^T\tilde{f}_{dis}. \tag{6.49}$$

Apply the following inequality,

$$-\tilde{f}_{dis}^T\dot{f}_{dis} \leq \frac{1}{2}\tilde{f}_{dis}^T\tilde{f}_{dis} + \frac{1}{2}\dot{f}_{dis}^T\dot{f}_{dis} \leq \frac{1}{2}\tilde{f}_{dis}^T\tilde{f}_{dis} + \frac{1}{2}\varsigma_0^2. \tag{6.50}$$

Then,

$$\begin{aligned}\dot{V}_2 \leq & -\sum_{i=1}^{N+1}\frac{k_i e_{1i}^2}{b_i^2 - e_{1i}^2} + \sum_{i=1}^{N+1}\frac{e_{1i}e_{2i}}{b_i^2 - e_{1i}^2} + e_2^T(U - f_{dis} - Kq \\ & - B\dot{\mu}) - \tilde{f}_{dis}^T D(e_2)B^{-1}\tilde{f}_{dis} + \frac{1}{2}\tilde{f}_{dis}^T\tilde{f}_{dis} + \frac{1}{2}\varsigma_0^2.\end{aligned} \tag{6.51}$$

Introduce the model-based controller U_0,

$$U_0 = -K_2 e_2 - \begin{bmatrix} \dfrac{e_{11}}{b_1^2 - e_{11}^2} \\ \dfrac{e_{12}}{b_2^2 - e_{12}^2} \\ \vdots \\ \dfrac{e_{1N+1}}{b_{N+1}^2 - e_{1N+1}^2} \end{bmatrix} + \hat{f}_{dis} + Kq + B\dot{\mu}, \tag{6.52}$$

where the gain matrix $K_2 = K_2^T > 0$. Apply $e_2^T\tilde{f}_{dis} \leq \frac{1}{2}e_2^T e_2 + \frac{1}{2}\tilde{f}_{dis}^T\tilde{f}_{dis}$. Substituting (6.52) into (6.51):

$$\begin{aligned}\dot{V}_2 \leq & -\sum_{i=1}^{N+1}\frac{k_i e_{1i}^2}{b_i^2 - e_{1i}^2} - e_2^T(K_2 - \frac{1}{2}I)e_2 \\ & - \tilde{f}_{dis}^T D(e_2)B^{-1}\tilde{f}_{dis} + \tilde{f}_{dis}^T\tilde{f}_{dis} + \frac{1}{2}\varsigma_0^2,\end{aligned} \tag{6.53}$$

where I denotes identity matrix. Obviously, it is unrealistic to assume that $B\dot{\mu}$ and K can be measured. Because of the complex nonlinear characteristics and strong coupling of the flexible system, it is impossible to obtain its dynamic model accurately. Due to the learning ability of NNs [12, 251], the global estimation scheme based on NN control is introduced to fit the uncertain system. Then, the corresponding controller is implemented to solve the constraint of the output signal.

$$U_1 = -K_2 e_2 - \begin{bmatrix} \dfrac{e_{11}}{b_1^2 - e_{11}^2} \\ \dfrac{e_{12}}{b_2^2 - e_{12}^2} \\ \vdots \\ \dfrac{e_{1N+1}}{b_{N+1}^2 - e_{1N+1}^2} \end{bmatrix} + \hat{f}_{dis} + \hat{W}^T S(Z), \tag{6.54}$$

where $\hat{W}_i$ is the NN weight estimation. $Z = [q,\ \dot{q},\ \mu,\ \dot{\mu}]^T$ is the input to radial basis function NN. Give the adaptive law:

$$\dot{\hat{W}}_i = -\Gamma_i (S_i(Z) e_{2,i} + \sigma_i \hat{W}_i), \tag{6.55}$$

where $\sigma_i > 0$ is a small constant. Γ_i is the constant gain matrix. $\hat{W}_i$ is the approximation of W_i^*.

$$W^{*T} S(Z) = Kq + B\dot{\mu} - \epsilon(Z), \tag{6.56}$$

where $\epsilon(Z)$ is the approximation error, $\|\epsilon\| \le \|\overline{\epsilon}\|$ and $\overline{\epsilon} > 0$ is a constant. Update the Lyapunov function (6.49),

$$V_3 = V_2 + \frac{1}{2}\sum_{i=1}^{N+1} \tilde{W}_i^T \Gamma_i^{-1} \tilde{W}_i. \tag{6.57}$$

Using the following inequality:

$$-\sum_{i=1}^{N+1} \sigma_i \tilde{W}_i^T \hat{W}_i \le -\sum_{i=1}^{N+1} \sigma_i \left(\frac{\|\tilde{W}_i\|^2}{2} - \frac{\|W_i^*\|^2}{2} \right). \tag{6.58}$$

Then,

$$\begin{aligned} \dot{V}_3 \le & -\sum_{i=1}^{N+1} \frac{k_i e_{1i}^2}{b_i^2 - e_{1i}^2} - e_2^T (K_2 - I) e_2 + \frac{1}{2}\|\overline{\epsilon}(Z)\|^2 \\ & - \tilde{f}_{dis}^T (D(e_2) B^{-1} - I) \tilde{f}_{dis} + \frac{1}{2}\varsigma_0^2 \\ & - \sum_{i=1}^{N+1} \frac{\sigma_i \|\tilde{W}_i\|^2}{2} + \sum_{i=1}^{N+1} \frac{\sigma_i \|W_i^*\|^2}{2} \\ \le & -a_1 V_3 + b_1, \end{aligned} \tag{6.59}$$

where

$$a_1 = \min\left(2\min(k_i), \frac{2\lambda_{\min}(K_2 - I)}{\lambda_{\max}(B)}, 2\lambda_{\max}(D(e_2)B^{-1} - I), \min_{i=1,2,\ldots,N+1} \left(\frac{\sigma_i}{\lambda_{\max}(\Gamma_i^{-1})} \right) \right),$$

$$b_1 = \frac{1}{2}\varsigma_0^2 + \frac{1}{2}\|\overline{\epsilon}(Z)\|^2 + \sum_{i=1}^{N+1} \frac{\sigma_i \|W_i^*\|^2}{2},$$

where λ_{max} and λ_{min} are the maximum and minimum eigenvalues. To ensure $a_1 > 0$,

$$\min k_i > 0,\ \lambda_{max}(K_2 - I) > 0,\ \lambda_{max}\left(\frac{\partial\Theta(e_2)}{\partial e_2^T}B^{-1} - I\right) > 0.$$

Multiplying (6.59) by $e^{a_1 t}$ and integrating it,

$$V_3 \leq \left(V_3(0) - \frac{b_1}{a_1}\right)e^{-a_1 t} + \frac{b_1}{a_1} \leq V_3(0) + \frac{b_1}{a_1}. \tag{6.60}$$

Thus, $\frac{1}{2}\|e_1\|^2 \leq V_3(0) + \frac{b_1}{a_1}$.

Theorem 6.1 As for the stand-alone tall building-like structure system (6.30), because the full-state information is measurable, by introducing adaptive NN control, for each compact set Ω_0 where $(x_1(0), x_2(0), \hat{W}_1(0)...\hat{W}_n(0)) \in \Omega_0$, the initial conditions are bounded and the closed-loop system is semi-globally uniformly bounded. e_1, e_2, $\tilde{f}_{dis}$ and $\tilde{W}$ will remain within Ω_{e_1}, Ω_{e_2}, $\Omega_{\tilde{f}_{dis}}$, Ω_W:

$$\Omega_{e_1} := \left\{e_1 \in \mathbb{R}^n \mid \|e_1\| \leq \sqrt{b_i^2(1 - e^{-X})}\right\}, \tag{6.61}$$

$$\Omega_{e_2} := \left\{e_2 \in \mathbb{R}^n \mid \|e_2\| \leq \sqrt{\frac{X}{\lambda_{min}(B)}}\right\}, \tag{6.62}$$

$$\Omega_{\tilde{f}_{dis}} := \left\{\tilde{f}_{dis} \in \mathbb{R}^n \mid \|\tilde{f}_{dis}\| \leq \sqrt{X}\right\}, \tag{6.63}$$

$$\Omega_{\tilde{W}} := \left\{\tilde{W} \in \mathbb{R}^{l \times n} \mid \|\tilde{W}\| \leq \sqrt{\frac{X}{\lambda_{min}(\Gamma^{-1})}}\right\}, \tag{6.64}$$

where $X = 2(V_3(0) + \frac{b_1}{a_1})$, $b_1 > 0$ and $a_1 > 0$.

6.4 Numerical Simulations

6.4.1 Simulation Parameters

Based on MATLAB, the NN control with output constraint is applied to the stand-alone tall building-like structure with an eccentric load, and compared with the full-state feedback (FSF) control and the NN control without output constraint. The effectiveness of the NN controller with output constraint is proved. Table 6.1, below, lists and characterizes the main parameters associated with the Smart Structure.

6.4.2 Simulation Results

The smart structure closed-loop response in FSF mode, NN control without output constraint, and NN control with output constraint are simulated. The simulation runs for 60 s, and the disturbance generates a sequence of sinusoidal reference with a frequency of 0.16 Hz and an amplitude of 0.5 V to the inertial load in order to perturb the Smart Structure. The vibration will exceed the maximum that the equipment can withstand under the uncontrolled conditions (without active control). Consider FSF control $U_{FSF} = -K_1 x - K_2\theta - K_3\dot{x} - K_4\dot{\theta}$, as represented in Figures 6.4 and 6.5, the amplitude of $\theta°$ which shows the angle of the pendulum is between −6 and 6 and x (cm)

Table 6.1 Parameters of the Quanser test platform.

Symbol	Description	Value	Unit
m_p	Mass of the cross beam	0.05	kg
L_p	Length of the cross beam	0.15	m
m_{rb}	Mass of rigid beam	0.072	kg
l_p	Length of rigid beam	0.292	m
J_p	Moment of inertia of bothrigid beams	4.09E-3	kg m^2
m_m	Mass of motor and fixture	0.60	kg
m_b	Mass of flexible beam	0.22	kg
L_b	Length of flexible beam	0.44	m
J_b	Moment of inertia of flexible beam	0.039	kg m^2
K_s	Stiffness of flexible beam	20.1	N/m
f_n	Natural frequency	0.90	Hz

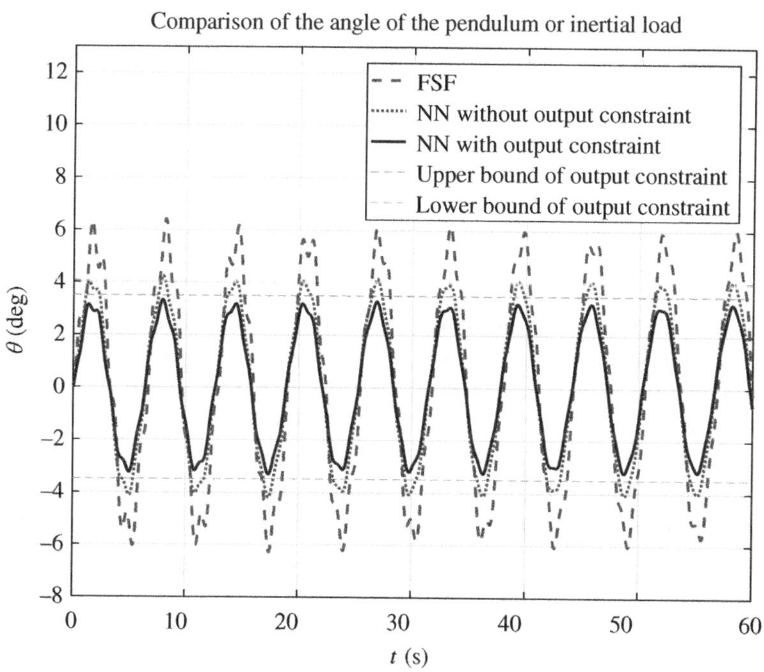

Figure 6.4 Simulation comparison of the angle of the pendulum.

which shows the simulated beam deflection converges to −0.4 to 0.4 in 60 s when the FSF control is employed, where control parameters are chosen as $K_1 = -1.1462,\ K_2 = 8.2734,\ K_3 = 8.5538$, and $K_4 = 1.2546$.

The amplitude of $\theta°$ is between −4 and 4 and x (cm) converges to −0.2 to 0.2 in 60 s when the NN control without output constraint is employed as $U_{NN} = -e_1 - K_2e_2 + \hat{f}_{dis} + \hat{W}^T S(Z)$, where control parameters are $K_1 = 15,\ K_2 = 2$. The parameters of NNs are set as $\sigma = 0.2,\ \Gamma = I_{256\times256}$. The center parameter of $S_i(Z)$ is chosen as either −1 or 1. The variance of centers is selected as $\eta = 1$.

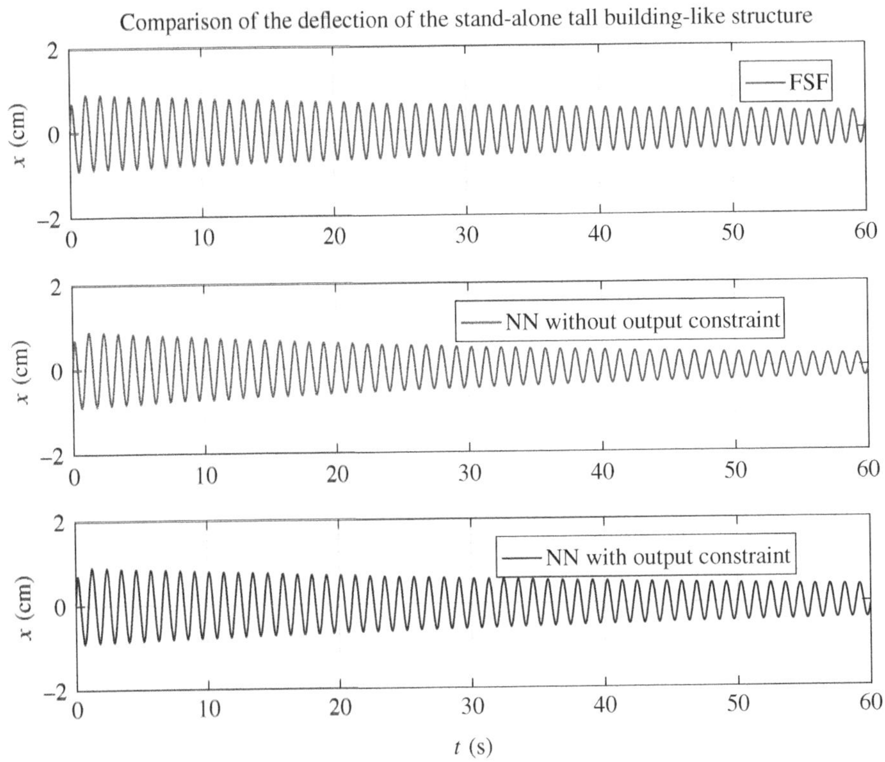

Figure 6.5 Simulation comparison of the deflection of the structure.

As shown in Figure 6.4, compared with the simulation results of FSF control, the peak value of the angle of the pendulum $\theta°$ is obviously reduced by 33.3%. As shown in Figure 6.5, it can be seen that in 60 s, the deflection of the flexible beam x (cm) converges to 0.2 cm, which is 50% less than the simulation results of FSF control. Therefore, the designed NN controller can obviously suppress the vibration of the flexible stand-alone tall building-like structure.

Considering output constraints, the NN control with output constraint is employed as Eq. (6.52), where the bound of output constraint is set as $\pm 3.5°$, control parameters are chosen as $K_1 = 35,\ K_2 = 1$. The amplitude of $\theta°$ is between -3 and 3 and x (cm) converges to -0.3 to 0.3 in 60 s. Comparison of control inputs is shown in Figure 6.6. As shown in Figure 6.4, compared with the simulation results of NN control without output constraint, the peak value of the angle of the pendulum $\theta°$ is obviously reduced by 25%. As shown in Figure 6.5, the deflection of the flexible beam x (cm) converges to 0.3 cm in 60 s, which is 25% less than the simulation results of FSF control. Therefore, the designed NN controller with output constraint can achieve obvious output constraint control and vibration suppression. The simulation comparison of the accelerations of the flexible structure is shown in Figure 6.7. Both control methods are successfully adopted to simulate of flexible structure system but NN control performance is more successful than FSF control according to acceleration criteria.

The frequency analysis of the deflections (simulation) of the flexible structure under the three controllers is shown in Figure 6.8. Given the same disturbance excitation, the dominant

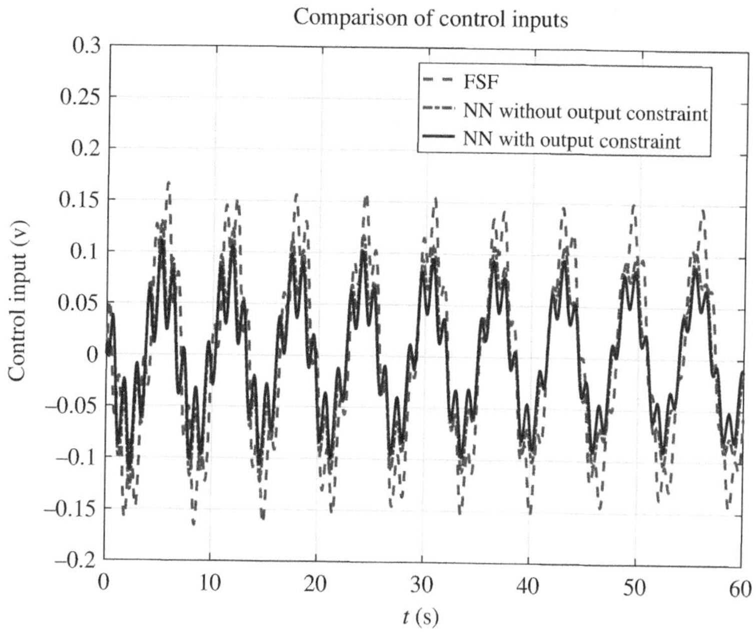

Figure 6.6 Simulation comparison of control inputs.

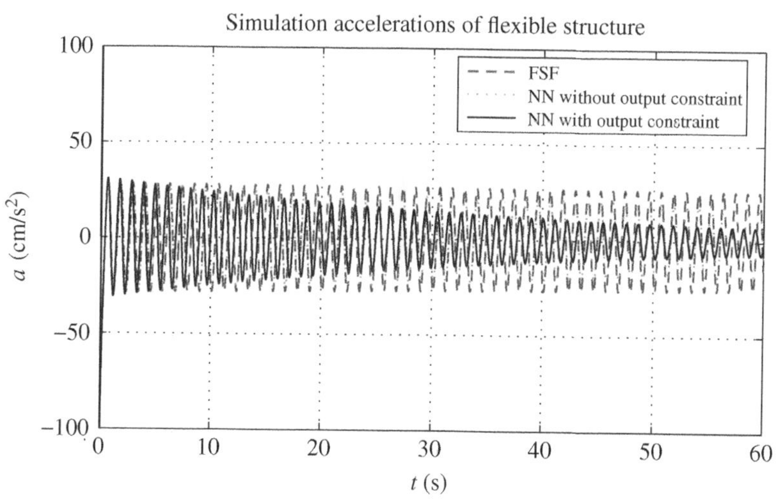

Figure 6.7 Simulation comparison of the accelerations of the flexible structure.

frequencies of the deflections are 2.8 Hz (FSF controller), 2.7 Hz (NN controller), and 2.7 Hz (NN controller with output constraint), respectively. Note that the dominant frequencies under the three controllers are far from the natural frequency of the smart structure system (0.9 Hz), it is thus evident to say that the three controllers are successful to avoid the building collapse from frequency resonance.

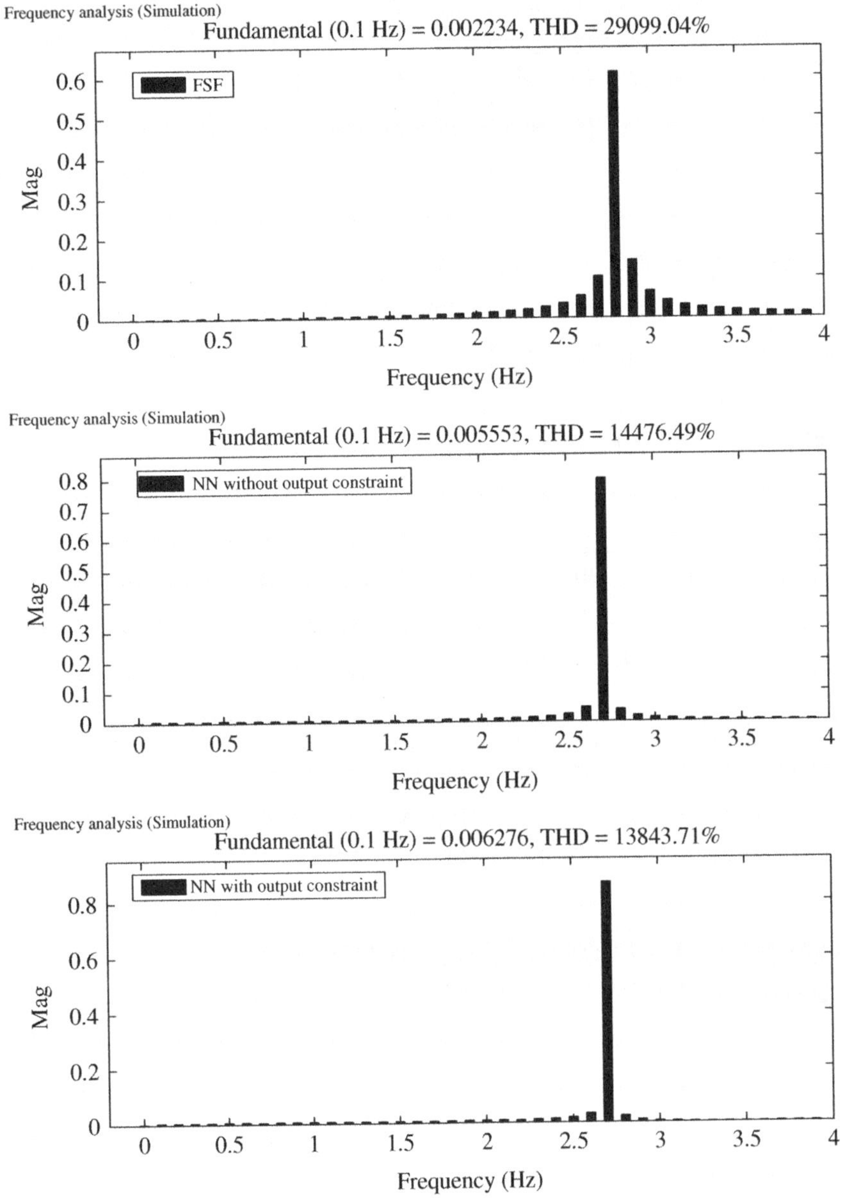

Figure 6.8 Frequency analysis of the deflections (simulation) of the flexible structure under the three controllers.

6.5 Experimental Investigation

6.5.1 Introduction of the Experimental Testbed

The Quanser Smart Structure is a stand-alone bench-scale model of a tall building instrumented with an eccentric load. This system shown in Figure 6.9 consists of a flexible beam mounted on a base support plate and a DC motor driving an eccentric load via a gearbox. It can be used to study control strategies of suppressing vibrations in tall buildings. A large pendulum with an inertial

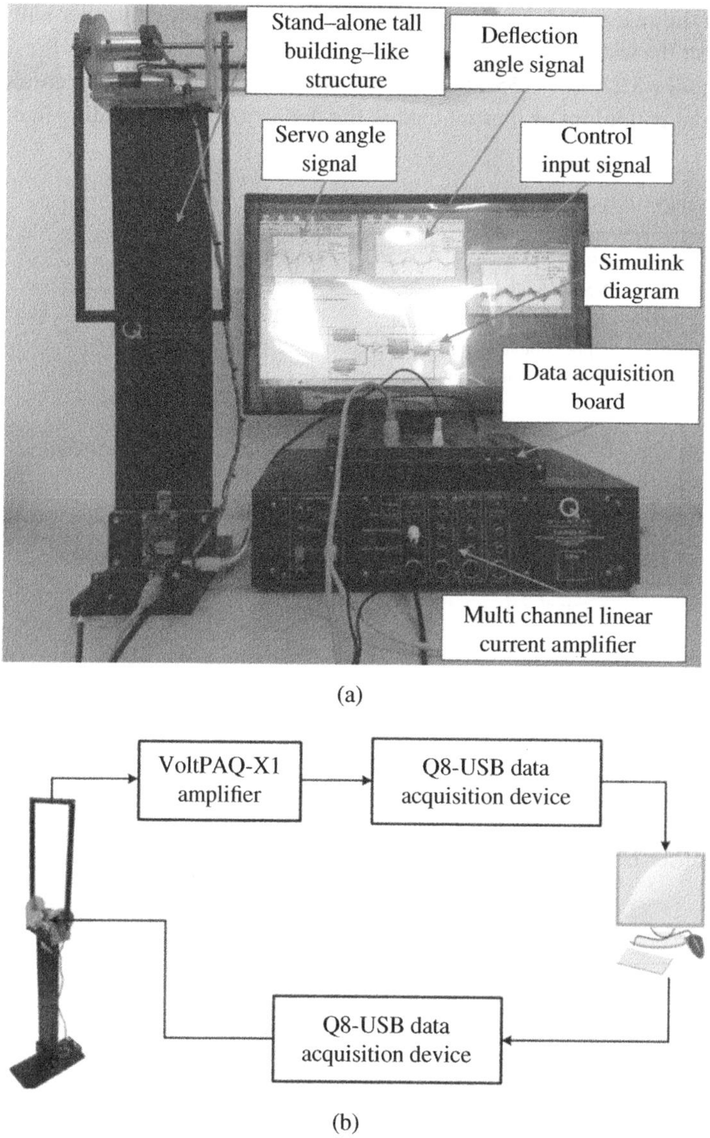

Figure 6.9 Flexible tall building-like structure with an eccentric load. (a) Laboratory setup. (b) Schematic showing details.

load is used to actively compensate the swaying motion of the structure. The system also includes a strain gage sensor to measure the flexible deflection and an optical encoder to measure the angle of the load. Detailed parameters are given in Table 5.1.

6.5.2 Experimental Results

In the experiment, the angle position signal of the pendulum θ and the linear motion of motor fixture x measured by the encoder and strain gauge can be observed in real time by the oscilloscope. A sinusoidal disturbance with a frequency of 0.16 Hz and an amplitude of 0.5 V is added to the flexible structure. Before start, the inertial load (rigid beam and cross beam) should be rotated until it is in the upright vertical position. The Simulink diagram contains a limit module which limits θ to less than 25°. When the swing angle of the pendulum reaches $\pm 25°$, the controller will automatically turns off to protect the test equipment.

FSF control with $K_1 = -1.15, K_2 = 8.27, K_3 = 8.55$, and $K_4 = 1.25$ is introduced. The experimental results including the angle of the pendulum or inertial load θ, the deflection of the building-like

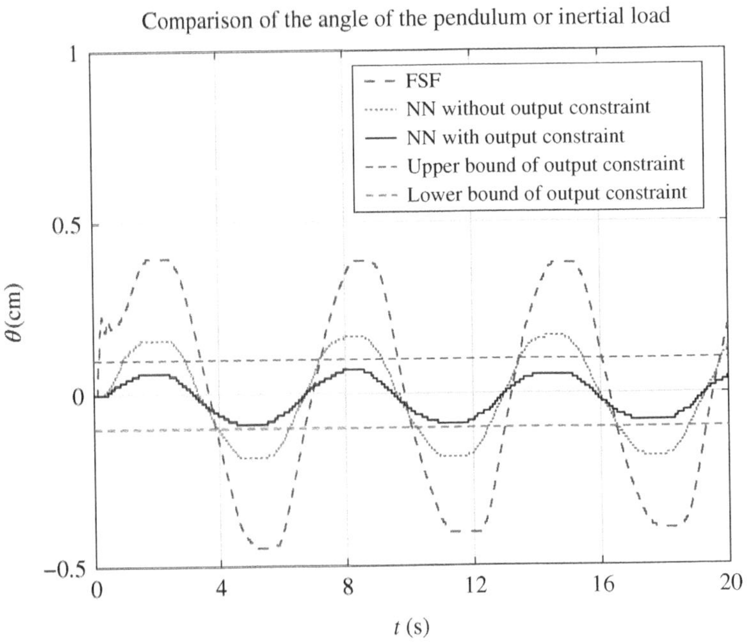

Figure 6.10 Experimental comparison of the angle of the pendulum.

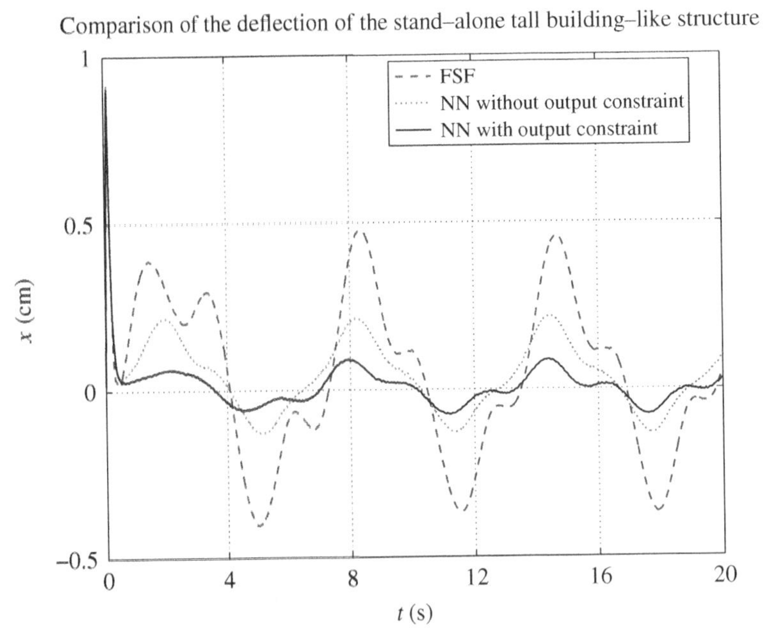

Figure 6.11 Experimental comparison of the deflection of the structure.

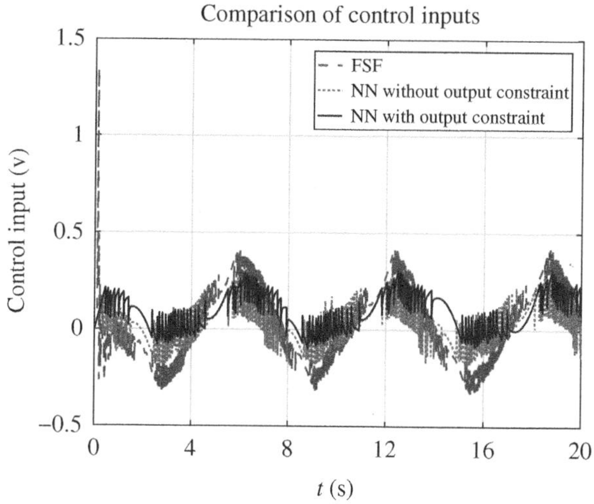

Figure 6.12 Experimental comparison of control inputs.

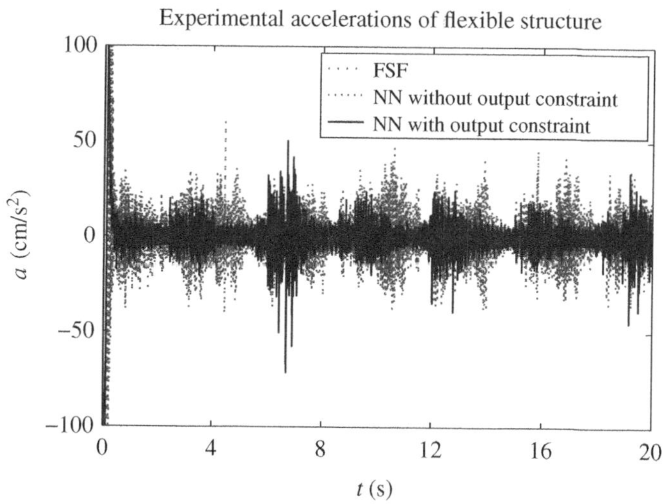

Figure 6.13 Experimental comparison of accelerations of flexible structure.

structure x, and control input V are shown in Figures 6.10–6.12. It can be observed that the amplitude of θ° under FSF control reaches 4, and the amplitude of the end displacement x (cm) is 0.5.

Therefore, NN controller without output constraint is applied on the Smart Structure. Parameters are chosen as $K_1 = 20$ and $K_2 = 2$. The experimental results including the angle of the inertial load θ, the deflection of the structure x, and control input V are shown in Figures 6.10–6.12. The amplitude of θ° under NN control without output constraint is 2, the amplitude of the end displacement x (cm) is 0.2. Compared with the experimental results of FSF control, the peak value of the angle of the pendulum θ° is obviously reduced by 50%, the deflection of the flexible beam x (cm) is 60% less than the experimental results of FSF control. Therefore, the control effect of the designed NN controller is better than that of the FSF controller.

Considering output constraints, the NN controller with output constraint is applied on the Smart Structure. The upper and lower bounds of θ are set to 1 and −1 cm, parameters are chosen as $K_1 = 35$

and $K_2 = 1.3$. The experimental results including the angle of the inertial load θ, the deflection of the structure x, and control input V are shown in Figures 6.10–6.12. The amplitude of $\theta°$ is constrained within 1. The amplitude of the end displacement x (cm) is 0.1. Compared with the experimental results of NN control without output constraint, the peak value of the angle of the pendulum $\theta°$ is obviously reduced by 50%. As shown in Figure 6.11, it can be seen that the deflection of the flexible beam x (cm) is 50% less than the experimental results of NN control without output constraint. The experimental acceleration results are shown in Figure 6.13. Therefore, the designed NN controller

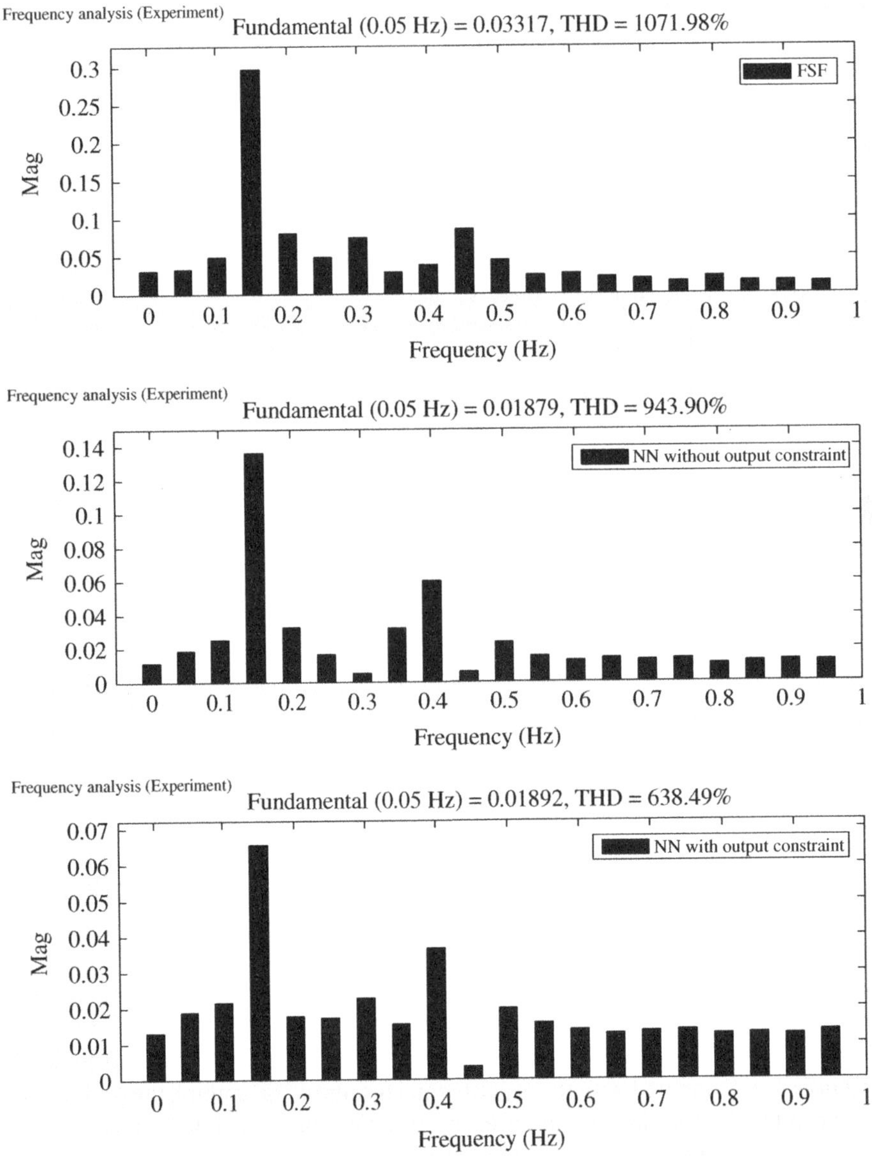

Figure 6.14 Frequency analysis of the deflections (experimental) of the flexible structure under the three controllers.

with output constraint can achieve obvious output constraint control and vibration suppression. And NN control performance is more successful than FSF control according to acceleration criteria.

Three controllers can successfully stabilize the smart structure. Compared with FSF controller, NN controller has strong anti-interference ability, and the amplitudes of θ, and x are smaller under the interference. Considering the output constraint, NN controller with output constraint is employed and θ is constrained to less than 1° successfully.

The frequency analysis of the deflections (experimental) of the flexible structure under the three controllers is shown in Figure 6.14. Given the same disturbance excitation, the dominant frequencies of the deflections are 0.15 Hz under all the three controllers. Note that the dominant frequencies under the three controllers are far from the natural frequency of the smart structure system (0.9 Hz), it is thus evident to say that the three controllers are successful to avoid the building collapse from frequency resonance.

6.6 Summary

In this chapter, the dynamic model of the tall building-like structure with an eccentric load is presented by using AMM. A RBFNN controller is designed to suppress the vibration of the flexible building. The constraint of output variables of smart structure is realized by using the barrier Lyapunov function. SGUUB of the closed-loop system is realized by Lyapunov direct method. A large number of simulations and experiments based on MATLAB and Quanser are carried out to verify the feasibility and effectiveness of the proposed control.

The presented results open a new route to vibration control of the tall buildings. In terms of computational demand, the proposed finite dimensional dynamic model effectively avoids the problems which are caused by the unmeasurable parameters in infinite dimensional distribution parameter systems. From a control design point of view, the proposed NN controller successfully deals with the challenge of system uncertainties while achieving vibration suppression. Moreover, the experimental investigation in this study may be very promising in applications of NN control in earthquake engineering.

The construction of super-high-rise structure brings great challenges to vibration control of the flexible structures. Future works include exploration of control algorithms that are suitable for a wide variety of tall building structures, the design of reinforcement learning control and the study on the vibration control of high-rise buildings with active mass damper.

7

Adaptive Vibration Control of a Flexible Structure Based on Hybrid Learning Controlled Active Mass Damping

After reading this chapter, you should be able to:

- Discusses an adaptive vibration control method for a single-floor building-like structure equipped with an active mass damper (AMD)
- Uses a hybrid learning control strategy to suppress vibrations caused by unknown time-varying disturbances such as earthquakes or strong winds
- Aim to bring new ideas and methods to the field of disaster reduction for engineering development

7.1 Introduction

The flexibility and complexity of the building structural system increase with the height. Coupled with the unpredictability of natural disasters such as earthquakes and strong winds, earthquake-resistant, and wind-resistant designs for ultra-high and high-rise building structures face new challenges [82, 90].

Under the excitation of complex environments (earthquakes, strong winds, etc.), the elastic deformations and vibrations of ultra-high and high-rise building structures cannot be ignored [61, 343]. These two types of building structures should be regarded as the flexible systems [269, 298]. Compared with the traditional rigid structure systems, the flexible structure systems are more widely applied in the field of manufacturing, medicine, and aerospace technology, such as flexible robotic manipulators, marine risers, bionic flapping-wing aircraft, and bridge structures [112, 119, 362]. As the flexible systems have very complicated dynamics and vibration characteristics [320, 329], such systems are thus usually described by distributed parameter systems based on partial differential equations (PDEs). Liu et al. [195] presented a PDEs system and employed the classical backstepping method, which avoided the problem of solving the Volterra mapping kernel function and greatly reduced the computation burden. The modeling of building-like structures is more complicated than that of industrial control system. This lies in the fact that the building-like structure is generally described by the PDEs model. Even if the precise finite element method (FEM) is adopted, the degree of freedom (DOF) of this type of model is infinite. It is noted that the control directly based on PDEs model requires more system information. Furthermore, the difficulty in control design will increase exponentially with the rising of uncertainty and disturbance problem. Therefore, it is of great theoretical and practical significance to establish simpler and more suitable models of flexible structures in complex environments.

Dynamic Modeling and Neural Network-Based Intelligent Control of Flexible Systems,
First Edition. Hejia Gao, Wei He, and Changyin Sun.

Motivated by the above reasons, it is imperative to transform the PDEs model of the flexible structures into the ordinary differential equations (ODEs) model [19]. At present, many modeling methods have been developed to analyze the vibration dynamics of the flexible structures [263]. Mode is the inherent property of the structural systems and can be used to analyze the vibration behavior. The vibration analysis of rectangular plates and stiffened panels subjected to point excitation force and enforced displacement at boundaries was conducted based on assumed mode method (AMM) by Cho et al. [41]. Considering the *n*th-order elastic mode, Gao et al. [80] derived the finite-dimensional ODEs model of a flexible beam with an unknown spatiotemporally disturbance in tangent coordinate system using AMM. As a type of colossal structures, the DOFs of high-rise buildings are infinite [253]. Even with relatively accurate finite element method, the model of the flexible building is very complicated. The AMM provides a new solution for the dimension reduction of high-rise building models.

The vibration control of flexible buildings under natural disasters has also attracted the attention of many researchers [73, 319]. Choi et al. [43] designed a magneto rheological (MR) damper-based intelligent passive control system using an electromagnetic induction device as a self-powered source. However, manufacture and installation of the device will lead to new challenges in control system design. To address these issues, Nguyen and Hong [215] proposed an economical friction damper device. Although the feasibility of the damping system in single-story frames was experimentally demonstrated, the rapidity, and robustness of vibration control need to be improved. Thenozhi and Wen [273] presented a model of building structures and designed an active mass damper (AMD) PD/PID control for a two-story building prototype. Based on a continuous-time system model, a H_2 control strategy was developed for controlling the lateral vibration of a jacket-type offshore platform by applying an AMD [172]. The performances of most of the reported control algorithms are directly related to the structural parameters. The uncertainties of the system parameters will lead to deterioration of control performance. However, due to the factors such as nonlinearity, time variability, and uncertainty in practical systems, it is hard to obtain the accurate mathematical model. It also should be noted that the nonconstant or time-invariant terms may exist as dynamic uncertainties and the system states are not usually all directly measurable in engineering applications. Vibration control of flexible buildings in the presence of system uncertainties has also been investigated by a number of researchers. Bani-Hani [11] developed a robust neural network (NN) approach for vibration mitigation of tall buildings under wind excitations, where stiffness uncertainty, time-delay, measurement noise, and sensor failure were addressed. Ding et al. [57] investigated a hybrid controller for low-frequency vibration isolation system in the presence of noises. Some control strategies have been successfully applied to the real high-rise buildings for vibration suppression in the presence of natural disasters. Aldawod et al. [6] developed a fuzzy logic controller for the Melbourne office tower building (76 stories, 306 m) to suppress vibration under wind excitation. In 1989, the world's first high-rise Kyobashi Seiwa building (11 stories) with active mass dampers was built in Tokyo, Japan. During the 1989 offshore earthquake, the seismic displacement response of the building was reduced by about 60%. In 1990, the building was subjected to a strong typhoon, and its wind-induced acceleration response was reduced by about 50%. AMD system is also installed in Nanjing TV Tower to reduce wind-induced vibration. In addition to the problems of uncertainties, time-varying and external disturbances, it is also critical to improve the vibration control performance as the requirements by safety, stability, and comfort of the buildings.

A number of advanced vibration control strategies have been successfully applied to the flexible systems [84]. These strategies provide new ideas for improving the immunity of the high-rise buildings for natural disasters. Liu et al. [196] proposed an adaptive NN control scheme for a flexible spacecraft, which effectively dealing with the parameter uncertainties and input nonlinearity. In [264], a finite-time sampled-data control based on fuzzy model was designed for nonlinear flexible spacecraft with stochastic actuator failures. Gao et al. [81] designed an intelligent NN control method to achieve vibration suppression of flexible robotic manipulators. Fang et al. [67] designed a backstepping fuzzy sliding mode control for a flexible cantilever beam to achieve vibration control performance. Jamshidifar et al. [140] proposed a robust and optimal vibration compensator using the linear parametric variable H_∞ control design techniques to realize the trajectory tracking and vibration control of redundant cable-driven parallel robots. In [198] and [197], the adaptive boundary control approach was developed for flexible systems, which successfully realize asymmetric input-output constraints.

In this chapter, dynamic modeling and hybrid learning (HL) control design are carried out for a single-floor building-like structure. Experiments on Quanser One-Floor AMD Structure are implemented to show the effectiveness of the HL strategy. The intensive experimental results show that, compared with the passive mode and the proportional-velocity (PV) position control, the proposed control policy can achieve smaller vibrations and smaller accelerations in a shorter time so as to guarantee the comfort and the safety of the residents under strong winds or earthquakes. The main contributions in this work are briefly summarized as:

1) A new finite-dimensional dynamic model for a single-floor building-like structure (with an AMD) is derived in this chapter. While this new model reduces the mathematical complexity, it can still ensure the acceptable modeling accuracy at a lower cost of computation.
2) An adaptive hybrid learning control based on full-order state observer is designed for vibration suppression of flexible structures. This method can approximate system uncertainties within any desired accuracy. Moreover, it is equipped with hybrid learning ability to ensure the convergene of multiple errors to zero through offline trainning of NNs.
3) An external disturbance observer is designed with utilization of the learning ability of the NN. Unlike the traditional disturbance observer technology, the designed observer avoids trial-and-error method to tunc parameters and it has significantly reduced the computational load in counteracting the unknown disturbance.

The remainder of the chapter is structured as follows. A new finite-dimensional dynamic model of the single-floor building-like structure is constructed by employing AMM in Section 7.2. A hybrid controller based on NN structure with DO is designed and the semi-global uniform ultimate boundedness (SGUUB) of the closed-loop system is proved in Section 7.3. Experimental results are given in Section 7.4 to verify the algorithm performance, followed by the conclusive remarks and the envision of the future work.

7.2 Dynamic Modeling

The dynamic modeling is considered in this section, and all the symbols and parameters used in the chapter are shown in Table 7.1 for clearance. The following notations are also used in this chapter: $(\dot{*}) = \frac{\partial *}{\partial t}$, $(\ddot{*}) = \frac{\partial^2(*)}{\partial t^2}$, $(*)' = \frac{\partial *}{\partial y}$, $(*)'' = \frac{\partial^2(*)}{\partial y^2}$, $(*)''' = \frac{\partial^3(*)}{\partial y^3}$, $(*)'''' = \frac{\partial^4(*)}{\partial y^4}$.

As illustrated in Figure 7.1, a schematic of the single-floor building-like structure with an AMD is represented. Based on energy analysis and Hamilton's principle, the motion equations are derived as follows

$$\rho \ddot{v}_l(y,t) + EIv_l''''(y,t) = 0, \tag{7.1}$$

$$\rho \ddot{v}_r(y,t) + EIv_r''''(y,t) = 0, \tag{7.2}$$

Table 7.1 Nomenclatures and symbols used in dynamic modeling.

Symbols	Parameters
H_f	Height of the building
M_f	Mass of the floor
M_c	Mass of the mass damper
$v_l(y, t)$	Elastic deflection of the building's left wall at the position y for time t
$v_r(y, t)$	Elastic deflection of the building's right wall at the position y for time t
EI	The bending stiffness of the flexible wall
ρ	The mass per unit length of the flexible wall
J_c	Cart rotor moment of inertia
d_{ex}	Unknown time-varying disturbance
$u(t)$	The input torque

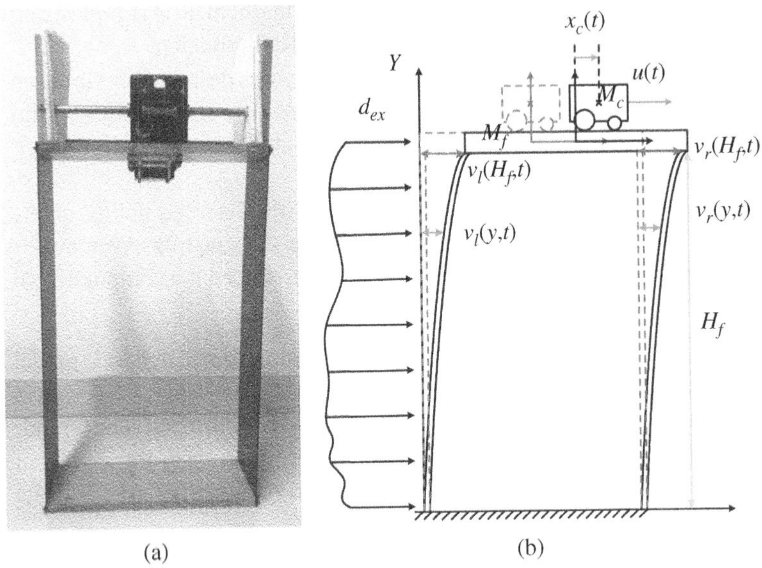

Figure 7.1 System dynamic modeling. (a) Stationary state. (b) System representation and notations.

and the boundary conditions are expressed as

$$EI[v_r'''(H_f, t) + v_l'''(H_f, t)] = M_f\ddot{v}(H_f, t) + J_c\ddot{x}_c(t), \tag{7.3}$$

$$J_c\ddot{x}_c(t) + M_c[\ddot{x}_c(t) + \ddot{v}(H_f, t)] = u(t), \tag{7.4}$$

$$v_l(H_f, t) = v_r(H_f, t) = v(H_f, t), \tag{7.5}$$

$$v_l(0, t) = v_l'(0, t) = v_l''(H_f, t) = 0, \tag{7.6}$$

where $v_l = v_r$ is employed. In the AMM modeling, the elastic vibration of the flexible beam is generally assumed to be of the form

$$v(y,t) = \sum_{i=1}^{\infty} \varpi_i(y)p_i(t), \tag{7.7}$$

where $\varpi_i(y) = \beta_1 \sin(\alpha y) + \beta_2 \cos(\alpha y) + \beta_3 \sinh(\alpha y) + \beta_4 \cosh(\alpha y)$ is the mode shape function, $p_i(t) = \exp(jwt)$ is the generalized coordinate.

In the constrained mode method, $\ddot{x}_c = 0$ is assumed. Based on (7.1), (7.3), (7.4), (7.5), and (7.6), one obtains

$$\beta_1 = -\beta_3,\ \beta_2 = -\beta_4, \tag{7.8}$$

$$M_f \varpi(H_f)\ddot{p}(t) = 2EI\varpi'''(H_f)p(t), \tag{7.9}$$

$$\varpi''(H_f) = 0, \tag{7.10}$$

$$EI\alpha^4 = -\rho j^2 \omega^2. \tag{7.11}$$

The derivatives for $\varpi(y)$ and $p(t)$,

$$\varpi'(y) = \alpha[\beta_1 \cos(\alpha y) - \beta_2 \sin(\alpha y) - \beta_1 \cosh(\alpha y) - C_2 \sinh(\alpha y)], \tag{7.12}$$

$$\varpi''(y) = \alpha^2[-\beta_1 \sin(\alpha y) - \beta_2 \cos(\alpha y) - \beta_1 \sinh(\alpha y) - C_2 \cosh(\alpha y)], \tag{7.13}$$

$$\varpi'''(y) = \alpha^3[-\beta_1 \cos(\alpha y) + \beta_2 \sin(\alpha y) - \beta_1 \cosh(\alpha y) - C_2 \sinh(\alpha y)], \tag{7.14}$$

$$\varpi''''(y) - \alpha^4[\beta_1 \sin(\alpha y) + \beta_2 \cos(\alpha y) - \beta_1 \sinh(\alpha y) - \beta_2 \cosh(\alpha y)], \tag{7.15}$$

$$\ddot{p}(t) = j^2\omega^2 \exp(jwt). \tag{7.16}$$

Substituting $p_i(t)$ and Eq. (7.9) into Eq. (7.10), one obtains

$$\beta_1[\sin(\alpha H_f) + \sinh(\alpha H_f)] + \beta_2[\cos(\alpha H_f) + \cosh(\alpha H_f)] = 0, \tag{7.17}$$

$$M_f\alpha[\beta_1(\sin(\alpha H_f) - \sinh(\alpha H_f)) + \beta_2(\cos(\alpha H_f) - \cosh(\alpha H_f))]$$
$$= 2\rho[-\beta_1(\cos(\alpha H_f) + \cosh(\alpha H_f)) + \beta_2(\sin(\alpha H_f) - \sinh(\alpha H_f))], \tag{7.18}$$

and define $s_1 = \sin(\alpha H_f) + \sinh(\alpha H_f)$, $s_2 = \sin(\alpha H_f) - \sinh(\alpha H_f)$, $c_1 = \cos(\alpha H_f) + \cosh(\alpha H_f)$, $c_2 = \cos(\alpha H_f) - \cosh(\alpha H_f)$, that is

$$\beta_1 s_1 + \beta_2 c_1 = 0, \tag{7.19}$$

$$M_f\alpha(\beta_1 s_2 + \beta_2 c_2) = 2\rho(-\beta_1 c_1 + \beta_2 s_2). \tag{7.20}$$

Therefore, $\varpi(y)$ can be further written as

$$\varpi(y) = \beta_1[\sin(\alpha y) - \sinh(\alpha y) + \gamma(\cos(\alpha y) - \cosh(\alpha y))], \tag{7.21}$$

where

$$\gamma = -\frac{\sin(\alpha H_f) + \sinh(\alpha H_f)}{\cos(\alpha H_f) + \cosh(\alpha H_f)}. \tag{7.22}$$

Design $F(\alpha) = \begin{bmatrix} A_1 & A_2 \\ A_3 & A_4 \end{bmatrix}$. Considering Eqs. (7.19) and (7.20), thus, $F(\alpha) \begin{bmatrix} \beta_1 \\ \beta_2 \end{bmatrix} = 0$, that is $|F(\alpha)| = 0$.

$$A_1 = s_1, \tag{7.23}$$

$$A_2 = c_1, \tag{7.24}$$

$$A_3 = M_f \alpha s_2 + 2\rho c_1, \tag{7.25}$$

$$A_4 = M_f \alpha c_2 - 2\rho s_2. \tag{7.26}$$

Therefore, $A_1 * A_4 - A_2 * A_3 = 0$. α is the solution of the following function.

$$s_1(M_f \alpha c_2 - 2\rho s_2) - c_1(M_f \alpha s_2 + 2\rho c_1) = 0. \tag{7.27}$$

By introducing the following orthogonal conditions

$$\int_0^{H_f} \varpi_i(y)\varpi_j(y)dy + \frac{M_f + M_c}{\rho} \varpi_i(H_f)\varpi_j(H_f) = \begin{cases} 0, & i \neq j \\ 1, & i = j \end{cases}$$

$$EI \int_0^{H_f} \varpi_i''(y)\varpi_j''(y)dy = \begin{cases} 0, & i \neq j \\ \omega_i^2 \rho, & i = j \end{cases}$$

We have

$$\beta_1 = 1 \bigg/ \left[H_f + \frac{\rho H_f^2}{(M_f + M_c)\alpha^2} \left(\frac{1 + \cosh\alpha \cos\alpha}{\sinh\alpha + \sin\alpha} \right)^2 \right]^{\frac{1}{2}}. \tag{7.28}$$

On the basis of the discretization, the dynamic model is obtained by satisfying

$$\frac{d}{dt}\frac{\partial(E_k - E_p)}{\partial \dot{q}_i} - \frac{\partial(E_k - E_p)}{\partial q_i} = f_i, \; i = 1, \dots, N. \tag{7.29}$$

where q_i are the generalized coordinates and f_i are the generalized forces.

The total kinetic energy of the single-floor building-like structure (with an AMD) is given by

$$\begin{aligned} E_k(t) &= \frac{1}{2}\int_0^{H_f} \rho[\dot{v}_l(y,t)]^2 dy + \frac{1}{2}\int_0^{H_f} \rho[\dot{v}_r(y,t)]^2 dy + \frac{M_f}{2}[\dot{v}(H_f,t)]^2 \\ &\quad + \frac{M_c}{2}[\dot{x}_c(t) + \dot{v}(H_f,t)]^2 + \frac{1}{2}J_c[\dot{x}_c(t)]^2, \\ &= \sum_{i=1}^{N} \dot{p}_i^2 \rho - \frac{1}{2}(M_f + M_c)\sum_{i=1}^{N} \dot{p}_i^2 \varpi_i^2(H_f) + M_c \dot{x}_c \sum_{i=1}^{N} \dot{p}_i \varpi_i(H_f) \\ &\quad - (M_f + M_c) \sum_{i,j=1, i\neq j}^{N} \dot{p}_i \dot{p}_j \varpi_i(H_f)\varpi_j(H_f) + \frac{1}{2}(M_c + J_c)\dot{x}_c^2. \end{aligned} \tag{7.30}$$

The potential energy is given by

$$E_p = \frac{1}{2}EI \int_0^{H_f} \{[v_l''(y,t)]^2 + [v_r''(y,t)]^2\} dy. \tag{7.31}$$

As a result of this procedure, the equation of motion for the smart structure can be expressed as

$$B\ddot{Q} + KQ + d_{ex} = U. \tag{7.32}$$

where $Q = (x_c, p_1, ..., p_N) \in \mathbb{R}^{N+1}$ is the generalized coordinates and $U \in \mathbb{R}^{N+1}$ is the joint torques. $B \in \mathbb{R}^{(N+1)\times(N+1)}$ is the inertia matrix and $K \in \mathbb{R}^{(N+1)\times(N+1)}$ is the stiffness matrix. $d_{ex} \in \mathbb{R}^{N+1}$ represents an external disturbance. The elements of B and K are defined as follows

$$B[1,1] = M_c + J_c,$$
$$B[1,j] = B[j,1] = \frac{1}{4}M_c\varpi_{j-1}(H_f),\ j = 2, ..., N+1,$$
$$B[j,j] = \rho - \frac{M_f + M_c}{2}\varpi_{j-1}^2(H_f),\ j = 2, ..., N+1,$$
$$B[j,n] = B[n,j] = \frac{1}{4}(M_f + M_c)\varpi_{j-1}(H_f)\varpi_{n-1}(H_f),\ j = 2, ..., N+1;\ n = j+1, ..., N+1,$$
$$K = \text{diag}[0\ 2\omega_1^2\rho\ 2\omega_2^2\rho\ ...\ 2\omega_N^2\rho].$$

7.3 Hybrid Learning Control

The purpose of the chapter is to design a control system that dampens the vibration of a bench-scale building-like tall structure using an AMD mounted on the top. It is conceptually similar to suppress vibrations caused by earthquakes or strong winds in real tall structures (e.g. high-rise buildings) using the AMD. The diagram of the HL control strategy is shown in Figure 7.2.

7.3.1 Disturbance Observer Design

Lemma 7.1 If a continuously differentiable function $f(t)$ satisfies $|f(t)| \leq \zeta$, $\forall t \in [t_1,\ t_2]$, where ζ is a positive constant, then $\dot{f}(t)$ is bounded, $\forall t \in [t_1,\ t_2]$ [179].

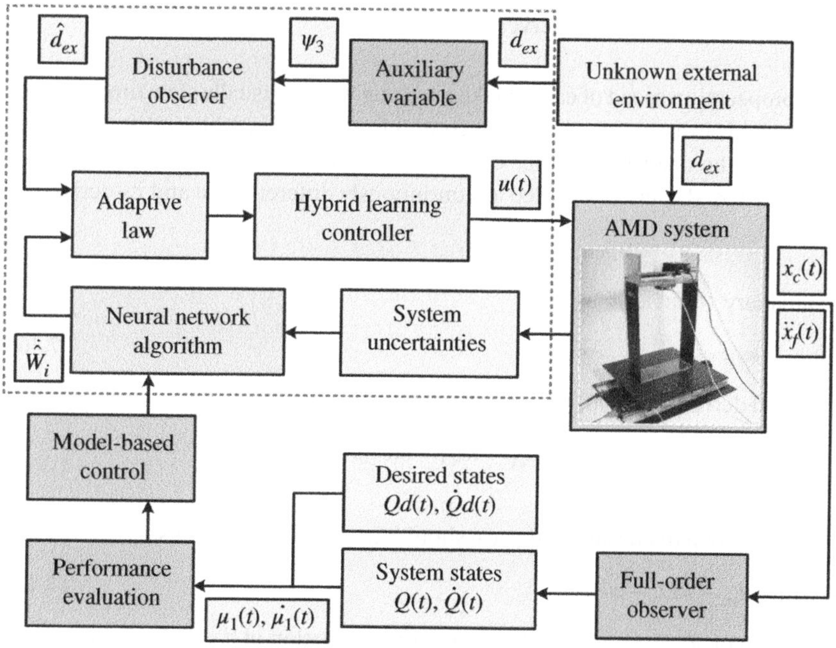

Figure 7.2 Diagram of NN control strategy.

The dynamic equation of the single-floor building-like structure (with an AMD) can be transformed into the following form by Eq. (7.32), as

$$\ddot{Q} = B^{-1}[u - d_{ex} - KQ]. \tag{7.33}$$

As for the system errors of vibration signals, we define two variables e_1 and e_2,

$$e_1 = Q - Q_d, \tag{7.34}$$

$$e_2 = \dot{Q} - \mu_1, \tag{7.35}$$

where $Q_d \in \mathbb{R}^{N+1}$ represents the desired position of the single-floor building-like structure without any external interference. $\mu_1 = -K_1e_1 + \dot{Q}_d$ is designed as a virtual variable with $K_1 = K_1^T > 0$.

Calculate the derivatives of e_1 and e_2 with respect to time t (by determining de_i/dt), the outcomes are

$$\dot{e}_1 = e_2 + \mu_1 - \dot{Q}_d = e_2 - K_1e_1, \tag{7.36}$$

$$\dot{e}_2 = \ddot{Q} - \dot{\mu}_1 = B^{-1}[u - d_{ex} - KQ] - \dot{\mu}_1. \tag{7.37}$$

The initial Lyapunov function F_1 is designed by relying on backstepping approach

$$F_1 = \frac{1}{2}e_1^Te_1. \tag{7.38}$$

Further, derive F_1 with respect to t and the outcome is

$$\dot{F}_1 = e_1^T(e_2 - K_1e_1) = e_1^Te_2 - e_1^TK_1e_1. \tag{7.39}$$

In practice, the unknown time-varying disturbances d_{ex} caused by earthquakes and strong winds exist. In order to suppress vibrations in tall structures and to protect against natural disasters, it is thus necessary to introduce a disturbance observer (DO).

Remark 7.1 As the propagation speed of earthquake or strong wind is usually slow time-varying in nature, it is thus reasonable to assume such speed is continuously differentiable and its derivative is bounded. Based on this, the disturbance force caused by such disaster, which is quadratically related to the propagation speed, can be regarded as continuously differentiable and its derivative is bounded accordingly.

Firstly, design an auxiliary variable ψ_3 as

$$\psi_3 = d_{ex} + \Phi(e_2), \tag{7.40}$$

where $\Phi(e_2) \in \mathbb{R}^{N+1}$. And derive Eq. (7.40) with respect to t,

$$\dot{\psi}_3 = \dot{d}_{ex} + o(e_2)\dot{e}_2 = \dot{d}_{ex} + o(e_2)B^{-1}[u - d_{ex} - KQ] - o(e_2)\dot{\mu}_1,$$

where $o(e_2) = \partial\Phi(e_2)/\partial e_2^T \in \mathbb{R}^{(N+1)\times(N+1)}$.

To approximate the unknown disturbances d_{ex}, we should first estimate ψ_3. Thus,

$$\dot{\hat{\psi}}_3 = o(e_2)B^{-1}[u - \hat{d}_{ex} - KQ] - o(e_2)\dot{\mu}_1. \tag{7.41}$$

where $\hat{\psi}_3$ is the estimation of ψ_3. According to Eq. (7.40), the estimation of d_{ex} is obtained as

$$\hat{d}_{ex} = \hat{\psi}_3 - \Phi(e_2). \tag{7.42}$$

Obviously, the following equation can be obtained as

$$\tilde{\psi}_3 = \hat{\psi}_3 - \psi_3 = \hat{d}_{ex} - d_{ex} = \tilde{d}_{ex}. \tag{7.43}$$

Then, derive $\tilde{d}_{ex}$ with respect to t as

$$\dot{\tilde{d}}_{ex} = \dot{\tilde{\psi}}_3 = \dot{\hat{\psi}}_3 - \dot{\psi}_3 = -o(e_2)B^{-1}\tilde{d}_{ex} - \dot{d}_{ex}. \tag{7.44}$$

Assume that the unknown time-varying disturbances d_{ex} caused by earthquakes and strong winds are slowly changing. According to Lemma 7.1, we have $\|\dot{d}_{ex}\| \leq \varsigma_0$, where the constant $\varsigma_0 > 0$. Due to the coupling term $e_1^T e_2$ in Eq. (7.39), the new Lyapunov function F_2 based on F_1 is designed as

$$F_2 = \frac{1}{2}e_1^T e_1 + \frac{1}{2}e_2^T B e_2 + \frac{1}{2}\tilde{d}_{ex}^T \tilde{d}_{ex}. \tag{7.45}$$

According to the following inequality,

$$-\tilde{d}_{ex}^T \dot{d}_{ex} \leq \frac{1}{2}\tilde{d}_{ex}^T \tilde{d}_{ex} + \frac{1}{2}\dot{d}_{ex}^T \dot{d}_{ex} \leq \frac{1}{2}\tilde{d}_{ex}^T \tilde{d}_{ex} + \frac{1}{2}\varsigma_0^2, \tag{7.46}$$

the derivative of F_2 with respect to t is expressed as

$$\begin{aligned} \dot{F}_2 &= e_1^T e_2 - e_1^T K_1 e_1 + e_2^T(u - d_{ex} - KQ - B\dot{\mu}_1) - \tilde{d}_{ex}^T(o(e_2)B^{-1}\tilde{d}_{ex} + \dot{d}_{ex}) \\ &\leq e_1^T e_2 - e_1^T K_1 e_1 + e_2^T(u - d_{ex} - KQ - B\dot{\mu}_1) - \tilde{d}_{ex}^T o(e_2)B^{-1}\tilde{d}_{ex} + \frac{1}{2}\tilde{d}_{ex}^T \tilde{d}_{ex} + \frac{1}{2}\varsigma_0^2. \end{aligned} \tag{7.47}$$

Design a model-based control u_0,

$$u_0 = -e_1 - K_2 e_2 + \hat{d}_{ex} + KQ + B\dot{\mu}_1, \tag{7.48}$$

where the gain matrix $K_2 = K_2^T > 0$.

Consider the inequality $e_2^T \tilde{d}_{ex} \leq \frac{1}{2}e_2^T e_2 + \frac{1}{2}\tilde{d}_{ex}^T \tilde{d}_{ex}$, and substitute Eq. (7.48) into Eq. (7.47), one obtains

$$\begin{aligned} \dot{F}_2 &\leq -e_1^T K_1 e_1 - e_2^T(K_2 - \frac{1}{2}I_{(N+1)\times(N+1)})e_2 - \tilde{d}_{ex}^T o(e_2)B^{-1}\tilde{d}_{ex} + \tilde{d}_{ex}^T \tilde{d}_{ex} + \frac{1}{2}\varsigma_0^2, \\ &\leq -e_1^T K_1 e_1 - e_2^T(K_2 - \frac{1}{2}I_{(N+1)\times(N+1)})e_2 - \tilde{d}_{ex}^T[o(e_2)B^{-1} - I_{(N+1)\times(N+1)}]\tilde{d}_{ex} + \frac{1}{2}\varsigma_0^2, \\ &\leq -a_1 F_2 + b_1, \end{aligned} \tag{7.49}$$

where

$$a_1 = \min\left(2\lambda_{\min}(K_1), \frac{\lambda_{\min}(2K_2 - I_{(N+1)\times(N+1)})}{\lambda_{\max}(B)}, 2\lambda_{\max}(o(e_2)B^{-1} - I_{(N+1)\times(N+1)}\right),$$

$$b_1 = \frac{1}{2}\varsigma_0^2.$$

It should be noted that the conditions $\lambda_{\min}(K_2 - \frac{1}{2}I) > 0$ and $\lambda_{\max}(\frac{\partial\Phi(e_2)}{\partial e_2^T}B^{-1} - I) > 0$ needs to be guaranteed to ensure $a_1 > 0$.

7.3.2 Hybrid Learning Control Design

Due to nonlinearity, time-variance and uncertainty in practical systems, it is hard to obtain the accurate mathematical model. The uncertainties of the system parameters will lead to deterioration of control performance. In addition, the traditional control methods usually eliminate the error based on feedback. Although these control structures are simple, the closed-loop dynamic quality is very sensitive to the change of control gains. For the vibration control of flexible structures under sudden disasters, these methods can not meet the control system requirements.

In this work, an adaptive HL control scheme is designed with utilization of the learning ability of radial basis function (RBF) NN. The approximation to system uncertainties will be ensured within the desired accuracy for every possible input, which also shows the universal approximation capability of the proposed controller. The HL capability lies that the tracking error, vibration error, approximation error, and observer error are simultaneously converged to zero through offline training of NN. This method avoids trial-and-error method to tune parameters and achieves the offline learning ability of the system. The HL control law has the following form

$$u_1 = -e_1 - K_2 e_2 + \hat{d}_{ex} + \hat{W}^T S(Z), \tag{7.50}$$

where $\hat{W}$ and $Z = [Q, \dot{Q}, \mu_1, \dot{\mu}_1]^T$ are the estimated weight vector and the inputs of RBFNN, respectively. Subsequently, the adaptive law can be designed as

$$\dot{\hat{W}}_i = -\Gamma_i (S_i(Z) e_{2,i} + \sigma_i \hat{W}_i), \tag{7.51}$$

where σ_i is a small positive constant and $\hat{W}^*$ is the NN weights. Introducing NNs to compensate for the system uncertainties,

$$W_i^{*T} S_i(Z) = KQ + B\dot{\mu}_1 - \epsilon(Z). \tag{7.52}$$

Remark 7.2 *The disturbance observer technology requires the assumption that the lumped disturbance is continuous, differentiable, and bounded. Such assumption is unfortunately always the case in real systems since nonconstant or time-invariant terms exist in dynamic uncertainties. The NN technique is still able to deal with such type of uncertainties, while the disturbance observer technology is unavailable. Therefore, the utilization of NNs is necessary in the controller design.*

The new Lyapunov function can be expressed as

$$F_3 = F_2 + \frac{1}{2} \sum_{i=1}^{n} \tilde{W}_i^T \Gamma_i^{-1} \tilde{W}_i. \tag{7.53}$$

Apply the approximation ability of NNs designed in Eq. (7.52) and derive F_3 with respect to t,

$$\begin{aligned} \dot{F}_3 \leq{} & e_1^T e_2 - e_1^T K_1 e_1 + e_2^T (u - d_{ex} - W^{*T} S(Z) - \epsilon(Z)) \\ & - \tilde{d}_{ex}^T o(e_2) B^{-1} \tilde{d}_{ex} + \frac{1}{2} \tilde{d}_{ex}^T \tilde{d}_{ex} + \frac{1}{2} \varsigma_0^2 + \sum_{i=1}^{n} \tilde{W}_i^T \Gamma_i^{-1} \dot{\hat{W}}_i. \end{aligned} \tag{7.54}$$

Substitute Eqs. (7.50) and (7.51) into Eq. (7.54),

$$\begin{aligned} \dot{F}_3 \leq{} & -e_1^T K_1 e_1 - e_2^T \left(K_2 - \frac{1}{2} I_{(N+1)\times(N+1)} \right) e_2 + e_2^T (\hat{W}^T S(Z) - W^{*T} S(Z) - \epsilon(Z)) \\ & - \tilde{d}_{ex}^T [o(e_2) B^{-1} - I_{(N+1)\times(N+1)}] \tilde{d}_{ex} + \frac{1}{2} \varsigma_0^2 - \sum_{i=1}^{n} \tilde{W}_i^T (S_i(Z) e_{2,i} + \sigma_i \hat{W}_i), \\ \leq{} & -e_1^T K_1 e_1 - e_2^T \left(K_2 - \frac{1}{2} I_{(N+1)\times(N+1)} \right) e_2 - e_2^T \epsilon(Z) \\ & - \tilde{d}_{ex}^T \left[o(e_2) B^{-1} - I_{(N+1)\times(N+1)} \right] \tilde{d}_{ex} + \frac{1}{2} \varsigma_0^2 - \sum_{i=1}^{n} \tilde{W}_i^T \sigma_i \hat{W}_i. \end{aligned} \tag{7.55}$$

Considering the following inequality

$$-\sum_{i=1}^{n} \sigma_i \tilde{W}_i^T \hat{W}_i \leq -\sum_{i=1}^{n} \sigma_i \left(\frac{\|\tilde{W}_i\|^2}{2} - \frac{\|W_i^*\|^2}{2} \right), \tag{7.56}$$

$\dot{F}_3$ can be rewritten as

$$\dot{F}_3 \leq -e_1^T K_1 e_1 - e_2^T [K_2 - I_{(N+1)\times(N+1)}] e_2 - \sum_{i=1}^{n} \frac{\sigma_i \|\tilde{W}_i\|^2}{2} + \frac{1}{2}\|\bar{\epsilon}\|^2$$
$$-\tilde{d}_{ex}^T [o(e_2)B^{-1} - I_{(N+1)\times(N+1)}]\tilde{d}_{ex} + \frac{1}{2}\varsigma_0^2 + \sum_{i=1}^{n} \frac{\sigma_i \|W_i^*\|^2}{2},$$
$$\leq -a_2 F_3 + b_2, \tag{7.57}$$

where

$$a_2 = \min(2\lambda_{\min}(K_1), \frac{2\lambda_{\min}(K_2 - I)}{\lambda_{\max}(B)}, 2\lambda_{\max}\left(o(e_2)B^{-1} - I\right), \min_{i=1,2,\ldots,N}\left(\frac{\sigma_i}{\lambda_{\max}(\Gamma_i^{-1})}\right)\Bigg),$$
$$b_2 = \frac{1}{2}\varsigma_0^2 + \frac{1}{2}\|\bar{\epsilon}\|^2 + \sum_{i=1}^{n} \frac{\sigma_i \|W_i^*\|^2}{2}.$$

It should be noted that the conditions $\lambda_{\min}(K_2 - I) > 0$ and $\lambda_{\max}(\frac{\partial \Phi(e_2)}{\partial e_2^T}B^{-1} - I) > 0$ need to be guaranteed to ensure $a_2 > 0$.

Multiplying Eq. (7.57) by $e^{a_2 t}$, and integrating

$$F_3 \leq \left(F_3(0) - \frac{b_2}{a_2}\right)e^{-a_2 t} + \frac{b_2}{a_2} \leq F_3(0) + \frac{b_2}{a_2} \tag{7.58}$$

The following conclusions can be obtained as

$$\frac{1}{2}e_1^T e_1 \leq F_3(0) + \frac{b_2}{a_2}, \tag{7.59}$$

$$\frac{1}{2}e_2^T B e_2 \leq F_3(0) + \frac{b_2}{a_2}, \tag{7.60}$$

$$\frac{1}{2}\tilde{d}_{ex}^T \tilde{d}_{ex} \leq F_3(0) + \frac{b_2}{a_2}, \tag{7.61}$$

$$\frac{1}{2}\sum_{i=1}^{n} \tilde{W}_i^T \Gamma_i^{-1} \tilde{W}_i \leq F_3(0) + \frac{b_2}{a_2}. \tag{7.62}$$

Theorem 7.1 An unknown time-varying disturbance (earthquake, strong wind, etc.) exists in the single-floor building-like structure (with an AMD). By designing a hybrid controller based on NN structure with disturbance observer, the closed-loop system is semi-global uniformly ultimately bounded only if the initial conditions are bounded. e_1, e_2, $\tilde{d}_{ex}$, and $\tilde{W}$ will remain within compact sets Ω_{e_1}, Ω_{e_2}, $\Omega_{\tilde{d}_{ex}}$, and Ω_W

$$\Omega_{z_1} := \left\{e_1 \in \mathbb{R}^{N+1} \mid \|e_1\| \leq \sqrt{X}\right\}, \tag{7.63}$$

$$\Omega_{z_2} := \left\{e_2 \in \mathbb{R}^{N+1} \mid \|e_2\| \leq \sqrt{\frac{X}{\lambda_{\min}(B)}}\right\}, \tag{7.64}$$

$$\Omega_{\tilde{d}_{ex}} := \left\{\tilde{d}_{ex} \in \mathbb{R}^{N+1} \mid \|\tilde{d}_{ex}\| \leq \sqrt{X}\right\}, \tag{7.65}$$

$$\Omega_{\tilde{W}} := \left\{\tilde{W} \in \mathbb{R}^{l\times(N+1)} \mid \|\tilde{W}\| \leq \sqrt{\frac{X}{\lambda_{\min}(\Gamma^{-1})}}\right\}, \tag{7.66}$$

where $X = 2(F_3(0) + \frac{b_1}{a_1})$, $b_1 > 0$ and $a_1 > 0$.

7.3.3 Full-order State Observer

Based on the above proposed HL controller, the semi-globally uniformly ultimate boundedness of the closed-loop system is guaranteed. However, in practical research and engineer applications, the system states are usually not directly measurable. It is essential to design a full-order state observer.

In this chapter, despite the flexible floor's acceleration $\ddot{v}(H_f, t)$ can be directly measured, its floor deflection $v(H_f, t)$ and velocity $\dot{v}(H_f, t)$ can not be measured. Such design of the system is consistent with the real full-scale building structure where floor deflections and velocities are difficult to measure directly. However, they can be estimated from acceleration measurements, since accelerometers provide an affordable and reliable way of sensing a building dynamic behavior.

Define the state vector $[x_c(t), v(H_f, t), \dot{x}_c(t), \dot{v}(H_f, t)]^T$ as S and the output vector $[x_c(t), \ddot{v}(H_f, t)]^T$ as O. The state-space representation can be written as

$$\dot{S} = T_1 S + T_2 U, \tag{7.67}$$

$$O = T_3 S + T_4 U, \tag{7.68}$$

where T_1, T_2, T_3, and T_4 are the state-space matrices. The system input U is equal to the linear cart's DC motor voltage V_m. Define $\hat{S}$ as the estimated state vector and $\hat{O}$ as the estimation of O, a full-order observer is designed as

$$\dot{\hat{S}} = T_1 \hat{S} + T_2 U + G(O - \hat{O}), \tag{7.69}$$

$$\hat{O} = T_3 \hat{S} + T_4 U, \tag{7.70}$$

where G is a full-order observer gain matrix. It can be seen that the estimated state vector $\hat{O}$ can be calculated based on the system input(s) and output(s). The observer is basically a replica of the actual plant with a corrective term $G(O - \hat{O})$ multiplied by the observer gain matrix.

Define the estimation error vector as follows:

$$S_e = S - \hat{S}. \tag{7.71}$$

The change of the estimation error can be expressed as

$$\dot{S}_e = (T_1 - GT_3) S_e. \tag{7.72}$$

Therefore, it is feasible to estimate the system states by a proper choice of G if the system is observable.

7.4 Simulation Verification and Comparative Analysis

The proposed HL control algorithm is implemented in the Matlab/Simulink environment considering an unexpected disturbance $d_{ex} = [0\ 0.1\sin(t) + d(t), 0.2\cos(t) + d(t), 0.3\sin(t) + d(t),\ 0.4\sin(t) + d(t),\ 0.5\cos(t) + d(t),\ 0.6\sin(t) + d(t),\ 0.7\sin(t) + d(t),\ 0.8\cos(t) + d(t),\ 0.9\sin(t) + d(t), 1.0\sin(t) + d(t)]^T$, where $d(t) \sim N(0, 1)$ is a white Gaussian noise of power 0 dBW. It should be noted that the flexible walls are initially deflected by a sudden external force (which is attained by setting the initial states not zero). In the meanwhile, an unexpected time-varying disturbance d_{ex} is also applied to the system to simulate the real environment. The comparative analysis between the PV position controller, NN controller and the HL controller is also investigated to further demonstrate the performance of the proposed algorithm.

The comparisons in different indexes of PV position control, NN control, and HL control are provided and illustrated in Figures 7.3–7.5. The vibration and the acceleration of the flexible floor as two important performance indexes are analyzed and discussed to understand the structure's safety and the user's comfort level.

Figure 7.3 shows the vibration of the flexible floor under PV position control, NN control, and HL control. The flexible vibration magnitude converges to a neighborhood of zero in just 0.7 s under HL algorithm. And for PV position controller, that even costs 7.0 s to stabilize the vibration amplitude within 0.03 cm. This value under the HL control is reduced by 90.0% and 46.2% compared to the PV position control and NN control, respectively.

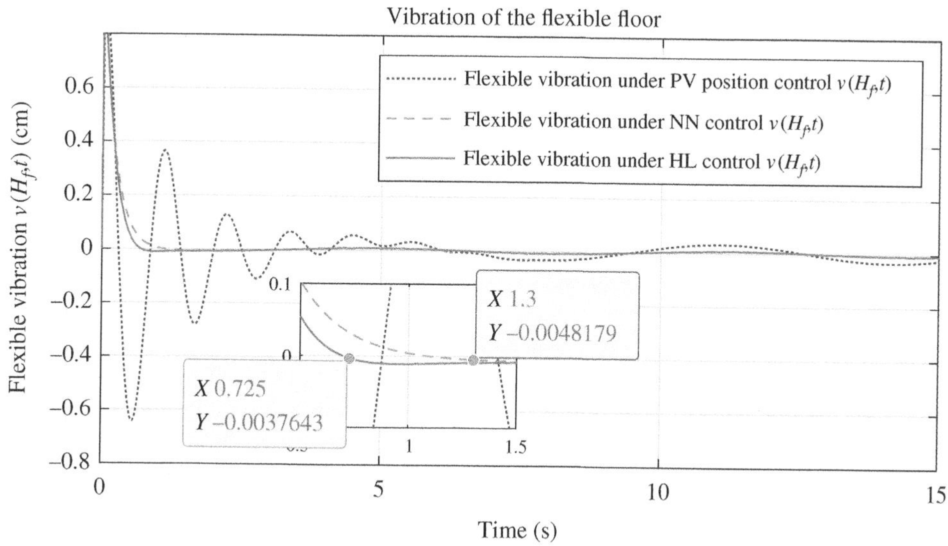

Figure 7.3 Vibration of the flexible floor in simulation verification.

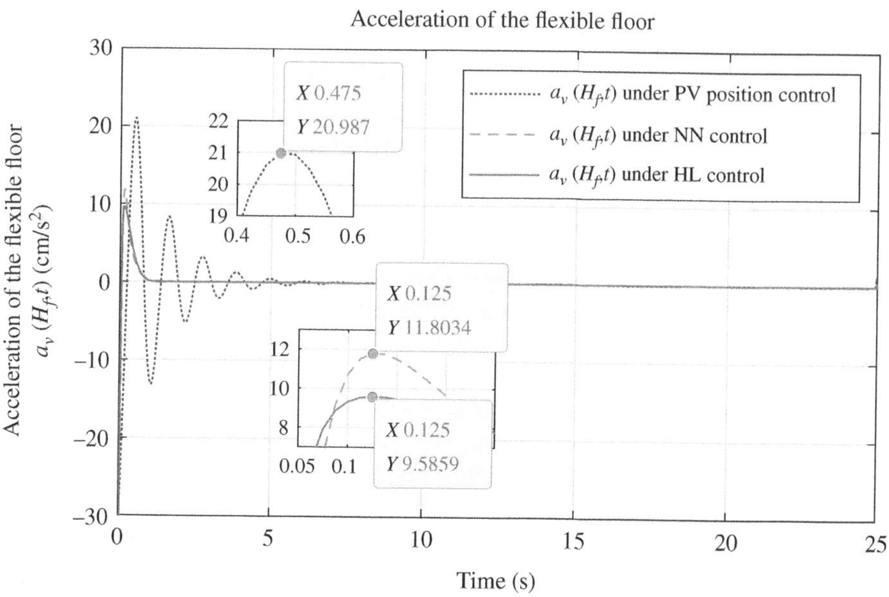

Figure 7.4 Acceleration of the flexible floor in simulation verification.

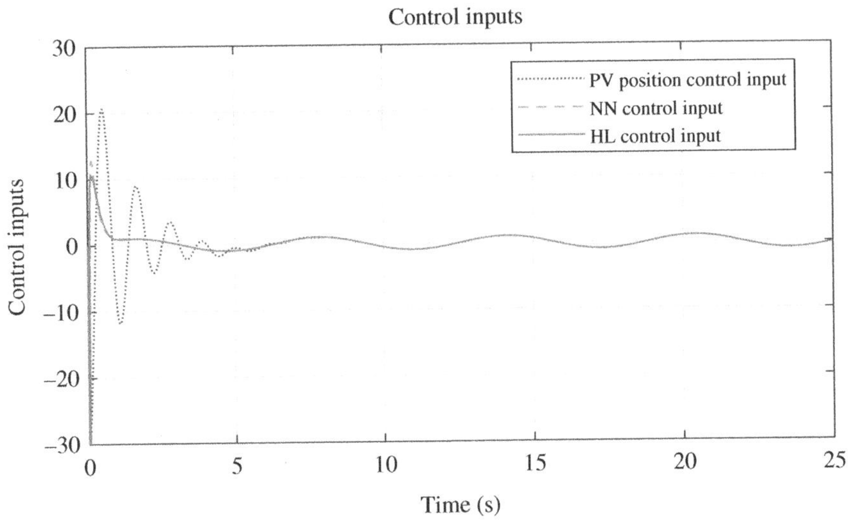

Figure 7.5 Control inputs in simulation verification.

Another important index is the floor acceleration, as shown in Figure 7.4. As revealed in Figure 7.4, the maximum accelerations of the flexible floor are 21.0, 11.8, and 9.6 cm/s^2, respectively, under PV position control, NN control, and HL control. This value under the HL control only accounts for 45.7% and 81.4% of the respective values under PV position control and NN control. Although the stabilization time of the floor acceleration under the NN control and the HL control are close to each other, the value under the HL control is reduced by 99.9% compared to the PV position control. Therefore, the proposed HL control strategy can achieve greater performance on the safety of the structure and the user's comfort.

7.5 Experimental Investigation

The Quanser One-Floor Active Mass Damper (QAMD) plant, as shown in Figure 7.6, consists of a single-floor building-like structure with a controllable linear cart on the top [234]. The cart is driven by a high-quality DC motor through a rack and pinion mechanism. The QAMD frame, which is made of steel, is flexible along its facade. The top floor is instrumented with an accelerometer to measure the acceleration of the "roof" relative to the ground ($\ddot{v}(H_f, t)$). The cart position ($x_c(t)$) is directly measured using an encoder. The encoder shaft meshes with the track via an additional pinion. Descriptions of the flexible structure and cart system are given in Table 7.2.

Extensive experiments were conducted on the QAMD plant to verify the effectiveness and the robustness of against the unknown time-varying disturbance. The proposed HL control algorithm was realized in the MATLAB/Simulink environment and was implemented on the QAMD plant. The essential performances of the safety of the structure and the user's comfort level were all analyzed and discussed. In the experimental studies, the external excitation problem is described as

$$d_{ex} = \begin{cases} |3\sin(5\pi t)|, & 0.8 < t \leq 1.2 \text{ s} \\ 0, & \text{otherwise}, \end{cases} \tag{7.73}$$

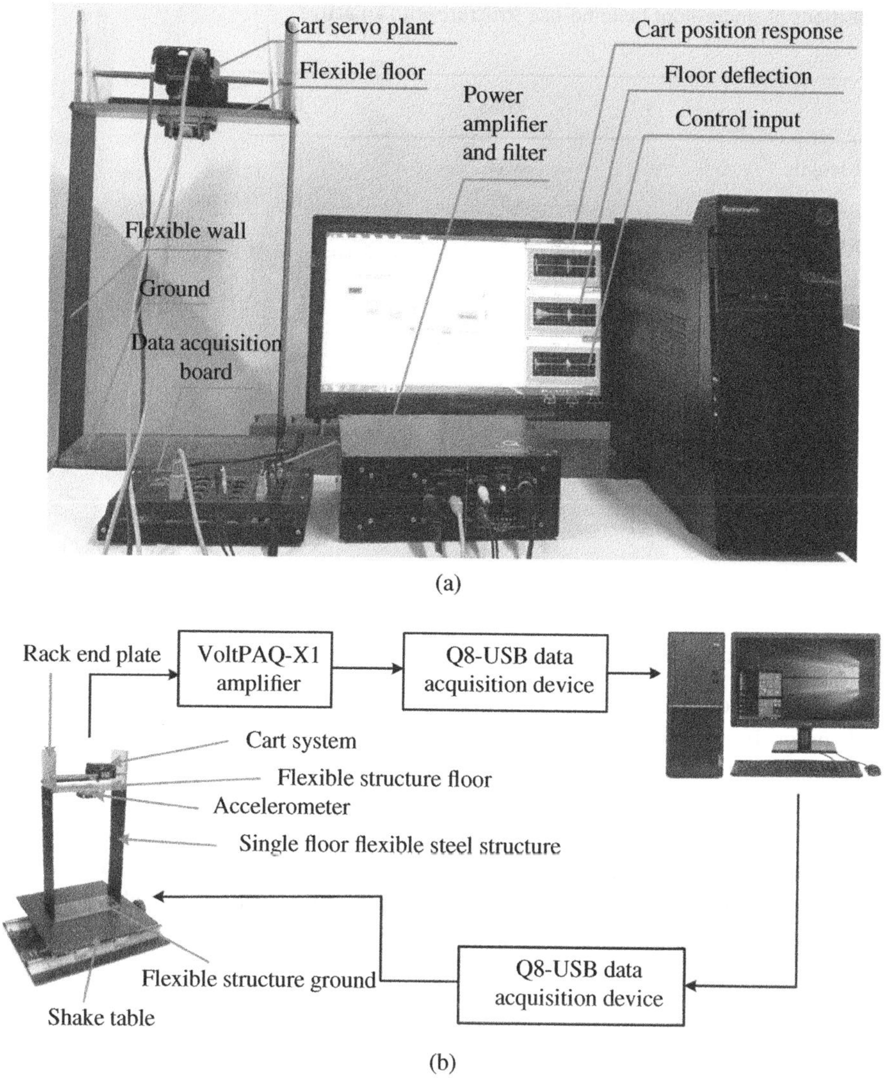

Figure 7.6 Single-floor building-like structure with an active mass damper. (a) Testbed setup. (b) Schematic showing details.

where the external excitation (earthquake, strong wind, etc.) occurs when $0.8 < t \leq 1.2$ s and the frequency of the disturbance is 5 Hz. The frequency of the sinusoidal excitation has been chosen according to the structure's natural frequency.

To further illustrate the effectiveness of the proposed HL control algorithm, both the passive mode and the PV position control algorithms were also employed on the QAMD plant. The control results of passive mode, PV position control and the proposed HL control are presented in Sections 7.5.1, 7.5.2, and 7.5.3, respectively. The analysis and the comparisons of the three approaches are given in Section 7.5.4.

Table 7.2 Specifications of single-floor building-like structure with an active mass damper.

Description	Value	Unit
Flexible structure length	0.32	m
Flexible structure height	0.5	m
Flexible structure depth	0.11	m
Structure floor height	0.53	m
Structure top floor mass	0.68	kg
Cart mass	0.39	kg
Cart rotor moment of inertia	3.90E-7	kg.m^2
Cart planetary gearbox gear ratio	3.71	
Cart motor pinion radius	6.35E-3	m
Cart travel	± 9.5	cm
Flexible structure natural frequency	1.57	Hz

7.5.1 Experimental Results of Passive Mode

In this part, relevant parameters of the flexible structure in passive mode are observed to reflect the strength and energy dissipation capacity of the building-like structure. To analyze and discuss the safety of the structure and the user's comfort level, cart position, vibration, and acceleration of the flexible floor are considered as three performance indexes. Considering that the structural responses during the period several seconds after the unexpected excitation can also reflect the performance of the structure, we select a time-window of 2 s after the unexpected excitation to analyze the performance. For convenience, we define the minimum peak magnitude of the vibration and the acceleration of the flexible floor during this time-window as the vibration response value (VRV) and the acceleration response value (ARV), respectively.

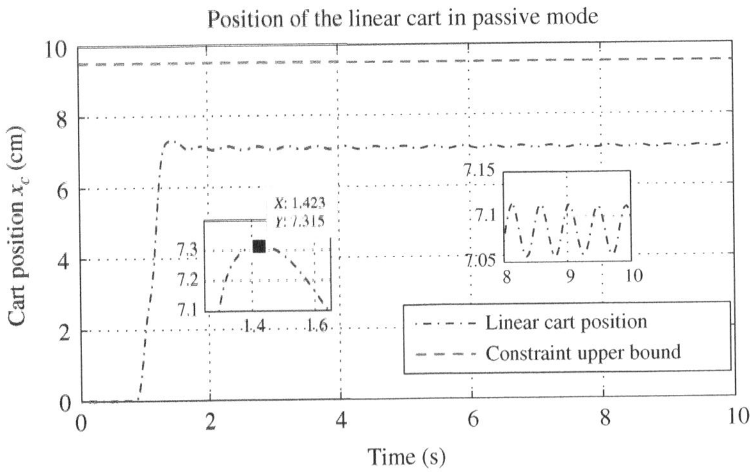

Figure 7.7 Position of the linear cart in passive mode.

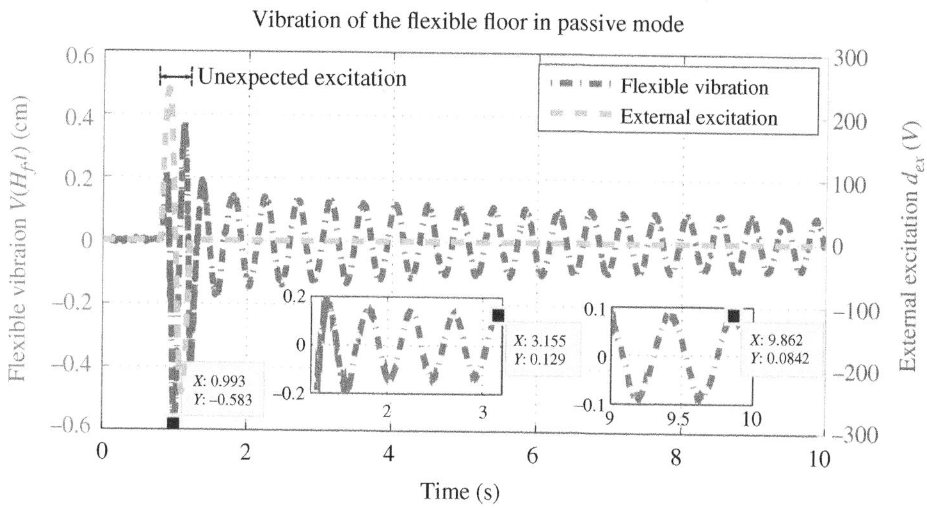

Figure 7.8 Vibration of the flexible floor in passive mode.

The position of the linear cart in passive mode in the presence of unknown external excitation is illustrated in Figure 7.7. When the unknown external excitation occurs, the maximum displacement of the linear cart is 7.315 cm. When the system stabilizes, the cart has an oscillation with an amplitude of 0.08 cm. Figure 7.8 shows the vibration of the flexible floor when excitation occurs in the passive mode. When the unexpected unknown excitation occurs, the vibration magnitude of the flexible floor reaches a maximum of 0.583 cm. The VRV of the flexible floor is reduced to 0.129 cm. Subsequently, only a slight reduction trend exists in the flexible vibration. When $t = 9.9$ s, the flexible vibration magnitude stabilizes at 0.0842 cm. Frequency analysis of the vibration in passive mode is provided in Figure 7.9. The dominate frequency of the vibration in passive mode is 2.2 Hz, which is far from the natural frequency of the flexible structure (1.57 Hz). Therefore, the flexible structure in the passive mode can successfully avoid the collapse caused by frequency resonance. The acceleration of the flexible floor in passive mode is shown in Figure 7.10. When an unexpected disturbance occurs, the maximum acceleration of the flexible floor is 4.431 cm/s^2. The

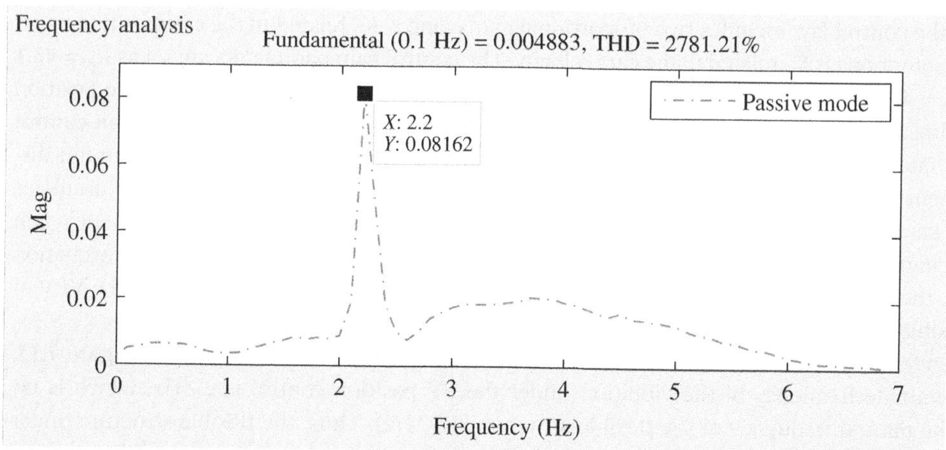

Figure 7.9 Frequency analysis of the vibration in passive mode.

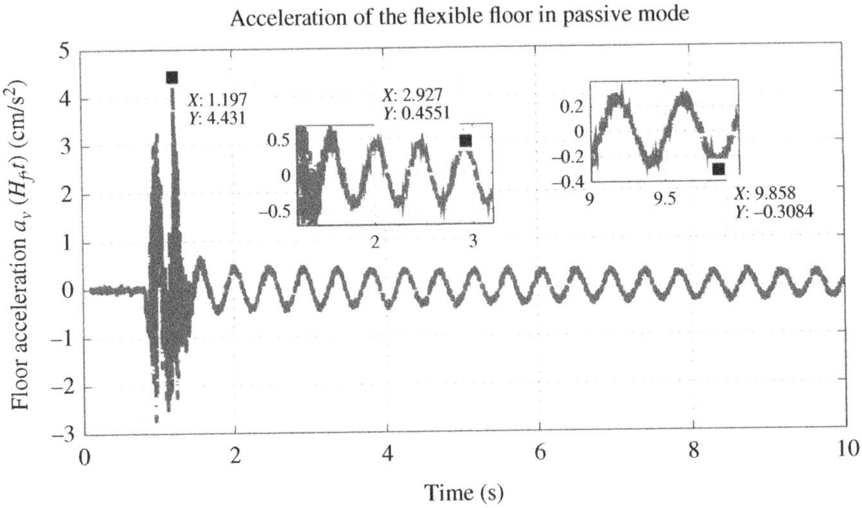

Figure 7.10 Acceleration of the flexible floor in passive mode.

ARV of the flexible floor is reduced to 0.4551 cm/s^2. Eventually, the acceleration of the flexible floor is stable within 0.3084 cm/s^2.

When the flexible system is in passive mode, the cart moves freely between the flexible floor single-story structural module and even crashes with the flexible structure. The movement of the cart also affects the system response in passive mode and this interaction increases vibration of the flexible floor in passive mode. Therefore, balancing the cart position while the flexible floor is experiencing vibration is of great importance.

7.5.2 Experimental Results of PV Position Controller

To balance the cart position while the flexible structure vibrates, a PV position controller is implemented on the QAMD platform. The PV position control law is expressed as

$$u_{pv} = K_p(x_{cd}(t) - x_c(t)) - K_v\dot{x}_c(t), \tag{7.74}$$

where the control law includes two proportional gains: one is K_p related to the cart position error, and the other one is K_v related to the cart velocity. The control gain parameters are set as $K_p = 45.1$ and $K_v = 4.3$. Besides, the expected value of the cart position is chosen as $x_{cd} = 0$. The position of the linear cart in the presence of unknown external excitation under the PV position control is illustrated in Figure 7.11. When the unknown external excitation occurs, the maximum displacement of the linear cart is 3.373 cm. When the system stabilizes, the cart position oscillates with a small magnitude of 0.016 cm. Figure 7.12 shows the vibration of the flexible floor when the excitation occurs under the PV position control. When the unexpected unknown excitation occurs, the vibration of the flexible floor reaches a maximum magnitude of 0.590 cm. The VRV of the flexible floor is reduced to 0.100 cm and the flexible vibration is stable at 0.030 cm after $t = 9.5$ s. Frequency analysis of the vibration under the PV position control is illustrated in Figure 7.13. The dominate frequency of the vibration under the PV position control is 2.2 Hz, which is far from the natural frequency of the flexible structure (1.57 Hz). Thus, the flexible structure under the PV position control has sufficient strength and energy dissipation ability to avoid frequency resonance. The acceleration of the flexible floor in passive mode is shown in Figure 7.14. When

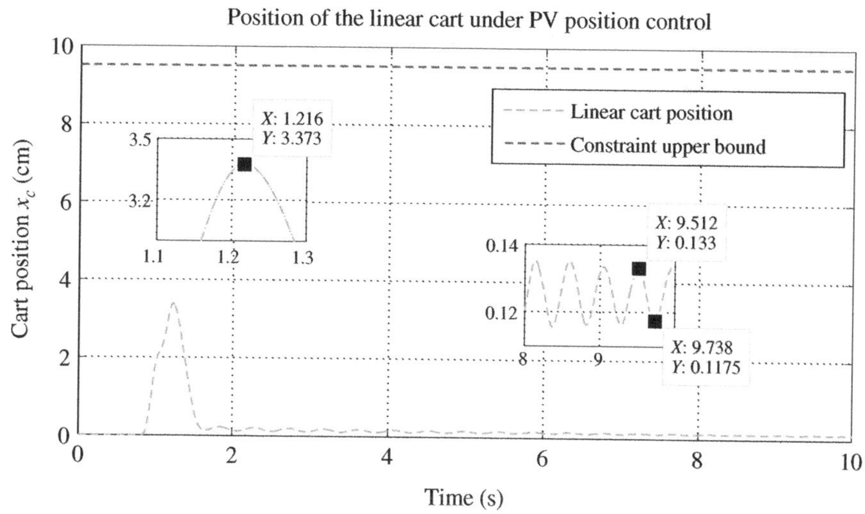

Figure 7.11 Position of the linear cart under PV position control.

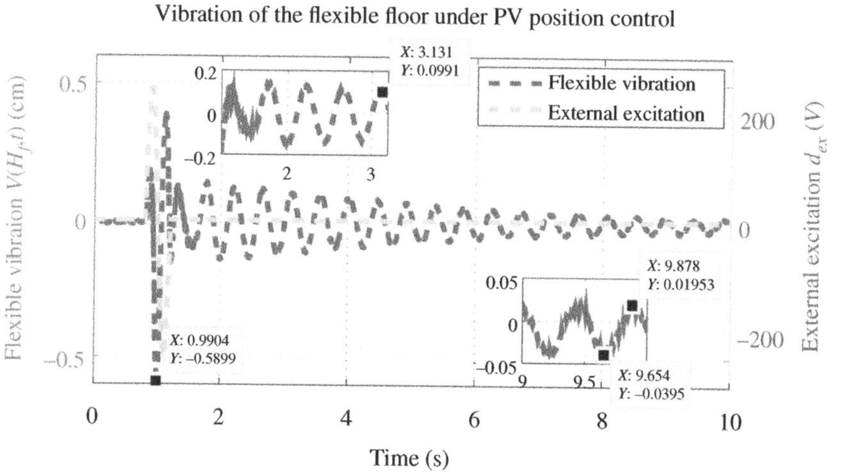

Figure 7.12 Vibration of the flexible floor under PV position control.

the unexpected disturbance occurs, the maximum acceleration of the flexible floor is 3.341 cm/s^2. The ARV of the flexible floor is reduced to 0.446 cm/s^2. Eventually, the acceleration of the flexible floor is stable within 0.144 cm/s^2. Under the PV position control, the system stabilization process is very slow, although both the vibration and the acceleration of the flexible floor gradually decrease. It is essential to design an effective intelligent control method to achieve faster vibration suppression.

7.5.3 Experimental Results of HL Controller

In order to achieve faster vibration suppression, the proposed HL control algorithm based on NN and DO is implemented on the QAMD platform. The HL control has the form described as Eq. (7.50) and the control parameters are selected as $K_1 = 40$ and $K_2 = 0.5$. The node number for

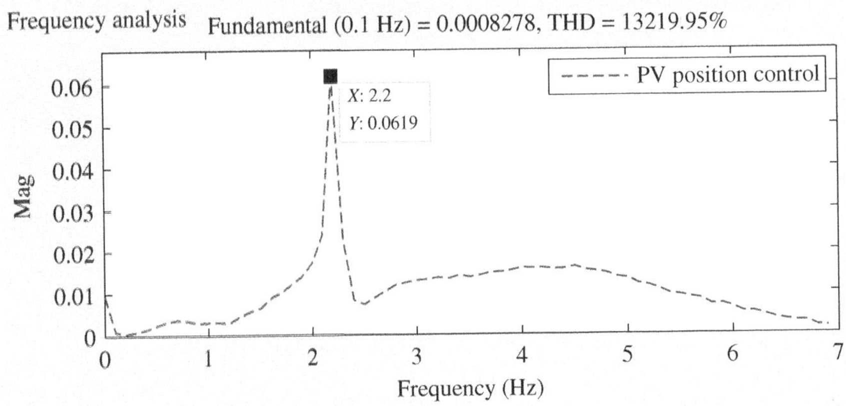

Figure 7.13 Frequency analysis of the vibration under PV position control.

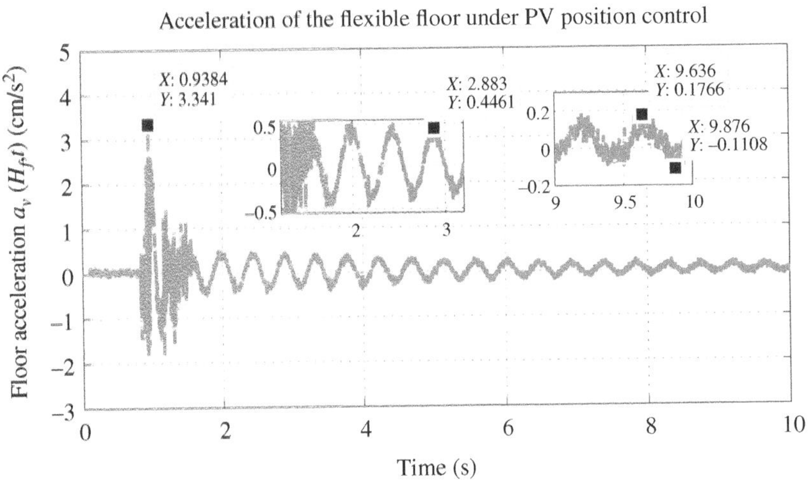

Figure 7.14 Acceleration of the flexible floor under PV position control.

each $S(Z)$ is 256 and the variance of centers is 2. The initial weights of NNs are chosen as $\hat{W}_i = 0$ $(i = 1, 2, ..., 256)$. The position of the linear cart in the presence of unknown external excitation under the HL control is illustrated in Figure 7.15. When the unknown external excitation occurs, the maximum displacement of the linear cart is 4.721 cm. When the system stabilizes, the cart position has an oscillation with a very small magnitude of 0.003 cm. Figure 7.16 shows the vibration of the flexible floor when the excitation occurs under the HL control. When the unexpected unknown excitation occurs, the vibration of the flexible floor reaches a maximum magnitude of 0.580 cm. The VRV of the flexible floor is reduced to 0.003 cm. In addition, the steady-state value of the vibration is also 0.003 cm and the flexible structure system is stable after $t = 2.2$ s. Frequency analysis of the vibration under the HL control is illustrated in Figure 7.17. The dominate frequency of the vibration under the HL control is 3.9 Hz, which is far from the natural frequency of the flexible structure (1.57 Hz). Thus, the flexible structure under the HL control can shift away its natural frequency from the resonance region under the disturbance excitation to maintain the structural health at a controllable embodied energy level. The acceleration of the flexible floor

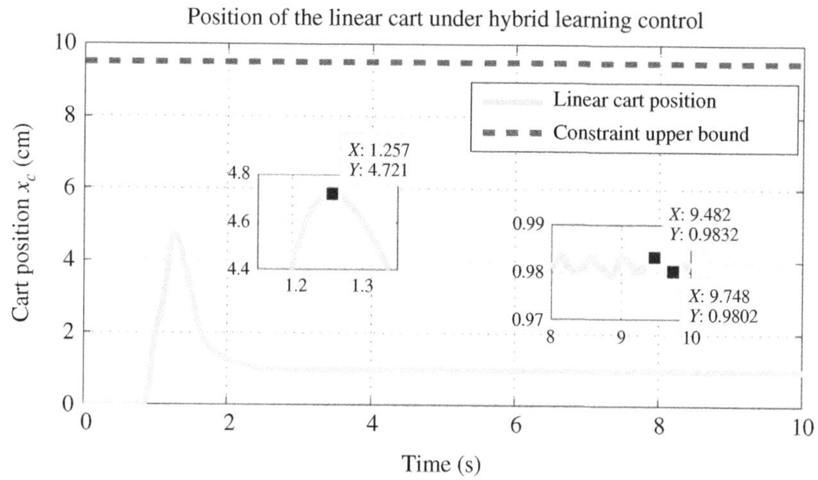

Figure 7.15 Position of the linear cart under hybrid learning control.

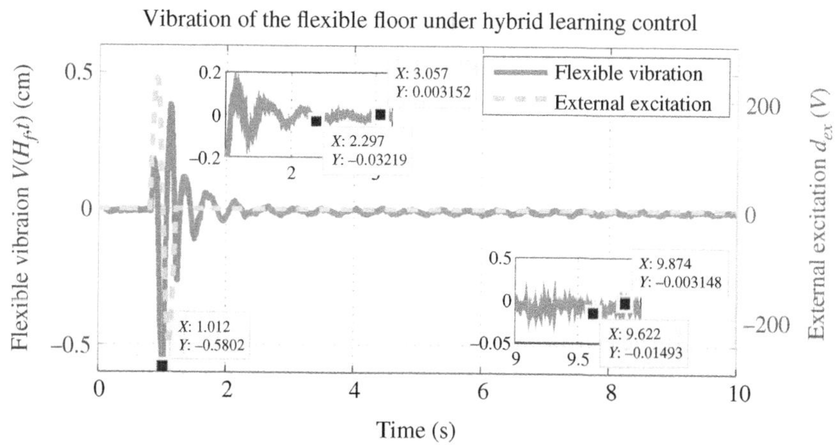

Figure 7.16 Vibration of the flexible floor under hybrid learning control.

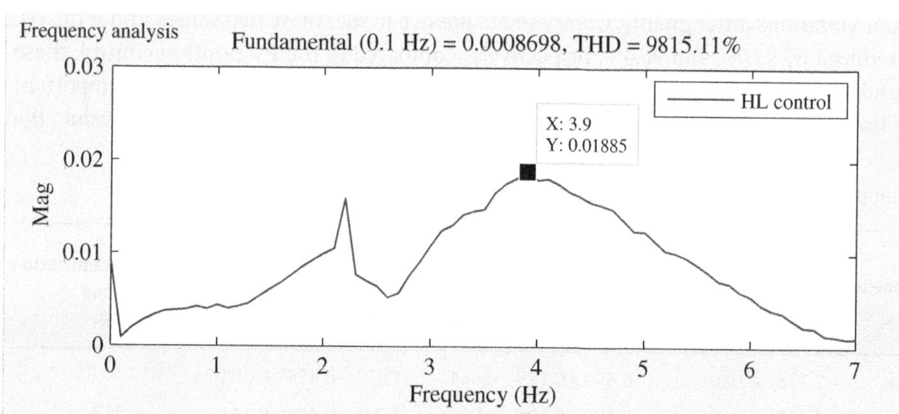

Figure 7.17 Frequency analysis of the vibration under hybrid learning control.

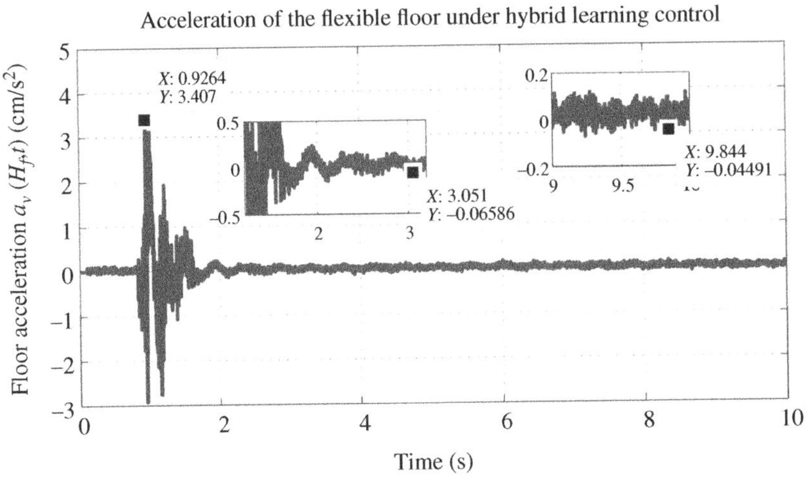

Figure 7.18 Acceleration of the flexible floor under hybrid learning control.

under the HL control is shown in Figure 7.18. When the unexpected disturbance occurs, the maximum acceleration of the flexible floor is 3.407 cm/s^2. The ARV of the flexible floor is reduced to 0.066 cm/s^2. Eventually, the acceleration of the flexible floor is stable within 0.045 cm/s^2.

7.5.4 Comparisons and Discussions

As can be seen from Figures 7.7–7.18, the experimental results in passive mode show that the flexible structure has adequate strength and energy dissipation capacity to limit the overall structural motion. The comparisons in different indexes of passive mode, PV position control and HL control have been listed in Table 7.3. The cart position is successfully confined to a range of 9.5 cm under the PV position control and the HL control. After the system is stable, the oscillation amplitude of the cart position under the HL control only accounts its respective values under the passive mode and the PV position control by 3.75% and 18.75%, respectively. Although the maximum floor vibrations under the passive mode, the PV position control and the HL control are close to each other (in the order of 0.583, 0.590, and 0.580 cm, respectively), the VRVs and the steady-state values of the floor vibrations differ greatly. Compared to passive mode, these two values under the HL control are reduced by 97.7% and 96.4%, respectively; compared to the PV position control, these two values under the HL control are reduced by 97.0% and 90.0%, respectively. Another important index is the floor acceleration. Similar trend has also been observed for the maximum value, the

Table 7.3 Comparison of passive mode, PV position control, and hybrid learning control.

Methods \ Indexes	Cart position (cm)		Floor vibration (cm)			Floor acceleration (cm/s^2)			Stabilization time (s)
	Max. value	Oscillation amplitude	Max. value	VRV	Steady-state	Max. value	ARV	Steady-state	
Passive mode	7.315	0.080	0.583	0.129	0.084	4.431	0.455	0.308	–
PV position	3.373	0.016	0.590	0.100	0.030	3.341	0.446	0.144	8.7
Hybrid learning	4.721	0.003	0.580	0.003	0.003	3.407	0.066	0.045	2.2

ARV and the steady-state value of the floor acceleration. The maximum value of the floor acceleration under the PV position and the HL control are close to each other, in the order of 3.341 and 3.407 cm/s^2, respectively. However, the ARV under the HL control is reduced by 85.2% compared to the PV position control. The steady-state value of the floor acceleration using the HL control is reduced by 85.4% and 68.8% compared to passive mode and the PV position control, respectively. Moreover, the stabilization time of the system under the HL control only accounts for 25.3% of the respective value under the PV position control. Therefore, the proposed HL control strategy can achieve greater performance on the safety of the structure and the user's comfort.

7.6 Summary

This chapter provides a solution for vibration control problem of single-floor building-like structures under strong winds or earthquakes based on NN methods. The problems caused by the infinite-dimensional characteristics of distributed parameter systems have been considered and a novel dimensional dynamic model described by ODEs has been derived via an AMM. A HL controller has been designed to deal with the dynamic uncertainties, unknown disturbances, and immeasurable states. The SGUUB of the closed-loop system has been proved by Lyapunov direct method. The comparative experimental results have shown that smaller magnitude and acceleration of vibrations can be fast achieved by the proposed HL control policy compared with the reported methods. It is further clearly observed that the proposed method guarantees the safety and comfort of residents under external disturbances. In the future, external disturbances with different frequency contents will be investigated to closely simulate the real-world disasters including strong winds and earthquakes. After optimizing the learning control strategy for single-deck buildings, the vibration control of double-deck buildings with active mass damper will also be further studied.

8

Reinforcement Learning Control of a Single-Floor Building-Like Structure with Active Mass Damper

After reading this chapter, you should be able to:

- Investigates a single-floor building-like structure equipped with an active mass damper (AMD)
- Optimal vibration control, while dealing with system uncertainties, is realized by the reinforcement learning technique
- Use proposed controller applying to the active mass damper can compensate the increase of the system vibration caused by external disturbances.

8.1 Introduction

The vibration of both high-rise and ultra high-rise building systems caused by external excitations will greatly reduce the comfort of the residents and even cause safety issues of the buildings. Effective vibration control can make building systems resistant to the unexpected natural disasters and ensure the buildings are immune to the danger of collapse. The literature review on recent research has shown that vibration control of tall building structures should be more emphasized on the performance optimization and practicality [329]. Passive control, which does not require an external power supply, has been widely used in civil engineering. It is known that the John Hancock Tower (60-story) in Boston, USA installs two 300-ton tuned mass damper (TMD) systems at the top so as to effectively control the responses of the structure to winds. In [156], a passive TMD system was designed based on a group of modal parameters to investigate the performance of mitigating the along-wind responses of a tall building. Moreover, passive control methods are primarily focused on the parameter selection and placement method of the respective dampers. However, the installation and maintenance as well as the replacement of the dampers are time-consuming and nontrivial work.

Due to the above-mentioned reasons, a number of innovative methods have been proposed for active vibration control of building structures [202, 275]. Active control can facilitate the performance of the building structures in terms of the controllability and stability. Apart from that, the active control can adapt to different external excitations, especially in emergency situations (when suddenly an earthquake happens) [126]. In [301], an active tendon control was studied for cable-stayed bridges based on a simple cable-supported cantilever beam model. An adaptive robust control algorithm, named time delay control (TDC), was proposed for an AMD system to effectively suppress the wind-induced vibration in [141]. Nagashima and Shinozaki investigated a variable gain feedback control for buildings with AMD systems in [213, 217]. Unfortunately, most

Dynamic Modeling and Neural Network-Based Intelligent Control of Flexible Systems,
First Edition. Hejia Gao, Wei He, and Changyin Sun.

of the current active control methods are based on the finite-dimension dynamic model without considering system uncertainties.

Actually, the complex flexible high-rise buildings, whose degrees of freedom are close to infinity, are distributed parameter systems with many dynamic uncertainties [177, 367]. Most importantly, the dynamic response cannot be predicted accurately [253, 283, 323]. As reinforcement learning control can guarantee the stability of the closed-loop systems in presence of uncertainties, it has thus been attracted the attention of many researchers [78, 159, 226]. The novel adaptive dynamic programming (ADP) RL algorithm was introduced for a single-link robotic arm to deal with the system nonlinearity [153]. Lewis and Vrabie combined reinforcement learning with adaptive optimal control to design a feedback controller for both the continuous- and the discrete-time dynamic systems. These works successfully solved the uncertainty issues caused by unknown disturbances and extraneous effects [166]. As can be seen that reinforcement learning may have the potential to address the practical issues encountered by the active control of vibrations in building structures.

This chapter proposes an active vibration control method based on reinforcement learning for a single-floor building-like structure equipped with an active mass damper. The stability of the closed-loop system is analyzed via Lyapunov direct method. The effectiveness of the proposed control algorithm is verified through the experiments which are conducted on a Quanser Active Mass Damper.

8.2 Problem Formulation

This chapter investigates a one-floor active mass damper system, which consists of one flexible floor, two flexible walls, and a linear cart system. The linear cart system includes a DC motor which provides the driven force, an accelerometer which measures the acceleration of the flexible floor relative to the ground, and an encoder which measures the cart position. The modeling assumptions are made as follows [139]: (1) Only lateral vibration, as shown in Figure 8.1a, is considered, the axial deformation, the shear deformation together with the twist deformation, as shown in Figure 8.1b, are neglected; (2) The length of the flexible wall is much larger than the width of its cross section. Therefore, the flexible wall can be regarded as an Euler–Bernoulli beam.

The flexible building structure under the unexpected disturbance is shown in Figure 8.1, where $d_{dis}(y,t)$ represents the unknown spatiotemporally time-varying disturbance. M_f and M_c are the mass of the flexible floor ("roof") and mass of the active mass damper ("linear cart"), respectively. $\omega_l(y,t)$ and $\omega_r(y,t)$ are the elastic deflection of the structure's left wall and right wall at the position y at time t. H_f represents the hight of the structure, $a_{w(H_f,t)}$ is the acceleration of the flexible floor, and x_c denotes the position of the active mass damper. EI is the bending stiffness of flexible walls. ρ is the mass per unit length of flexible walls, J_c represents the cart rotor moment of inertia, and $\tau(t)$ is the control input.

8.2.1 Dynamic Modeling

The single-floor AMD system is a distributed parameter system which is usually described by a set of motion equations expressed by partial differential equations (PDEs) and boundary conditions expressed by ordinary differential equations (ODEs). The motion equations of the flexible building system can be expressed as

$$\rho\ddot{\omega}_l(y,t) + EI\omega_l''''(y,t) = 0, \tag{8.1}$$

$$\rho\ddot{\omega}_r(y,t) + EI\omega_r''''(y,t) = 0, \tag{8.2}$$

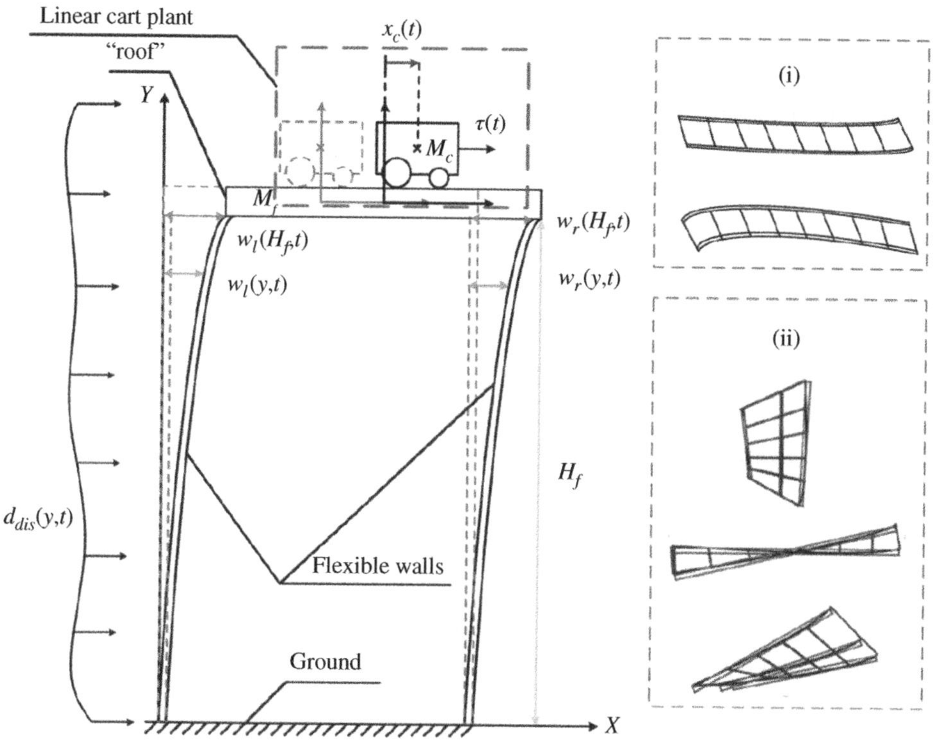

Figure 8.1 One-floor active mass damper system.

and boundary conditions are

$$M_f\ddot{\omega}(H_f,t) - J_c\ddot{x}_c(t)$$
$$- EI[\omega_l'''(H_f,t) + \omega_r'''(H_f,t)] = 0, \tag{8.3}$$

$$\omega_l(0,t) = \omega_l'(0,t) = \omega_l''(H_f,t) = 0, \tag{8.4}$$

$$\omega_l(H_f,t) = \omega_r(H_f,t) = \omega(H_f,t), \tag{8.5}$$

$$J_c\ddot{x}_c(t) + M_c[\ddot{x}_c(t) + \ddot{\omega}(H_f,t)] = \tau(t). \tag{8.6}$$

In order to solve the vibration control problem of uncertainty systems, this section introduces assume mode method (AMM) to approximate the system dynamics with infinite-dimensional properties. Based on AMM, the elastic deflection of the structure's walls, $\omega(y,t)$, can be described as

$$\omega(y,t) = \sum_{i=1}^{\infty} \gamma_i(y)p_i(t), \tag{8.7}$$

where $\gamma_i(y)$ is the mode shape function, $p_i(t)$ represents ith mode. The solution for $\gamma_i(y)$ can be seen in [246]. According to Lagrange method, the second-order continuous-time model of the one-floor AMD system is described as follows

$$H\ddot{G} + BG + d_{dis} = T, \tag{8.8}$$

where $G = (x_c, p_1, \dots p_N) \in \mathbb{R}^{N+1}$ is the generalized coordinates, and $T = (\tau, 0, \dots 0) \in \mathbb{R}^{N+1}$ is the joint torques. $H \in \mathbb{R}^{(N+1)\times(N+1)}$ is the inertia matrix and $B \in \mathbb{R}^{(N+1)\times(N+1)}$ is the stiffness matrix, where,

$$H[1,1] = M_c + J_c,$$
$$H[1,j] = H[j,1] = \frac{1}{4}M_c\gamma_{j-1}(H_f),\ j = 2,\dots,N+1,$$
$$H[j,j] = \rho - \frac{M_f + M_c}{2}\gamma_{j-1}^2(H_f),\ j = 2,\dots,N+1,$$
$$H[j,n] = H[n,j] = \frac{1}{4}(M_f + M_c)\gamma_{j-1}(H_f)\gamma_{n-1}(H_f),$$
$$j = 2,\dots,N+1;\ n = j+1,\dots,N+1.$$
$$B = \mathrm{diag}[0\ 2v_1^2\rho\ 2v_2^2\rho\ \dots\ 2v_N^2\rho],$$

where v_i is the natural frequency of the mode $p_i(t)$.

8.2.2 Model Discretization

Consider a long-run cost function

$$C(t) = \int_0^\infty e^{-\frac{m-t}{\psi}}\varphi(m)dm, \tag{8.9}$$

where constant ψ is employed to discount future costs and function $\varphi(t) = (G - G_r)^T D(G - G_r) + \tau^T R\tau$ denotes immediate cost, matrices D and R are positive definite, and $G_r = [x_{cr}, p_{1r}, p_{2r}, \dots, p_{Nr}]^T$ is the desired value.

The following Radial basis function (RBF) NN can be used to approximate a continuous function $L(Z) : \mathbb{R}^k \to \mathbb{R}$:

$$L_{nn}(Z) = WS(Z), \tag{8.10}$$

where vector $Z = [Z_1, Z_2, \dots, Z_k]^T \in \Omega \subset \mathbb{R}^k$ represents the input of NN, vector $W = [w_1, w_2, \dots, w_l] \in \mathbb{R}^l$ with NN nodes number $l > 1$ denotes the weight of NN. $S(Z) = [s_1(Z), s_2(Z), \dots, s_l(Z)]$ is the radial basis function, where $s_i(Z) = \exp[-(Z - \mu_i)^T(Z - \mu_i)/\hbar_k^2]$, $\mu_i = [\mu_{i1}, \mu_{i2}, \dots, \mu_{ik}]$ is the center of receptive field and $\hbar_k$ is the width of the Gaussian function.

$$L(Z) = W^* S(Z) + \epsilon, \forall Z \in \Omega_Z, \tag{8.11}$$

where W^* is the optical weight and ϵ is the bounded approximation error.

8.3 Reinforcement Learning Control

In this section, the back-stepping method and the iterative design method are employed to gradually design control input signal [186, 241]. In addition, a Lyapunov candidate is used to derive a semi-global uniformly ultimately bounded (SGUUB) property. The RL control design diagram based on actor-critic structure is shown as Figure 8.2.

8.3.1 Control Design

Define error variables e_1 and e_2 as $e_1(t) = G(t) - G_r(t)$, $e_2(t) = \dot{G}(t) - \alpha_1(t)$, where $\alpha_1(t) = \dot{G}_r - K_1 e_1$ is a virtual variable with $K_1 = K_1^T \in \mathbb{R}^{(N+1)\times(N+1)} > 0$. Accordingly,

$$\dot{e}_1 = e_2 - K_1 e_1,\ \dot{e}_2 = H^{-1}[\tau - BG] - \dot{\alpha}_1. \tag{8.12}$$

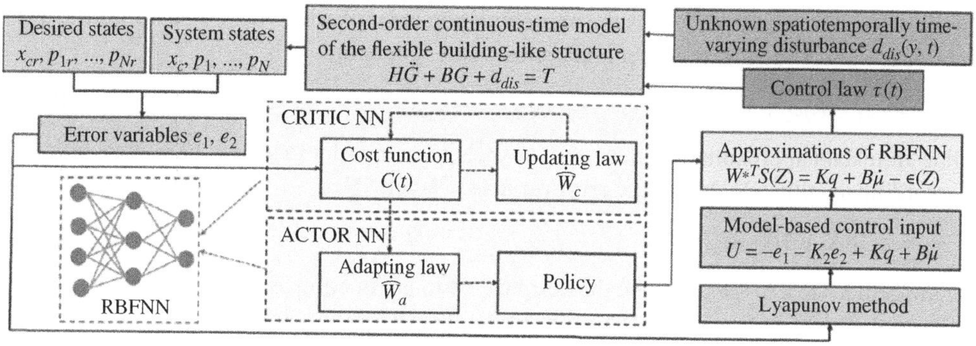

Figure 8.2 The RL control design diagram.

Consider $C = W_c^{*T}S_c(Z_c) + \epsilon_c$ and $\hat{C} = \hat{W}_c^T S_c(Z_c)$ with $Z_c = e_1$. Calculating Eq. (8.9), one obtains

$$\begin{aligned} \varrho(t) &= \varphi(t) - \frac{1}{\psi}\hat{C}(t) + \dot{\hat{C}}(t) \\ &= \varphi(t) - \frac{1}{\psi}\hat{C}(t) + \nabla\hat{C}(t)\dot{Z}_c, \end{aligned} \tag{8.13}$$

∇ is the gradient of Z_c. Design the updating law for the critic NN as

$$\dot{\hat{W}}_c = -\sigma_c \frac{\partial E_c}{\partial W_c}, \tag{8.14}$$

where $E_c = \frac{1}{2}\varrho^T\varrho$. Combining (8.13) and (8.14),

$$\begin{aligned} \dot{\hat{W}}_c &= -\sigma_c \varrho(t) \frac{\partial \varrho}{\partial W_c} \\ &= -\sigma_c \varrho(t) \frac{\partial[\varphi(t)\ \ 1/\psi\hat{C}(t) + \dot{\hat{C}}(t)]}{\partial W_c} \\ &= -\sigma_c \varrho(t)\left[-\frac{1}{\psi}\frac{\partial \hat{C}}{\partial W_c} + \frac{\partial}{\partial W_c}\left(\frac{\partial \hat{C}}{\partial Z_c}\right)\right] \\ &= -\sigma_c(\varphi(t) + W_c^T\Lambda)\Lambda, \end{aligned} \tag{8.15}$$

where constant $\sigma_c > 0$ represents the learning rate, and $\Lambda = -(S_c/\psi) + \nabla S_c \dot{Z}_c$. Initially, design a Lyapunov function as $F_1 = \frac{1}{2}e_1^T e_1$. We subsequently obtain its time derivative,

$$\dot{F}_1 = -e_1^T K_1 e_1 + e_1^T e_2. \tag{8.16}$$

As can be seen from the above Eq. (8.16), there exists a coupling term. Regarding the present problem, F_1 can be adapted as $F_2 = F_1 + (1/2)e_2^T H e_2$ and, through it,

$$\dot{F}_2 = -e_1^T K_1 e_1 + e_1^T e_2 + e_2^T(\tau - BG - H\dot{\alpha}_1). \tag{8.17}$$

Therefore, the model-based control strategy can be designed as the form of

$$\tau_1 = -e_1 - K_2 e_2 + BG + H\dot{\alpha}_1, \tag{8.18}$$

where $K_2 \in \mathbb{R}^{(N+1)\times(N+1)}$ and the minimum eigenvalue of $K_2 > 0$ (that is $\lambda_{\min}(K_2)$). Along with this comes the problem of system uncertainty (namely BG and H are not available). Concerning the fact that NNs can approximate continuous functions with any accuracy, we introduce a NN to approach the system uncertainty.

$$\tau_2 = -e_1 - K_2 e_2 + W_a^{*T} S_a(Z_a) + \epsilon_a, \tag{8.19}$$

where $Z_a = [G^T, \dot{G}^T, G_r^T, \dot{G}_r^T, \ddot{G}_r^T]$. ϵ_a is the approximation error of actor NN. $\hat{\diamond} = \tilde{\diamond} + \diamond^*$ is the estimated value, $\tilde{\diamond}$ and $\diamond^*$ denote the error and optical value. Therefore,

$$\tau_3 = -e_1 - K_2 e_2 + \hat{W}_a^T S_a(Z_a). \tag{8.20}$$

Define an instant approximation error as $\varsigma_a = \tilde{W}_a^T S_a(Z_a)$. In order to adjust the control input in time according to critic NN, design the error variable of actor NN as

$$e_a = \varsigma_a + K_C(\hat{C}(t) - C_d(t)), \tag{8.21}$$

where the gain $K_C \in \mathbb{R}^{N+1} > 0$. Define the desired cost-to-go function as $C_d(t) = 0$.

Similarly, define E_a as $E_a = \frac{1}{2} e_a^T e_a$. Thus, the updating law for the actor NN can be obtained as

$$\dot{\hat{W}}_a = -\sigma_a \frac{\partial E_a}{\partial W_a} = -\sigma_a(\varsigma_a + K_C \hat{C}) S_a, \tag{8.22}$$

where $\sigma_a > 0$ is the learning rate. As ς_a is unavailable, a new updating law would be

$$\dot{\hat{W}}_a = -\sigma_a(\hat{W}_a^T S_a(Z_a) + K_C \hat{C}) S_a. \tag{8.23}$$

8.3.2 Stability Analysis

First of all, define a Lyapunov function as

$$F_c = \frac{1}{2} \tilde{W}_c^T \tilde{W}_c, \tag{8.24}$$

Combining (8.15) and (8.24), the following equation subsequently is obtained,

$$\dot{F}_c = -\sigma_c \tilde{W}_c^T (\varphi(t) + W_c^T \Lambda) \Lambda. \tag{8.25}$$

Gradient descent update is used to minimize the error to the "optimal" weights W^*. When time is close to infinity, we have $\varrho(t) \leq \jmath$, where $\jmath$ is a small constant. So $\varphi(t) \leq \frac{1}{\psi}\varrho - \dot{\varrho} + \jmath$,

$$\begin{aligned} \varphi(t) &\leq W_c^{*T} \frac{S_c}{\psi} + \frac{\epsilon_c}{\psi} - \nabla(W_c^{*T} S_c(Z_c) + \epsilon_c) \dot{Z}_c + \jmath \\ &\leq -W_c^{*T} \Lambda + \varepsilon_c, \end{aligned} \tag{8.26}$$

where $\varepsilon_c = \epsilon_c + \nabla \epsilon_c \dot{Z}_c + \jmath$, and $\|\varepsilon_c\| \leq \varepsilon_{c,\max}$. According to Eqs. (8.25) and (8.26),

$$\begin{aligned} \dot{F}_c &\leq -\sigma_c \tilde{W}_c^T (\tilde{W}_c^T S_c + \varepsilon_c) \Lambda \\ &\leq -\sigma_c \Lambda^T \Lambda \tilde{W}_c^T \tilde{W}_c - \sigma_c \tilde{W}_c^T \varepsilon_c \Lambda \\ &\leq \frac{-\sigma_c \Lambda^T \Lambda}{2} \tilde{W}_c^T \tilde{W}_c + \frac{\sigma_c}{2} \varepsilon_c^T \varepsilon_c, \end{aligned} \tag{8.27}$$

Therefore, we propose a new Lyapunov function,

$$F = \frac{1}{2} e_1^T e_1 + \frac{1}{2} e_2^T H e_2 + \frac{1}{2} \tilde{W}_a^T \tilde{W}_a + \frac{1}{2} \tilde{W}_c^T \tilde{W}_c. \tag{8.28}$$

Combining (8.23) and (8.28),

$$\begin{aligned} \dot{F} &= e_1^T \dot{e}_1 + e_2^T H \dot{e}_2 + \tilde{W}_a^T \dot{\tilde{W}}_a + \tilde{W}_c^T \dot{\tilde{W}}_c \\ &\leq -e_1^T K_1 e_1 - e_2^T K_2 e_2 + e_2^T (\tilde{W}_a^T S_a - \varepsilon_a) \\ &\quad -\sigma_a \tilde{W}_a^T S_a (\hat{W}_a^T S_a(Z_a) + K_C \hat{C}) \\ &\quad -\sigma_c \tilde{W}_c (-W_c^{*T} \Lambda + \varepsilon_c) \Lambda. \end{aligned} \tag{8.29}$$

According to $\hat{C} = W_c^{*T} S_c(Z_c) + \tilde{W}_c^T S_c(Z_c)$,

$$\hat{C}^T\hat{C} \leq 2(W_c^{*T}S_c)^T W_c^{*T} S_c + 2(\tilde{W}_a^T S_c)^T \tilde{W}_a^T S_c, \tag{8.30}$$

substituting (8.30) into (8.29),

$$\begin{aligned}
\dot{F} \leq & -e_1^T K_1 e_1 - e_2^T (K_2 - E) e_2 - \frac{\sigma_a - 1}{2} \|\tilde{W}_a\|^2 \|S_a\|^2 \\
& - \frac{\sigma_c \Lambda^T \Lambda - 2\sigma_a K_C^2 \|S_c\|^2}{2} \|\tilde{W}_c\|^2 \\
& + \frac{\sigma_a}{2} \|W_a^*\|^2 \|S_a\|^2 + \sigma_a K_C^2 \|W_c^*\|^2 \|S_c\|^2 \\
& + \frac{1}{2} \|\epsilon_a\|^2 + \frac{1}{2} \|\epsilon_{c,max}\|^2 \\
\leq & -\kappa F + \lambda,
\end{aligned} \tag{8.31}$$

where E presents an identity matrix,

$$\kappa = \min\left(\lambda_{\min}(K_1), \lambda_{\min}(K_2 - E), \frac{\sigma_a - 1}{2} b_S^2, \frac{\sigma_c b_\Lambda^2 - 2\sigma_a K_C^2 \|S_c\|^2}{2}\right)$$

$$\lambda = \frac{\sigma_a}{2} \|W_a^*\|^2 \|S_a\|^2 + \sigma_a K_C^2 \|W_c^*\|^2 \|S_c\|^2 + \frac{1}{2} \|\epsilon_a\|^2 + \frac{1}{2} \|\epsilon_{c,max}\|^2,$$

where $b_S \leq \|S_a\|$ and $b_\Lambda \leq \|\Lambda\|$. Additionally, the following conditions need to be guaranteed to ensure $\kappa > 0$.

$$\lambda_{\min}(K_1) > 0, \lambda_{\min}(K_2 - E) > 0, \frac{\sigma_a - 1}{2} b_S^2 > 0,$$
$$\frac{\sigma_c b_\Lambda^2 - 2\sigma_a K_C^2 \|S_c\|^2}{2} > 0.$$

Theorem 8.1 Regarding the AMD system with the initial conditions are bounded, the closed-loop system signals e_1, e_2, $\tilde{W}_a$, and $\tilde{W}_c$ are semi-globally uniformly ultimately bounded using RL strategy. These error signals will eventually remain within the compact sets Ω_{e_1}, Ω_{e_2}, $\Omega_{\tilde{W}_a}$, and $\Omega_{\tilde{W}_c}$, respectively.

$$\Omega_{e_1} = \{e_1 \in \mathbb{R}^{N+1} \mid \|e_1\| \leq \sqrt{D}\}, \tag{8.32}$$

$$\Omega_{e_2} = \{e_2 \in \mathbb{R}^{N+1} \mid \|e_2\| \leq \sqrt{\frac{D}{\lambda_{\min}(H)}}\}, \tag{8.33}$$

$$\Omega_{\tilde{W}_a} = \{\tilde{W}_a \in \mathbb{R}^{l\times(N+1)} \mid \|\tilde{W}_a\| \leq \sqrt{D}\}, \tag{8.34}$$

$$\Omega_{\tilde{W}_c} = \{\tilde{W}_c \in \mathbb{R}^{l} \mid \|\tilde{W}_c\| \leq \sqrt{D}\}, \tag{8.35}$$

where $D = 2(F(0) + \lambda/\kappa)$, λ and κ are positive constants.

8.4 Experimental Investigation

In order to verify the feasibility of the RL scheme, a large number of experiments have been carried out on the Quanser AMD platform, which connects the FLEXGAGE module with the physical model such that the experimental process can be monitored in real time. The platform consists of

Table 8.1 Specifications of one-floor AMD system.

Description	Value	Unit
Structure length	0.32	m
Flexible structure height	0.5	m
Structure depth	0.11	m
Structure total height	0.53	m
Structure top floor mass	0.68	kg
Cart mass	0.39	kg
Cart rotor moment of inertia	3.90E-7	kg m^2
Cart planetary gearbox gear ratio	3.71	
Cart motor pinion radius	6.35E-3	m

a computer, AMD structure, SRV02 servo, data acquisition card, power amplifier, and emergency stop button. The control algorithm is designed using the Simulink module in MATLAB. The data acquisition card and power amplifier can be used to communicate with the SRV02 servo device. The parameters of the Quanser AMD system are listed in Table 8.1. The Active Mass Damper-One-Floor (AMD-1) plant is fully described in Section 8.2 and illustrated in Figure 8.3. The purpose of the AMD-1 laboratory is to design a RL control system that first excites the vibration mode of the one-storey structure and then dampens the structure oscillation.

In the experiment, the feedback signals are the analog signal of the elastic deflection of the flexible floor and the displacement of the AMD. The analog signal is transmitted to the data acquisition card through the power amplifier for analog-to-digital conversion, and the converted digital signal is sent to the computer. The controlled inputs are subsequently adjusted according to the received

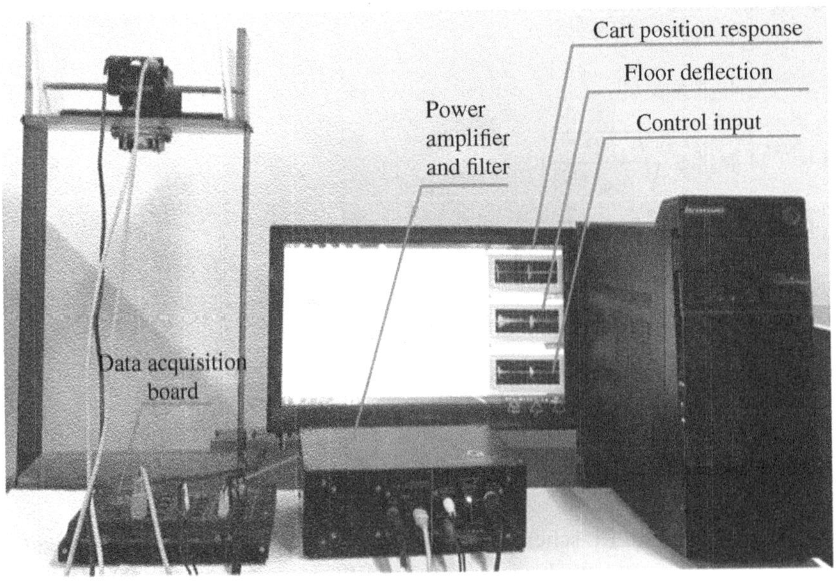

Figure 8.3 Experimental setup of one-floor AMD system.

Table 8.2 Comparison of passive mode, PV and RL control.

Modes	Passive mode	PV control	RL control
Max. cart displacement	1.752 cm	1.004 cm	1.036 cm
Max. displacement of floor	0.216 cm	0.215 cm	0.215 cm
Max. acceleration of floor	1.018 cm/s^2	1.287 cm/s^2	1.198 cm/s^2
Stable range of displacement	0.053 cm	0.049 cm	0.014 cm
Stable range of acceleration	0.225 cm/s^2	0.240 cm/s^2	0.057 cm/s^2

feedback signal, and then transferred to the servo device through the data acquisition board for digital-to-analog conversion.

In the following results analysis, the vibration control performances of passive mode, PV controller, and RL controller have been compared. The experimental time is set to 10 s. A sine wave with amplitude of 1 and frequency of 2.5 Hz is set as the excitation disturbance signal and is magnified 40 times before the arc-angle conversion. As three main indexes reflecting the residents' comfort, the acceleration and the displacement of the flexible floor ($a_w(H_f, t)$, $w(H_f, t)$), together with the position of the linear cart x_c, have been analyzed and discussed. The results are shown in Figures 8.4–8.6, respectively. In addition, passive mode, PV and RL controlled active mode experimental results are compared and summarized in Table 8.2. The gains of the PV controller are set as $K_p = 48.0$ and $K_v = 4.3$, and the gains of the RL controller are set as $K_1 = 23$, $K_2 = 0.3$, and $K_C = 10$. As actor-critic structures are employed to realize RL control, there are 256 nodes and 4 nodes used for $S_a(Z_a)$ and $S_c(Z_c)$, respectively. Moreover, $S_a(Z_a)$ and $S_c(Z_c)$ are set with center parameters as either −1 or 1 and variances as 1. The initial values of the weights W_a and W_c are set as zero. Learning rates σ_a and σ_c are selected as 10 and 1, respectively.

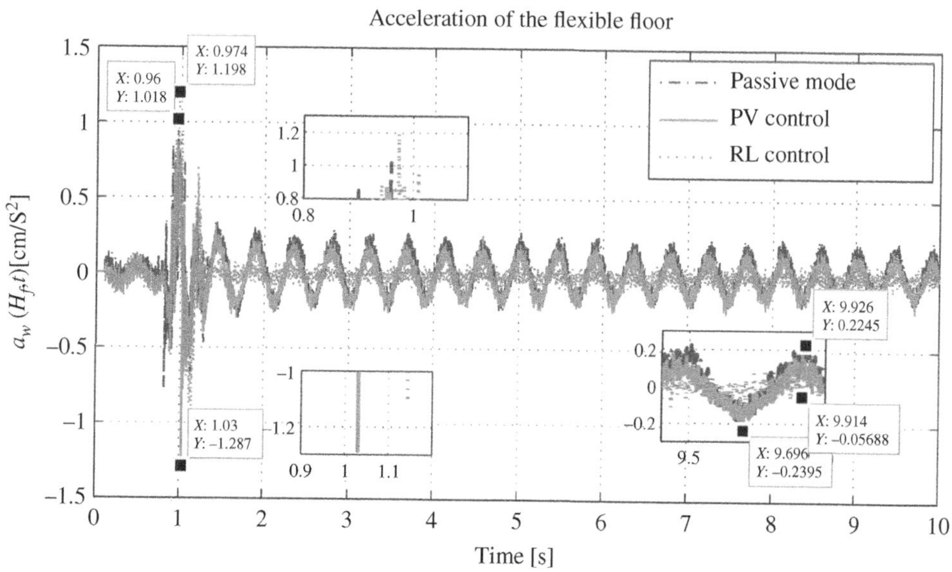

Figure 8.4 Comparison results of the acceleration of the flexible floor.

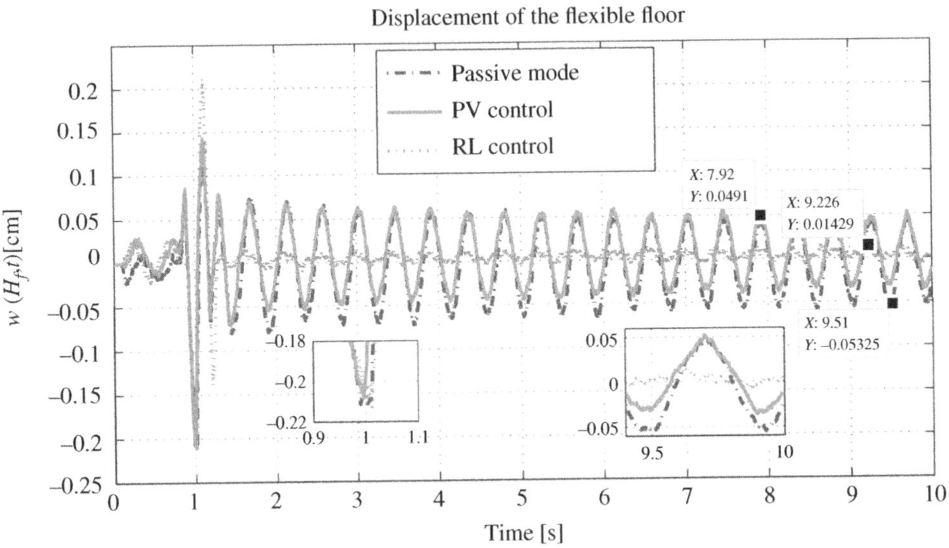

Figure 8.5 Comparison results of the displacement of the flexible floor.

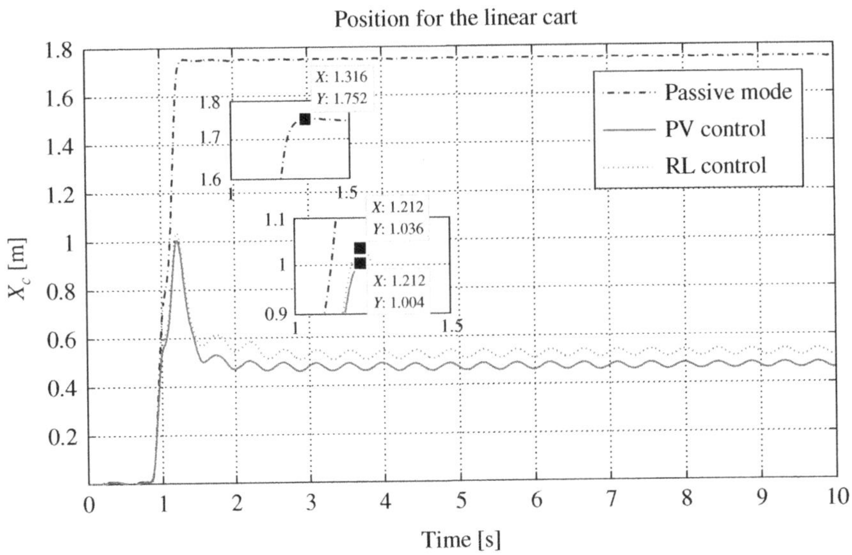

Figure 8.6 Comparison results of the position of the linear cart.

As can be seen from Figures 8.4–8.6, the cart displacement is successfully controlled within the desired range of magnitude (16 cm) in passive mode and active mode. The maximum floor displacements of the three control strategies are close to each other, in the order of 0.216, 0.215, and 0.215 cm, respectively. However, the stable range of displacement and acceleration under RL control only accounts 28.6% and 23.7%, respectively, for their respective values under PV control. Moreover, the maximum floor acceleration is reduced by 6.9% using RL control compared to PV control. While the passive mode achieves smallest maximum acceleration among the three controllers, the stable range of displacement and acceleration under RL control are only 26.4% and 25.3%, respectively, of those values under passive mode.

8.5 Summary

A single-floor building-like structure equipped with an active mass damper is investigated in this study. In order to achieve its vibration suppression, a reinforcement learning control approach has been proposed and the system uncertainty issues of such building structure have been resolved through neural networks. The experimental results have shown the effectiveness of this reinforcement learning control algorithm, and the final performance has been compared with those of passive mode and PV control.

In future works, control design of the multi-floor building-like structure will be considered. Multiple constraint problems such as constraints, saturation, and backlash will be studied. In addition, excitations caused by wind and earthquakes which may have wider range of frequency are further included so as to more practically simulate real-world engineering problems.

9

Disturbance Observer-Based Neural Network Control of a Flexible Flapping-Wing System

After reading this chapter, you should be able to:

- Develops the visualization model of the rigid-flexible coupled bionic flapping wing by the advanced system-level modeling software MapleSim
- Proposed a novel neural network controller based on disturbance observer technology to compensate the system uncertainties
- Suppress the vibration of the flapping wing while accurately track the desired trajectory according to the proposed method

9.1 Introduction

With the extensive application of flexible structures, its control theory and method has become a forward-looking research direction, which has attracted wide attention from academia and industry [33, 147, 211]. At present, the problem of control theory and method for flexible structures has developed into a common scientific issue with extreme challenge, such as intelligent control problem of nonlinear systems in complex environment [34], adaptive control problem of uncertain multi input multi output (MIMO) nonlinear systems [40, 155, 180], tracking control problem for hypersonic flight vehicle or multiagent systems [25, 322], vibration control problem for flexible systems [334, 351].

For a control engineer, few areas of control theory are as physically motivated–but also as impenetrable–as control of partial differential equations (PDEs) [307]. In control of PDEs, the challenge comes from the infinite-dimensional dynamic model of the distributed parameter system. In order to facilitate the analysis of dynamic uncertainty and the design of controller, how to realize the conversion from PDE to ordinary differential equation (ODE) and design an intelligent control method with learning ability is the core issue.

Flexible structure is a distributed parameter system described by a hybrid PDE-ODE model [306, 311]. It is difficult to obtain the exact solution of the dynamic model, and the complexity of the dynamic equation also brings great difficulty to the controller design. The discretization method is used to transform PDE into ODE. The common discretization methods include finite element method (FEM) [56], finite segment method (FSM) [100], lumped parameter method (LPM) [210, 263], and assumed mode method (AMM) [72, 161, 265]. In [20], the dynamical equations of a two-link flexible manipulator moving in the vertical plane are derived.

Dynamic Modeling and Neural Network-Based Intelligent Control of Flexible Systems,
First Edition. Hejia Gao, Wei He, and Changyin Sun.

With the popularization and application of intelligent technology, intelligent control theory and methods are widely used in nonlinear systems [136, 180, 261], including unmanned vehicles [336], service robots [182] and space robots. In order to meet the basic performance requirements such as rapidity, accuracy and robustness of nonlinear systems in complicated environment, many scholars have carried out in-depth research from NN control method [118, 321], fuzzy control [194, 296, 365], robust control method [177, 291, 293, 341], model predictive control [304] and adaptive dynamic programming method [268]. In [108], in order to deal with input deadzone and output constraint of a robotic manipulator system, an adaptive NN tracking control was proposed. In [257], an adaptive dynamic programming method was introduced for continuous-time complex-valued nonlinear systems.

In this chapter, we discuss the dynamic modeling of the flexible beam with the unknown spatiotemporally varying disturbances as shown in Figure 9.1 by using AMM. Subsequently, we investigate a neural network control for the flexible beam to achieve vibration suppression. Compared to the existing works, the main contributions of the chapter include:

1) A n-dimensional ODE model of the flexible beam system with the unknown spatiotemporally varying disturbance is derived based on AMM.
2) NN control with both full-state feedback and output feedback is proposed to compensate for the dynamic uncertainty of the flexible beam system. And a disturbance observer (DO) is used to estimate unknown spatiotemporally varying disturbance.
3) With the proposed NN control, it is proved by Lyapunov's stability and numerical simulation that the elastic vibration of the flexible beam system with unknown spatiotemporally varying disturbance will eventually converge to zero with a small neighborhood.

The rest of the chapter is arranged as follows. The problem formulation is developed in Section 9.2. Neural network controller with full-state and output feedback are designed in Section 9.3. Simulation results are given in Section 9.4. Conclusions are obtained in Section 9.5.

9.2 Problem Formulation

9.2.1 Dynamic Modeling

Figure 9.1 shows an Euler–Bernoulli beam model with unknown spatiotemporally varying disturbance. The left boundary of the beam is fixed at origin. L is the length of the flexible beam. M is the mass of the tip payload. ρ denotes the uniform mass per unit length of the flexible beam. EI represents the bending stiffness of the flexible beam. $w(x,t)$ indicates the elastic deflection with respect to frame XOY at the position x for time t. $\tau(t)$ is the control input. $f(x,t)$ is the distributed spatiotemporally varying disturbance.

Remark 9.1 For clarity, notations $(\cdot)' = \partial(\cdot)/\partial x$, $(\cdot)'' = \partial^2(\cdot)/\partial x^2$, $(\cdot)''' = \partial^3(\cdot)/\partial x^3$, $(\cdot)'''' = \partial^4(\cdot)/\partial x^4$ and $(\dot{\cdot}) = \partial(\cdot)/\partial t$ are used throughout this chapter.

Hamilton's principle which is used to derive the model of the beam system is expressed as

$$\int_{t_1}^{t_2} \delta[E_k(t) - E_p(t) + W(t)]dt = 0, \tag{9.1}$$

where t_1 and t_2 are two time instants, and δ denotes the variational operator.

The kinetic energy of the beam $E_k(t)$ can be represented as

$$E_k(t) = \frac{1}{2}M[\dot{\omega}(L,t)]^2 + \frac{1}{2}\rho\int_0^L [\dot{\omega}(x,t)]^2 dx, \tag{9.2}$$

where x and t represent the independent spatial and time variables, respectively.

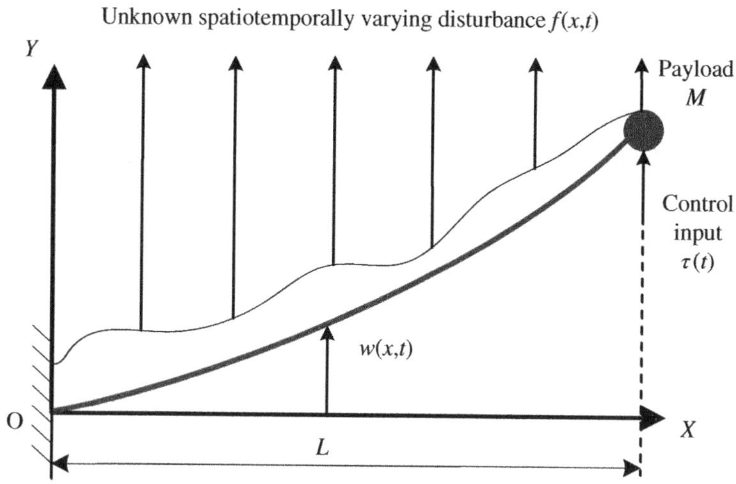

Figure 9.1 A typical flexible cantilevered beam system.

The potential energy $E_p(t)$ due to the bending can be obtained from

$$E_p(t) = \frac{1}{2}EI\int_0^L [\omega''(x,t)]^2 dx, \tag{9.3}$$

The virtual work done by the control input force $u(t)$ which produces a transverse force for vibration suppression can be written as

$$\delta W(t) = \tau(t)\delta\omega(L,t) + \int_0^L f(x,t)\delta\omega(x,t)dx, \tag{9.4}$$

Applying the variation operator and integrating the above equations by parts respectively, we obtain

$$\begin{aligned}
\int_{t_1}^{t_2} \delta E_k(t)dt &= M\int_{t_1}^{t_2} \dot{\omega}(L,t)\delta\dot{\omega}(L,t)dt + \rho\int_{t_1}^{t_2}\int_0^L \dot{\omega}(x,t)\delta\dot{\omega}(x,t)dxdt \\
&= M\dot{\omega}(L,t)\delta\omega(L,t)\,|_{t_1}^{t_2} - M\int_{t_1}^{t_2} \ddot{\omega}(L,t)\delta\omega(L,t)dt \\
&\quad + \rho\int_0^L [\dot{\omega}(x,t)\delta\omega(x,t)\,|_{t_1}^{t_2}]dx - \rho\int_0^L\int_{t_1}^{t_2} \ddot{\omega}(x,t)\delta\omega(x,t)dtdx \\
&= -M\int_{t_1}^{t_2} \ddot{\omega}(L,t)\delta\omega(L,t)dt - \rho\int_0^L\int_{t_1}^{t_2} \ddot{\omega}(x,t)\delta\omega(x,t)dtdx,
\end{aligned} \tag{9.5}$$

$$\begin{aligned}
\int_{t_1}^{t_2} \delta E_p(t)dt &= \int_{t_1}^{t_2}\int_0^L [EI\omega''(x,t)\delta\omega''(x,t)]dxdt, \\
&= \int_{t_1}^{t_2} [EI\omega''(x,t)\delta\omega'(x,t)\,|_0^L - EI\omega'''(x,t)\delta\omega(x,t)\,|_0^L \\
&\quad + \int_0^L EI\omega''''(x,t)\delta\omega(x,t)dx
\end{aligned} \tag{9.6}$$

$$\int_{t_1}^{t_2} \delta W(t)dt = \int_{t_1}^{t_2} \tau(t)\delta\omega(L,t)dt + \int_{t_1}^{t_2}\int_0^L f(x,t)\delta\omega(x,t)dxdt. \tag{9.7}$$

Based on the energy analysis of the flexible beam system [87], the governing equations can be obtained via Hamilton's principle

$$\rho \ddot{w}(x,t) + EIw''''(x,t) - f(x,t) = 0, \tag{9.8}$$

where $\forall (x,t) \in (0,L) \times [0,\infty)$. The boundary conditions of the flexible beam system also can be obtained

$$w(0,t) = w'(0,t) = w''(L,t) = 0, \tag{9.9}$$

$$M\ddot{w}(L,t) - EIw'''(L,t) = \tau(t). \tag{9.10}$$

where $\forall t \in [0,\infty)$.

9.2.2 Model Discretization

In order to discretize the dynamic model described by PDEs into ODEs, we introduce a AMM which uses the linear sum of assumed modes to describe the vibration of the elastic body to achieve this purpose. The elastic vibration of a flexible beam can be represented by

$$w(x,t) = \sum_{i=1}^{\infty} \phi_i(x) q_i(t). \tag{9.11}$$

where $\phi_i(x)$ are the modes shape functions or the eigen-functions, and $q_i(t)$ are the generalized coordinates which corresponds to a DOF of the flexible beam system.

Considering that the energy occupying ratio of the high order mode is too low and has little influence on the structure vibration, i.e.,

$$w(x,t) = \sum_{i=1}^{N} \phi_i(x) q_i(t), \ 0 \le x \le L. \tag{9.12}$$

where N is the number of the modes.

We regard flexible beams as Euler–Bernoulli beams. The elastic deflection $w(x,t)$ is assumed to be small. Therefore, $w(x,t)$ can be represented by

$$w(x,t) = \phi(x) q(t). \tag{9.13}$$

Substituting Eq. (9.13) into Eq. (9.8), we have

$$\frac{\phi''''}{\phi} \frac{EI}{\rho} = -\frac{\ddot{q}}{q}. \tag{9.14}$$

Define a constant k, and we can obtain two ODEs instead of Eq. (9.14),

$$\ddot{q}(t) = -kq(t), \tag{9.15}$$

$$\phi''''(x) = \frac{\rho}{EI} k\phi(x). \tag{9.16}$$

and the boundary conditions Eqs. (9.9) and (9.10) can be rewritten as

$$\phi(0) = \phi'(0) = \phi''(L) = 0, \tag{9.17}$$

$$M\phi(L)\ddot{q}(t) - EI\phi'''(L)q(t) = \tau(t). \tag{9.18}$$

Obviously, $q(t)$ and $\phi(x)$ of Eqs. (9.15) and (9.16) should be such that $w(x,t) = \phi(x)q(t)$ satisfies the boundary conditions. $q(t)$ is assumed to be harmonic with frequency h and thus $k = h^2$. Thus, Eq. (9.16) can be rewritten as

$$\phi''''(x) = \left(\frac{\beta}{L}\right)^4 \phi(x). \tag{9.19}$$

Consider the general solution of Eq. (9.19)

$$\phi(x) = \kappa_1 \cos\frac{cx}{L} + \kappa_2 \cosh\frac{cx}{L} + \kappa_3 \sin\frac{cx}{L} + \kappa_4 \sinh\frac{cx}{L}. \tag{9.20}$$

Substituting the general solution (9.20) into boundary conditions, we have

$$\begin{cases} \kappa_1 + \kappa_2 = 0, \\ \kappa_3 + \kappa_4 = 0, \\ -\kappa_1 \cos c + \kappa_2 \cosh c - \kappa_3 \sin c + \kappa_4 \sinh c = 0, \\ \kappa_1(\sin c + \frac{M}{\rho L}\cos c) + \kappa_2(\sinh c + \frac{M}{\rho L}\cosh c) \\ +\kappa_3(\frac{M}{\rho L}\sin c - \cos c) + \kappa_4(\cosh c + \frac{M}{\rho L}\sinh c) = 0. \end{cases}$$

Therefore, we can obtain the following equation about β,

$$1 + \cosh c \cos c + \frac{Mc}{\rho L}(\sinh c \cos c - \cosh c \sin c) = 0. \tag{9.21}$$

Consider the boundary conditions, $\phi(x)$ can be further written as

$$\phi(x) = \kappa_2 \left[\cosh\frac{cx}{L} - \cos\frac{cx}{L} - \gamma\left(\sinh\frac{cx}{L} - \sin\frac{cx}{L}\right)\right]. \tag{9.22}$$

where

$$\kappa_2 = \begin{cases} \sqrt{L}/L, & M = 0 \\ 1\Big/\left[L + \frac{\rho L^2}{Mc_i^2}\left(\frac{1+\cosh c_i \cos c_i}{\sinh c_i \sin c_i}\right)^2\right]^{\frac{1}{2}}, & M \neq 0 \end{cases} \tag{9.23}$$

$$\gamma = \frac{\cosh c + \cos c}{\sinh c + \sin c}. \tag{9.24}$$

Defining the generalized coordinates vector as

$$Q = [q_1\ q_2\ \dots\ q_N]^T. \tag{9.25}$$

we can rewrite the kinetic energy E_k into the following compact form

$$E_k = \frac{1}{2}\dot{Q}^T M_A \dot{Q}, \tag{9.26}$$

considering the following equation

$$\int_0^L \dot{w}^2(x,t)dx = \sum_{i=1}^{N} \dot{q}_i^2(t)\left[1 - \frac{M}{\rho}\phi_i^2(L)\right] + 2\sum_{i,j=1\ i\neq j}^{N} \dot{q}_i \dot{q}_j \left[-\frac{M}{\rho}\phi_i(L)\phi_j(L)\right], \tag{9.27}$$

where $M_A \in R^{N\times N}$ is the symmetric and positive definite inertia matrix of the system which is given by

$$M_A = \begin{bmatrix} \gamma_A^1 & b_A^{12} & \cdots & b_A^{1N} \\ b_A^{12} & \gamma_A^2 & \cdots & b_A^{2N} \\ \vdots & \vdots & \vdots & \vdots \\ b_A^{1N} & b_A^{2N} & \cdots & \gamma_A^N \end{bmatrix}, \tag{9.28}$$

where $\gamma_A^i = \rho[1 - \frac{M}{\rho}\phi_i^2(L)] + M\phi_i^2(L) = \rho$, $b_A^{ij} = 0$.

We can rewrite the kinetic energy E_p into the following compact form

$$E_p = \frac{1}{2}\sum_{i=1}^{N} q_i^2 w_i^2 \rho = \frac{1}{2} Q^T K_A Q, \tag{9.29}$$

where K_A, the stiffness matrix of the system, is given by

$$K_A = \text{diag}[\omega_1^2 \rho \ \omega_2^2 \rho \ \dots \ \omega_N^2 \rho]. \tag{9.30}$$

By using the Euler–Lagrange's equations

$$M_A \ddot{Q} + K_A Q = u + f_{dis}, \tag{9.31}$$

where $u \in \mathbb{R}^N = [0, 0, \dots, \tau(t)]^T, f_{dis} \in \mathbb{R}^N = [f(l,t), f(2l,t), \dots, f(L,t)]^T$.

9.3 Disturbance Observer-Based Neural Network Control

For the system (9.31), we designed a NN control strategy and a disturbance observer in this section. In addition, we theoretically prove the stability of the controller by Lyapunov direct method.

In the process of the control design, we adopt the backstepping method which is to decompose the complex nonlinear system into subsystems may not exceed the number of system order. Subsequently, by using the iterative design method, through coordinate transformation, introduce a virtual stabilization function, design the control input starting from the first subsystem without control input signal, and gradually backstepping to the last subsystem with control input, and then the real control input signal is obtained.

9.3.1 Control Design

Transform the Eq. (9.31), we have the description of the system dynamics:

$$\ddot{Q} = M_A^{-1}[u + f_{dis} - K_A Q]. \tag{9.32}$$

Given the errors of system states z_1 and z_2 as follows

$$z_1 = Q - Q_d, \tag{9.33}$$

$$z_2 = \dot{Q} - \mu_1. \tag{9.34}$$

where μ_1 is the virtual control to z_1. We choose $\mu_1 = -K_1 z_1 + \dot{Q}_d, K_1 = K_1^T > 0$. Then the derivative of the error variables can be described as

$$\dot{z}_1 = z_2 + \alpha_1 - \dot{Q}_d = z_2 - K_1 z_1. \tag{9.35}$$

$$\dot{z}_2 = \ddot{Q} - \dot{\mu}_1 = M_A^{-1}[u + f_{dis} - K_A Q] - \dot{\mu}_1. \tag{9.36}$$

Design V_1 as the initial Lyapunov function:

$$V_1 = \frac{1}{2} z_1^T z_1, \tag{9.37}$$

Further, $\dot{V}_1$ can be obtained:

$$\dot{V}_1 = z_1^T (z_2 - K_1 z_1) = z_1^T z_2 - z_1^T K_1 z_1. \tag{9.38}$$

To compensate for the disturbance f_{dis}, we use a nonlinear DO to estimate it. Define the following auxiliary variable as:

$$z_3 = f_{dis} + \Theta(z_2). \tag{9.39}$$

where $\Theta(z_2) \in \mathbb{R}^N$ is a function vector of z_2 to be designed. Taking the time derivative of z_3, we have

$$\begin{aligned}\dot{z}_3 &= \dot{f}_{dis} + D(z_2)\dot{z}_2 \\ &= \dot{f}_{dis} + D(z_2)M_A^{-1}[u + f_{dis} - K_A Q] - D(z_2)\dot{\mu}_1,\end{aligned} \tag{9.40}$$

where $D(z_2) = \partial\Theta(z_2)/\partial z_2^T \in \mathbb{R}^{N\times N}$. Generally, we design $\Theta(z_2)$ as a linear function of z_2 so that $D(z_2)$ is a constant matrix and the DO can be easily designed and implemented. Since z_2 can be obtained by the state feedback, we can get the approximate value of the disturbance f_{dis} if we have a appropriate way to estimate the z_3. To get the approximate value, $\dot{\hat{z}}_3$ is presented as follows:

$$\dot{\hat{z}}_3 = D(z_2)M_A^{-1}[u + \hat{f}_{dis} - K_A Q] - D(z_2)\dot{\mu}_1, \tag{9.41}$$

where $\hat{z}$ is the estimation of z_3. From Eq. (9.39), we have the estimation of the disturbance f_{dis}

$$\hat{f}_{dis} = \hat{z}_3 - \Theta(z_2). \tag{9.42}$$

It is easy to obtain that

$$\tilde{z}_3 = \hat{z}_3 - z_3 = \hat{f}_{dis} - f_{dis} = \tilde{f}_{dis}. \tag{9.43}$$

Taking its time derivative, we have

$$\dot{\tilde{f}}_{dis} = \dot{\tilde{z}}_3 = D(z_2)M_A^{-1}\tilde{f}_{dis} - \dot{f}_{dis}. \tag{9.44}$$

Lemma 9.1 [333] For a continuously differentiable function $\Psi(t)$, if $\Psi(t)$ satisfies $|\Psi(t)| \leq \psi, \forall t \in [0, \infty)$, where ψ is a positive constant, then $\dot{\Psi}(t)$ is bounded $\forall t \in [0, \infty)$.

Assumption 9.1 [109] Based on the actual application environment, the disturbance $f_{dis}(t)$ exerted by humans and environment is assumed to be continuous and bounded.

According to Lemma 9.1 and Assumption 9.1, we can obtain

$$\|\dot{f}_{dis}\| \leq \varsigma_0, \tag{9.45}$$

where ς_0 is an unknown positive constant. Then, aiming at removing the coupling term $z_1^T z_2$ in Eq. (9.38), the Lyapunov function as follows can be employed:

$$V_2 = \frac{1}{2}z_1^T z_1 + \frac{1}{2}z_2^T M_A z_2 + \frac{1}{2}\tilde{f}_{dis}^T \tilde{f}_{dis}, \tag{9.46}$$

Apply the following inequality,

$$-\tilde{f}_{dis}^T \dot{f}_{dis} \leq \frac{1}{2}\tilde{f}_{dis}^T \tilde{f}_{dis} + \frac{1}{2}\dot{f}_{dis}^T \dot{f}_{dis} \leq \frac{1}{2}\tilde{f}_{dis}^T \tilde{f}_{dis} + \frac{1}{2}\varsigma_0^2. \tag{9.47}$$

Further, $\dot{V}_2$ can be obtained:

$$\begin{aligned}\dot{V}_2 &= z_1^T \dot{z}_1 + z_2^T M_A \dot{z}_2 + \tilde{f}_{dis}^T \dot{\tilde{f}}_{dis} \\ &= z_1^T z_2 - z_1^T K_1 z_1 + z_2^T(u + f_{dis} - K_A Q - M_A\dot{\mu}_1) \\ &\quad + \tilde{f}_{dis}^T(D(z_2)M_A^{-1}\tilde{f}_{dis} - \dot{f}_{dis}) \\ &\leq z_1^T z_2 - z_1^T K_1 z_1 + z_2^T(u + f_{dis} - K_A Q - M_A\dot{\mu}_1) \\ &\quad + \tilde{f}_{dis}^T D(z_2)M_A^{-1}\tilde{f}_{dis} + \frac{1}{2}\tilde{f}_{dis}^T \tilde{f}_{dis} + \frac{1}{2}\varsigma_0^2.\end{aligned} \tag{9.48}$$

Then, the model-based controller u_0 is proposed

$$u_0 = -z_1 - K_2 z_2 - \hat{f}_{dis} + K_A Q + M_A \dot{\mu}_1, \tag{9.49}$$

where the gain matrix $K_2 = K_2^T > 0$. Apply the following inequality,

$$z_2^T \tilde{f}_{dis} \leq \frac{1}{2} z_2^T z_2 + \frac{1}{2} \tilde{f}_{dis}^T \tilde{f}_{dis}. \tag{9.50}$$

Substituting Eq. (9.49) into Eq. (9.48)

$$\begin{aligned} \dot{V}_2 &\leq -z_1^T K_1 z_1 - z_2^T (K_2 + \frac{1}{2} I_{N\times N}) z_2 + \tilde{f}_{dis}^T D(z_2) M_A^{-1} \tilde{f}_{dis} + \frac{1}{2}\varsigma_0^2 \\ &\leq -a_1 V_2 + b_1, \end{aligned} \tag{9.51}$$

where $I_{N\times N}$ denotes identity matrix and

$$a_1 = \min\left(2\lambda_{\min}(K_1), \frac{2\lambda_{\min}(K_2 + \frac{1}{2} I_{N\times N})}{\lambda_{\max}(M_A)}, 2\lambda_{\max}\left(D(z_2) M_A^{-1}\right)\right)$$

$$b_1 = \frac{1}{2}\varsigma_0^2.$$

To ensure $a_1 > 0$, the linear function vector $\Theta(z_2)$ are chosen to satisfy that

$$\lambda_{\max}\left(\frac{\partial \Theta(z_2)}{\partial z_2^T} M_A^{-1}\right) > 0. \tag{9.52}$$

Since $M_A^{-1}(Q)$ is known, the $\Theta(z_2)$ is not hard to be designed as a linear function of z_2.

Since $M_A \dot{\mu}_1$ and K_A are uncertain, the control law (9.49) may be infeasible. Therefore, neural network is introduced to deal with the dynamic uncertainty via the online estimation.

$$u_1 = -z_1 - K_2 z_2 - \hat{f}_{dis} + \hat{W}^T S(Z), \tag{9.53}$$

where $\hat{W}_i$ is the estimation of neural network weight. $Z = [Q, \dot{Q}, \mu_1, \dot{\mu}_1]^T$ denotes the inputs to the adaptive radial basis function neural network (RBFNN).

The adaptive law $\dot{\hat{W}}_i$ is proposed

$$\dot{\hat{W}}_i = -\Gamma_i (S_i(Z) z_{2,i} + \sigma_i \hat{W}_i), \tag{9.54}$$

where, because the system needs strong robustness, a small constant $\sigma_i > 0$ is introduced. $\hat{W}_i^T$ is the approximation of neural network weights W_i^{*T}.

$$W_i^{*T} S_i(Z) = K_A Q + M_A \dot{\mu}_1 - \epsilon(Z). \tag{9.55}$$

We propose the Lyapunov function candidate as follows:

$$V_3 = V_2 + \frac{1}{2}\sum_{i=1}^{n} \tilde{W}_i^T \Gamma_i^{-1} \tilde{W}_i. \tag{9.56}$$

Then, applying the approximation Eq. (9.55), $\dot{V}_3$ can be derived

$$\begin{aligned} \dot{V}_3 \leq\; & z_1^T z_2 - z_1^T K_1 z_1 + z_2^T (u + f_{dis} - W^{*T} S(Z) - \epsilon(Z)) \\ & + \tilde{f}_{dis}^T D(z_2) M_A^{-1} \tilde{f}_{dis} + \frac{1}{2} \tilde{f}_{dis}^T \tilde{f}_{dis} + \frac{1}{2}\varsigma_0^2 + \sum_{i=1}^{n} \tilde{W}_i^T \Gamma_i^{-1} \dot{\hat{W}}_i, \end{aligned} \tag{9.57}$$

Substituting Eqs. (9.53) and (9.54) into the $\dot{V}_3$

$$\dot{V}_3 \leq -z_1^T K_1 z_1 - z_2^T (K_2 + \frac{1}{2} I_{N\times N}) z_2 + z_2^T (\hat{W}^T S(Z) - W^{*T} S(Z) - \epsilon(Z))$$
$$+ \tilde{f}_{dis}^T D(z_2) M_A^{-1} \tilde{f}_{dis} + \frac{1}{2}\varsigma_0^2 - \sum_{i=1}^{n} \tilde{W}_i^T (S_i(Z) z_{2,i} + \sigma_i \hat{W}_i)$$
$$\leq -z_1^T K_1 z_1 - z_2^T (K_2 + \frac{1}{2} I_{N\times N}) z_2 - z_2^T \epsilon(Z)$$
$$+ \tilde{f}_{dis}^T D(z_2) M_A^{-1} \tilde{f}_{dis} + \frac{1}{2}\varsigma_0^2 - \sum_{i=1}^{n} \tilde{W}_i^T \sigma_i \hat{W}_i, \tag{9.58}$$

Using the following inequality:

$$-\sum_{i=1}^{n} \sigma_i \tilde{W}_i^T \hat{W}_i \leq -\sum_{i=1}^{n} \sigma_i \left(\frac{\|\tilde{W}_i\|^2}{2} - \frac{\|W_i^*\|^2}{2} \right), \tag{9.59}$$

We have $\dot{V}_3$

$$\dot{V}_3 \leq -z_1^T K_1 z_1 - z_2^T K_2 z_2 + \frac{1}{2}\|\bar{\epsilon}\|^2 + \frac{1}{2}\varsigma_0^2 + \tilde{f}_{dis}^T D(z_2) M_A^{-1} \tilde{f}_{dis}$$
$$- \sum_{i=1}^{n} \frac{\sigma_i \|\tilde{W}_i\|^2}{2} + \sum_{i=1}^{n} \frac{\sigma_i \|W_i^*\|^2}{2}$$
$$\leq -a_2 V_3 + b_2, \tag{9.60}$$

where

$$a_2 = \min \left(2\lambda_{\min}(K_1), \frac{2\lambda_{\min}(K_2)}{\lambda_{\max}(M_A)}, 2\lambda_{\max}(D(z_2) M_A^{-1}), \min_{i=1,2,\ldots,N} \left(\frac{\sigma_i}{\lambda_{\max}(\Gamma_i^{-1})} \right) \right)$$
$$b_2 = \frac{1}{2}\varsigma_0^2 + \frac{1}{2}\|\bar{\epsilon}\|^2 + \sum_{i=1}^{n} \frac{\sigma_i \|W_i^*\|^2}{2},$$

where $\lambda_{\max}$ and $\lambda_{\min}$, respectively are the maximum and minimum eigenvalues. To ensure $a_1 > 0$, the linear function vector $\Theta(z_2)$ are chosen to satisfy that

$$\lambda_{\max} \left(\frac{\partial \Theta(z_2)}{\partial z_2^T} M_A^{-1} \right) > 0. \tag{9.61}$$

Multiplying Eq. (9.60) by $e^{a_2 t}$, and integrating, we have

$$V_3 \leq \left(V_3(0) - \frac{a_2}{b_2} \right) e^{-a_2 t} + \frac{a_2}{b_2} \leq V_3(0) + \frac{a_2}{b_2}. \tag{9.62}$$

Thus

$$\frac{1}{2}\|z_1\|^2 \leq V_3(0) + \frac{a_2}{b_2}. \tag{9.63}$$

Theorem 9.1 For the dynamic model of the flexible beam in Eq. (9.31), by employing the control input and the adaptive law, because the full-state information is known, for each compact set Ω_0 where $(x_1(0), x_2(0), \hat{W}_1(0)\ldots\hat{W}_n(0)) \in \Omega_0$, the initial conditions are bounded and the closed-loop system is semiglobally uniformly bounded. z_1, z_2, $\tilde{f}_{dis}$ and $\tilde{W}$ will remain within Ω_{z_1}, Ω_{z_2}, $\Omega_{\tilde{f}_{dis}}$, Ω_W:

$$\Omega_{z_1} := \left\{ z_1 \in R^n \mid \|z_1\| \leq \sqrt{X} \right\}, \tag{9.64}$$

$$\Omega_{z_2} := \left\{ z_2 \in R^n \mid \|z_2\| \leq \sqrt{\frac{X}{\lambda_{\min}(M_A)}} \right\}, \tag{9.65}$$

$$\Omega_{\tilde{f}_{dis}} := \left\{ \tilde{f}_{dis} \in R^n \mid \|\tilde{f}_{dis}\| \leq \sqrt{X} \right\}, \tag{9.66}$$

$$\Omega_{\tilde{W}} := \left\{ \tilde{W} \in R^{l\times n} \mid \|\tilde{W}\| \leq \sqrt{\frac{X}{\lambda_{\min}(\Gamma^{-1})}} \right\}, \tag{9.67}$$

where $X = 2(V_3(0) + \frac{b_1}{a_1})$, $b_1 > 0$ and $a_1 > 0$.

9.3.2 Stability Analysis

However, the control law with full-state feedback (9.53) needs to satisfy that both Q and $\dot{Q}$ can be obtained by measurement. In practice, if $\dot{Q}$ is immesurable, a high-gain observer can be introduced to estimate $\dot{Q}$ by employing separation principle and the certainty equivalence property.

According to [13], $\frac{\pi_{k+1}}{\varepsilon^k}$ converges asymptotically to $Q^{(k)}$. As a result of the high-gain $\frac{1}{\varepsilon}$, ξ_k converges to zero with a small time constant if Q and its k derivatives are bounded. Therefore, the observer $\frac{\pi_{k+1}}{\varepsilon^k}$ is appropriate to estimate the system unmeasurable states up to the nth order. Therefore,

$$Q = \pi_1,\ \hat{\dot{Q}} = \frac{\pi_2}{\varepsilon},\ \hat{z}_2 = \frac{\pi_2}{\varepsilon} - \mu_1, \tag{9.68}$$

where

$$\varepsilon\dot{\pi}_1 = \pi_2,\ \varepsilon\dot{\pi}_2 = -\bar{\lambda}_1\pi_2 - \pi_1 + Q. \tag{9.69}$$

Rewrite the control input (9.53) and adaptive law (9.54) by backstepping.

$$u_3 = -z_1 - K_2\hat{z}_2 - \hat{f}_{dis} + \hat{W}^T S(\hat{Z}), \tag{9.70}$$

$$\dot{\hat{W}}_i = -\Gamma_i(S_i(\hat{Z})\hat{z}_{2,i} + \sigma_i\hat{W}_i), \tag{9.71}$$

We propose the Lyapunov function candidate as follows:

$$V_4 = \frac{1}{2}z_1^T z_1 + \frac{1}{2}z_2^T M_A z_2 + \frac{1}{2}\tilde{f}_{dis}^T\tilde{f}_{dis} + \frac{1}{2}\sum_{i=1}^{n}\tilde{W}_i^T\Gamma_i^{-1}\tilde{W}_i, \tag{9.72}$$

Then, $\dot{V}_4$ can be derived:

$$\begin{aligned}
\dot{V}_4 \leq\ & z_1^T z_2 - z_1^T K_1 z_1 + z_2^T(u + f_{dis} - W^{*T}S(Z) - \epsilon(Z)) \\
& + \tilde{f}_{dis}^T D(z_2) M_A^{-1}\tilde{f}_{dis} + \frac{1}{2}\tilde{f}_{dis}^T\tilde{f}_{dis} + \frac{1}{2}\varsigma_0^2 + \sum_{i=1}^{n}\tilde{W}_i^T\Gamma_i^{-1}\dot{\hat{W}}_i \\
\leq & -z_1^T K_1 z_1 - z_2^T(K_2 - I_{N\times N})z_2 - z_2^T K_2\tilde{z}_2 + z_2^T(\hat{W}^T S(\hat{Z}) - W^{*T}S(Z)) \\
& + \frac{1}{2}\|\bar{\epsilon}(Z)\|^2 + \tilde{f}_{dis}^T D(z_2) M_A^{-1}\tilde{f}_{dis} + \frac{1}{2}\varsigma_0^2 - \sum_{i=1}^{n}\tilde{W}_i^T S_i(\hat{Z})\hat{z}_{2,i} \\
& - \sum_{i=1}^{n}\sigma_i\frac{\|\tilde{W}_i\|^2}{2} + \sum_{i=1}^{n}\sigma_i\frac{\|W_i^*\|^2}{2}.
\end{aligned} \tag{9.73}$$

From the lemma from the reference [13], we have

$$\xi_2 = \frac{\pi_2}{\varepsilon} - \dot{Q} = -\varepsilon\psi^{(2)}, \tag{9.74}$$

$$\tilde{z}_2 = \hat{z}_2 - z_2 = \frac{\pi_2}{\varepsilon} - \mu_1 - \dot{Q} + \mu_1 = \xi_2. \tag{9.75}$$

where ε is an any small positive constant, $\psi = \pi_2 + \overline{\lambda}_1 \pi_1$. For $\forall t > t^* > 0$ and $h_2 > 0$, then $\|\xi_2\| \leq \varepsilon h_2$. Therefore, $\dot{Q}$ can be estimated by $\frac{\pi_2}{\varepsilon}$.

$$\hat{Q} = \frac{\pi_2}{\varepsilon}, \tag{9.76}$$

$$\hat{z}_2 = \frac{\pi_2}{\varepsilon} - \mu_1. \tag{9.77}$$

Using the following inequality:

$$\|S_i(\hat{Z})\|^2 \leq l_i, \tag{9.78}$$

$$\hat{W}_i^T S_i(\hat{Z}) = W_i^{*T} \epsilon S_{ti} + W_i^{*T} S_i(Z) + \tilde{W}_i^T S_i(\hat{Z}), \tag{9.79}$$

$$\sum_{i=1}^{n} z_{2,i} W_i^{*T} \epsilon S_{ti} \leq \sum_{i=1}^{n} \frac{\|W_i^*\|^2 \epsilon^2 \|S_{ti}\|^2}{2} + \frac{1}{2} z_2^T z_2 \tag{9.80}$$

$$-\sum_{i=1}^{n} \tilde{W}_i^T S_i(\hat{Z}) \tilde{z}_{2,i} \leq \sum_{i=1}^{n} \frac{\sigma_i \|\tilde{W}_i\|^2}{4} + \sum_{i=1}^{n} \frac{2\|S_i(\hat{Z})\|^2}{\sigma_i} \frac{1}{2} \tilde{z}_2^T \tilde{z}_2, \tag{9.81}$$

where ϵ and l_i are positive and S_{ti} is bounded.

We can obtain:

$$\begin{aligned}
\dot{V}_4 \leq& -z_1^T K_1 z_1 - z_2^T (K_2 - I_{N\times N}) z_2 - z_2^T K_2 \tilde{z}_2 + \sum_{i=1}^{n} z_{2,i} W_i^{*T} \epsilon S_{ti} \\
&+ \frac{1}{2}\|\overline{\epsilon}(Z)\|^2 + \tilde{f}_{dis}^T D(z_2) M_A^{-1} \tilde{f}_{dis} + \frac{1}{2}\varsigma_0^2 - \sum_{i=1}^{n} \tilde{W}_i^T S_i(\hat{Z}) \tilde{z}_{2,i} \\
&- \sum_{i=1}^{n} \sigma_i \frac{\|\tilde{W}_i\|^2}{2} + \sum_{i=1}^{n} \sigma_i \frac{\|W_i^*\|^2}{2} \\
\leq& -z_1^T K_1 z_1 - z_2^T \left(K_2 - \frac{3}{2} I_{N\times N}\right) z_2 - z_2^T K_2 \tilde{z}_2 + \frac{1}{2}\|\overline{\epsilon}(Z)\|^2 \\
&+ \tilde{f}_{dis}^T D(z_2) M_A^{-1} \tilde{f}_{dis} + \frac{1}{2}\varsigma_0^2 - \sum_{i=1}^{n} \frac{\sigma_i \|\tilde{W}_i\|^2}{4} \\
&+ \sum_{i=1}^{n} \frac{2\|S_i(\hat{Z})\|^2}{\sigma_i} \frac{1}{2} \tilde{z}_2^T \tilde{z}_2 + \sum_{i=1}^{n} \left(\frac{\sigma_i}{2} + \frac{\epsilon^2 \|S_{ti}\|^2}{2}\right) \|W_i^*\|^2.
\end{aligned} \tag{9.82}$$

Using Eqs. (9.75), (9.78) and $-z_2^T K_2 \tilde{z}_2 \leq \frac{1}{2} z_2^T z_2 + \frac{1}{2}(K_2 \tilde{z}_2)^T (K_2 \tilde{z}_2)$, we further have:

$$\begin{aligned}
\dot{V}_4 \leq& -z_1^T K_1 z_1 - z_2^T (K_2 - 2I_{N\times N}) z_2 + \frac{1}{2} K_2^T K_2 \xi_2^T \xi_2 + \frac{1}{2}\|\overline{\epsilon}(Z)\|^2 \\
&+ \tilde{f}_{dis}^T D(z_2) M_A^{-1} \tilde{f}_{dis} + \frac{1}{2}\varsigma_0^2 - \sum_{i=1}^{n} \frac{\sigma_i \|\tilde{W}_i\|^2}{4} \\
&+ \sum_{i=1}^{n} \frac{2l_i}{\sigma_i} \frac{1}{2} \xi_2^T \xi_2 + \frac{1}{2} \sum_{i=1}^{n} (\sigma_i + \epsilon^2 \|S_{ti}\|^2) \|W_i^*\|^2 \\
\leq& -z_1^T K_1 z_1 - z_2^T (K_2 - 2I_{N\times N}) z_2 + \frac{1}{2}\|\overline{\epsilon}(Z)\|^2 + \frac{1}{2}\varsigma_0^2 \\
&+ \lambda_{\max}\left(K_2^T K_2 + \mathrm{diag}[\frac{2l_i}{\sigma_i}]\right) \frac{1}{2} \xi_2^T \xi_2 + \tilde{f}_{dis}^T D(z_2) M_A^{-1} \tilde{f}_{dis} \\
&- \sum_{i=1}^{n} \frac{\sigma_i}{4} \|\tilde{W}_i\|^2 + \frac{1}{2} \sum_{i=1}^{n} (\sigma_i + \epsilon^2 \|S_{ti}\|^2) \|W_i^*\|^2.
\end{aligned} \tag{9.83}$$

Applying $\frac{1}{2}\xi_2^T\xi_2 \le \frac{1}{2}\epsilon^2 h_2^2$, we have:

$$\begin{aligned}\dot{V}_4 \le& -z_1^T K_1 z_1 - z_2^T(K_2 - 2I_{N\times N})z_2 + \frac{1}{2}\|\bar{\epsilon}(Z)\|^2 + \frac{1}{2}\varsigma_0^2 \\ &+ \lambda_{\max}\left(K_2^T K_2 + \mathrm{diag}\left[\frac{2l_i}{\sigma_i}\right]\right)\frac{1}{2}\varepsilon^2 h_2^2 + \tilde{f}_{dis}^T D(z_2) M_A^{-1}\tilde{f}_{dis} \\ &-\sum_{i=1}^{n}\frac{\sigma_i}{4}\|\tilde{W}_i\|^2 + \frac{1}{2}\sum_{i=1}^{n}(\sigma_i + \epsilon^2\|S_{ti}\|^2)\|W_i^*\|^2 \\ \le& -a_3 V_3 + b_3,\end{aligned} \tag{9.84}$$

where a_3 and b_3 are two constants defined as:

$$\begin{aligned} a_3 =& \min\left(2\lambda_{\min}(K_1), \frac{2\lambda_{\min}(K_2 - 2I)}{\lambda_{\max}(M_A)}, 2\lambda_{\min}(D(z_2)M_A^{-1}),\right. \\ &\left.\min_{i=1,2\dots n}\left(\frac{\sigma_i}{2\lambda_{\max}(\Gamma_i^{-1})}\right)\right), \\ b_3 =& \frac{1}{2}\sum_{i=1}^{n}(\epsilon^2\|S_{ti}\|^2 + \sigma_i)\|W_i^*\|^2 + \lambda_{\max}\left(K_2^T K_2 + \mathrm{diag}\left[\frac{2l_i}{\sigma_i}\right]\right)\frac{1}{2}\varepsilon^2 h_2^2 \\ &+\frac{1}{2}\|\bar{\epsilon}\|^2 + \frac{1}{2}\varsigma_0^2. \end{aligned}$$

To guarantee that $a_3 > 0$, the control parameters K_1, K_2, and the linear function vector $\Theta(z_2)$ require to meet two conditions as follows, respectively.

$$\lambda_{\min}(K_1) > 0, \lambda_{\min}(K_2 - 2I) > 0, \lambda_{\max}\left(\frac{\partial\Theta(z_2)}{\partial z_2^T}M_A^{-1}\right) > 0. \tag{9.85}$$

Multiplying (9.84) by $e^{a_3 t}$, and integrating, we have

$$V_4 \le \left(V_4(0) - \frac{a_3}{b_3}\right)e^{-a_3 t} + \frac{a_3}{b_3} \le V_4(0) + \frac{a_3}{b_3}. \tag{9.86}$$

Thus

$$\frac{1}{2}\|z_1\|^2 \le V_4(0) + \frac{a_3}{b_3}. \tag{9.87}$$

Theorem 9.2 For the dynamic model in Eq. (9.31), by employing the control input and the adaptive law, because the output information is known, for each compact set Ω_0 where $(x_1(0), x_2(0), \hat{W}_1(0)\dots\ \hat{W}_n(0)) \in \Omega_0$, the initial conditions are bounded and the closed-loop system are semiglobally uniformly bounded. $z_1, z_2, \tilde{f}_{dis}$, and $\tilde{W}$ will remain within Ω_{z_1}, Ω_{z_2}, $\Omega_{\tilde{f}_{dis}}$, Ω_W:

$$\Omega_{z_1} := \left\{z_1 \in R^n \mid \|z_1\| \le \sqrt{X}\right\}, \tag{9.88}$$

$$\Omega_{z_2} := \left\{z_2 \in R^n \mid \|z_2\| \le \sqrt{\frac{X}{\lambda_{\min}(M_A)}}\right\}, \tag{9.89}$$

$$\Omega_{\tilde{f}_{dis}} := \left\{\tilde{f}_{dis} \in R^n \mid \|\tilde{f}_{dis}\| \le \sqrt{X}\right\}, \tag{9.90}$$

$$\Omega_{\tilde{W}} := \left\{\tilde{W} \in R^{l\times n} \mid \|\tilde{W}\| \le \sqrt{\frac{X}{\lambda_{\min}(\Gamma^{-1})}}\right\}, \tag{9.91}$$

where $X = 2(V_4(0) + \frac{b_3}{a_3})$, $b_3 > 0$ and $a_3 > 0$.

9.3.3 Simulation Results Without Control

The flexible beam system is vibrating without control force, namely, the control torque $u(t) = 0$. The parameters of the flexible beam are chosen as $N = 10$ and the initial status are selected as $w(L, t) = 1$ m. In Figure 9.2, we can see that the vibration amplitude of the end deflection of the flexible beam system without disturbance is varying periodically. Subsequently, the spatiotemporally varying disturbance is introduced as $f_{dis} = [\sin(t),\ \cos(t),\ \sin(t),\ \sin(t),\ \cos(t),\ \sin(t),\ \sin(t),\ \cos(t),\ \sin(t),\ \sin(t)]^T$. When spatiotemporally varying disturbance exists, the vibration amplitude is increasing which may cause serious damage to this system.

9.3.4 Simulation Results for PD Control

In order to suppress the vibration of flexible beam system, PD control is introduced. The parameter of the flexible beam is chosen as $N = 10$ and the initial status is set as $\omega(L, t) = 1$ m. The control torque is given as $u = -K_p(q_{10} - q_{10d}) - K_d\dot{q_{10}}$ and $\tau \in \mathbb{R}^n = [0, 0, ..., u]^T$, where K_p and K_d represent proportional gain and differential gain, respectively. q_{10} and q_{10d} are the tenth system state and the desired value, respectively. When the disturbance does not exist, the control parameters are set to $K_p = 30,\ K_d = 2$. The dark gray lines in Figures 9.3 and 9.4 represent the end vibration $w(L, t)$ of the flexible beam system and the PD control input, respectively. We can see that the end deflection is suppressed to a small neighborhood with zero in 6 s.

Subsequently, we introduce the spatiotemporally varying disturbance $f_{dis} = [\sin(t),\ \cos(t),\ \sin(t),\ \sin(t),\ \cos(t),\ \sin(t),\ \sin(t),\ \cos(t),\ \sin(t),\ \sin(t)]^T$. In order to suppress the divergent vibration, we employ a disturbance observer and choose the PD control parameters $K_p = 30,\ K_d = 2$. The red lines in Figures 9.3–9.5 show the end vibration $w(L, t)$, the PD control input and the disturbance observer error, respectively. We can get the conclusion that the real elastic vibration of the flexible beam system under PD control converges to a small neighborhood of zero within 7 s.

9.3.5 Simulation Results for Full-State Feedback

In order to reduce the elastic vibration within a shorter time, a NN control strategy for the flexible beam system is introduced. For neural network control strategy, the parameters of the flexible beam are chosen as $N = 10$ and $\omega(L, t) = 1$ m. The parameters of the RBFNN are set as $\hat{W}_i = 0\ (i = 1, 2, ..., 256)$, $\sigma = 0.0002$ and $\Gamma = 100I_{256\times256}$. The NN control parameters with full-state feedback are set as $K_1 = 4 * I_{10\times10}$, $K_2 = 30 * I_{10\times10}$. The dark gray lines in Figures 9.6

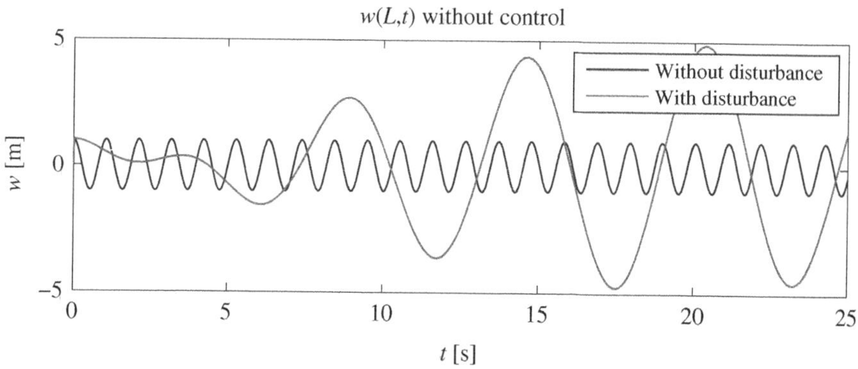

Figure 9.2 $w(L, t)$ without control.

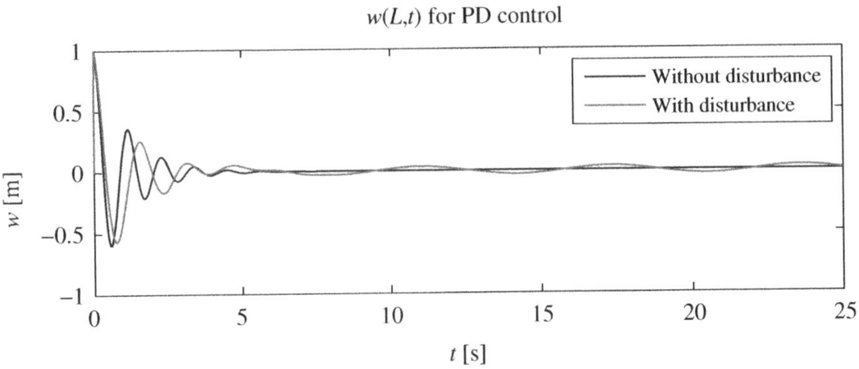

Figure 9.3 $w(L,t)$ with PD control.

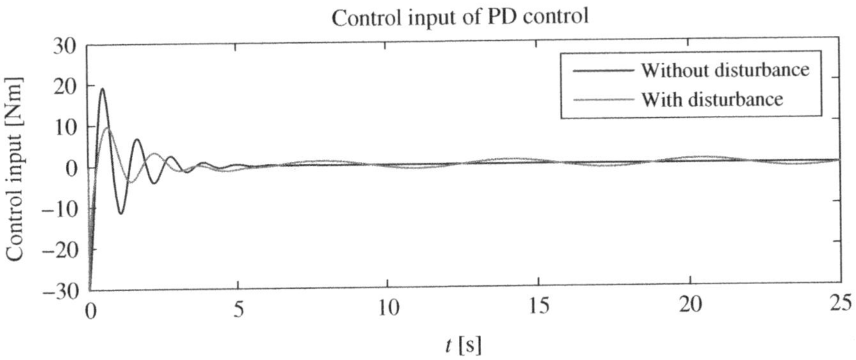

Figure 9.4 Control input of PD control.

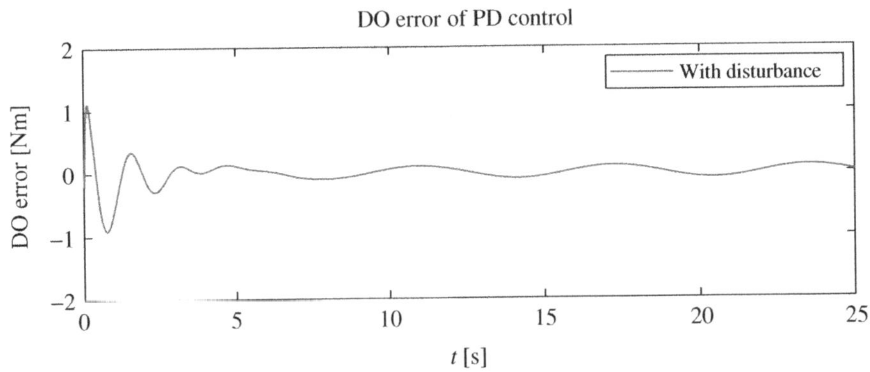

Figure 9.5 DO error of PD control.

and 9.7 represent the end vibration $w(L,t)$ of the flexible beam system and the NN control input with full-state feedback, respectively. We can see that the end deflection is suppressed to a small neighborhood with zero in 1 s. Compared with PD control, the time is reduced by 6 times by NN control with full-state feedback.

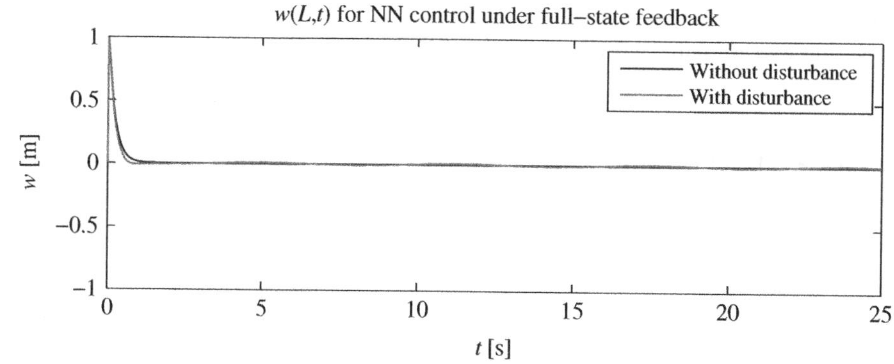

Figure 9.6 $w(L,t)$ with NN control under full-state feedback.

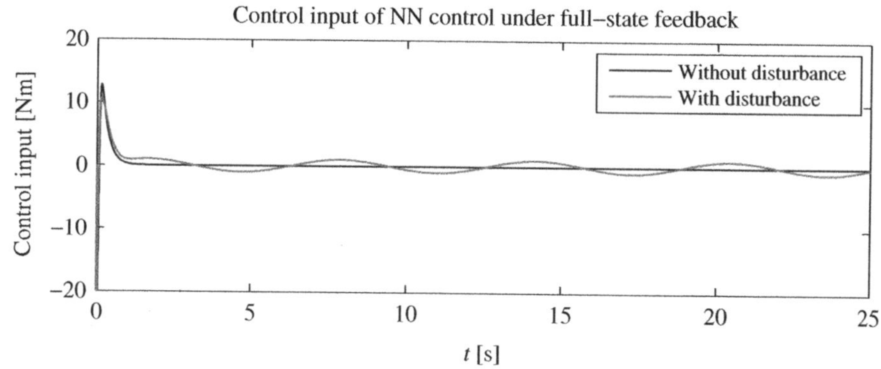

Figure 9.7 Control input of NN control under full-state feedback.

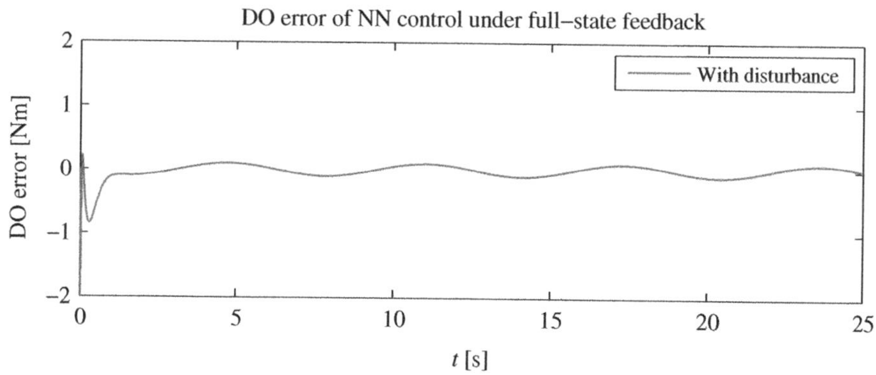

Figure 9.8 DO error of NN control under full-state feedback.

Subsequently, we introduce the spatiotemporally varying disturbance $f_{dis} = [\sin(t), \cos(t), \sin(t), \sin(t), \cos(t), \sin(t), \sin(t), \cos(t), \sin(t), \sin(t)]^T$. In order to estimate the spatiotemporally varying disturbance, we employ a disturbance observer and choose the NN control parameters $K_1 = 4 * I_{10\times10}$, $K_2 = 30 * I_{10\times10}$. The red lines in Figures 9.6–9.8 show the end vibration $w(L,t)$, the NN control input and the disturbance observer error, respectively. We can see that the real elastic vibration of the flexible beam system under NN control with full-state feedback converges to a small neighborhood of zero within 1 s. Compared with PD control, the time is reduced by 6 times by NN control with full-state feedback.

9.3.6 Simulation Results for Output Feedback

In this part, we introduce a NN controller with output feedback for the flexible beam system. The parameters of the flexible beam are chosen as $N = 10$ and $\omega(L, t) = 1$ m. The parameters of the RBFNN are set as $\hat{W}_i = 0$ $(i = 1, 2, \ldots, 256)$, $\sigma = 0.0002$ and $\Gamma = 100I_{256\times256}$. The NN control parameters with output feedback are set as $K_1 = 4 * I_{10\times10}$, $K_2 = 30 * I_{10\times10}$. The dark gray lines in Figures 9.9 and 9.10 represent the end vibration $w(L, t)$ of the flexible beam system and the NN control input with output feedback, respectively. We can see that the end deflection is suppressed to a small neighborhood with zero in 1 s. Compared with PD control, the time is reduced by 6 times by NN control with output feedback.

Subsequently, we introduce the spatiotemporally varying disturbance $f_{dis} = [\sin(t), \cos(t), \sin(t), \sin(t), \cos(t), \sin(t), \sin(t), \cos(t), \sin(t), \sin(t)]^T$. In order to estimate the spatiotemporally varying disturbance, we employ a disturbance observer and choose the NN control parameters $K_1 = 4 * I_{10\times10}$, $K_2 = 30 * I_{10\times10}$. The red lines in Figures 9.9–9.11 show the end vibration $w(L, t)$, the NN control input and the disturbance observer error, respectively. The real elastic vibration of the system under NN control with output feedback converges to a small neighborhood of zero within

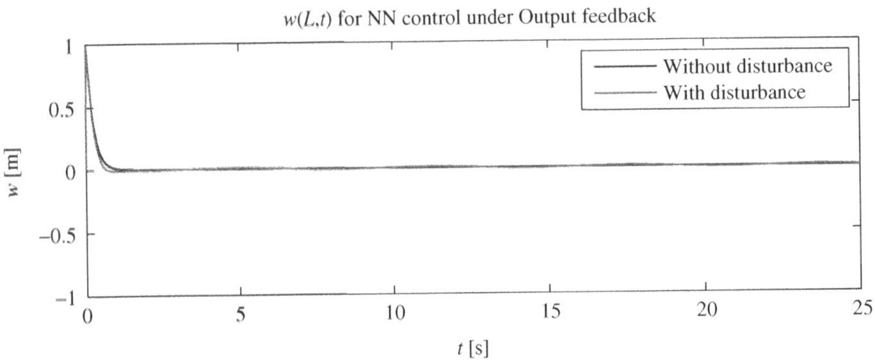

Figure 9.9 $w(L, t)$ with NN control under output feedback.

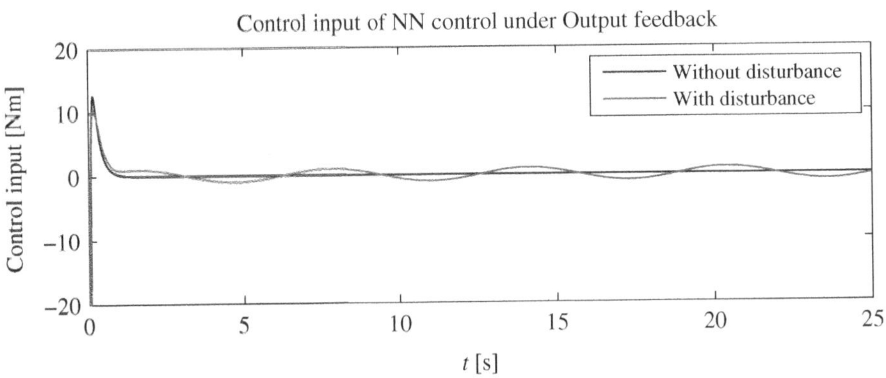

Figure 9.10 Control input of NN control under output feedback.

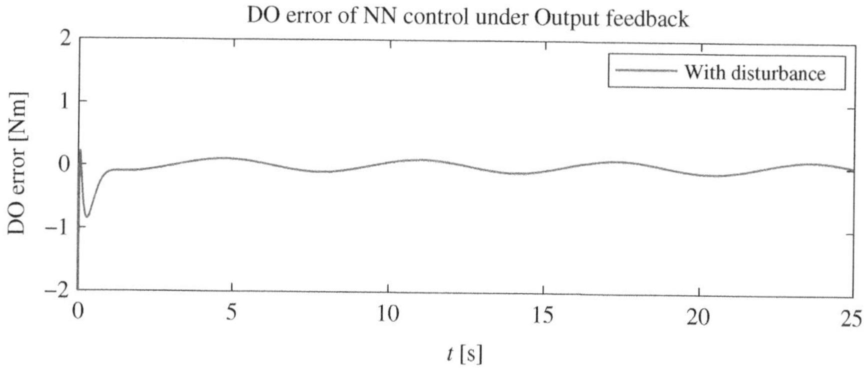

Figure 9.11 DO error of NN control under output feedback.

1 s. Compared with PD control, the time is reduced by 6 times by NN control with output feedback. Neural network control can make the flexible beam system stable in a short time and greatly reduce the transient overshoots which will bring irreparable consequences to practical projects.

9.4 Summary

In this chapter, considering the nth order elastic mode, the dynamics of a flexible beam equation with unknown spatiotemporally varying disturbance are derived in tangent coordinates system using AMM. Subsequently, a disturbance observer and a NN control are proposed to approximate the dynamic uncertainty and achieve vibration suppression. It is proved by Lyapunov's stability and extensive numerical simulation that the elastic vibration of the flexible beam system will converge to zero with a small neighborhood. In future work, we will continue to study the vibration control of flexible systems. Considering three-dimensional space, the advanced adaptive intelligent control method of flexible systems like nonhomogeneous flexible wing with bounded input disturbances, flexible robotic manipulator and standalone tall building-like structure with an eccentric load in complex environment will be studied. This work will provide a theoretical basis for further works.

10

Adaptive Finite-Time Control of a Bionic Flexible Flapping-Wing Aircraft with Actuator Failures

After reading this chapter, you should be able to:

- They are inspired by the flight of birds or insects and have many benefits for low energy consumption and high flexibility
- It is a novel control method that uses fuzzy neural network and nonsingular fast terminal sliding mode to achieve tracking and vibration control of the flexible wings
- They are used to test the effectiveness of the proposed controller under different scenarios of system uncertainties and ac tuator failures

10.1 Introduction

All flying animals in nature fly though flapping wing flight without exception. Inspired by the flight of birds or insects, the bionic flapping wing robotic aircraft has attracted special attention of many researchers in recent years [229, 335, 367]. The aircraft can stabilize the fuselage in the horizontal position or glide in the sky through flapping wings. The flexible wings of the aircraft have great advantages such as light weight, high flexibility, low energy consumption, and so on [145]. The bionic flapping-wing aircraft can generate lift force in high efficiency. In addition, with the advantage of low energy consumption, the bionic flapping-wing aircraft is suitable for flight missions without energy replenishment under long distance conditions [237]. The bionic flapping-wing aircraft is therefore widely used in both the military tasks (low-altitude surveillance, urban combat, accurate delivery, etc.) and the civilian applications (disaster monitoring and relief, field exploration, etc.) [93, 228]. Compared with fixed wings and rotary wings, flapping wings have the advantage in flight efficiency [133, 230, 241], which shows vital importance of study of the bionic flexible flapping wings. The flexible flapping wing is essentially an infinite-dimensional distributed parameter system with complex dynamic characteristics, which causes enormous challenges to system dynamic analysis and modeling [26, 175, 205]. The existing modeling methods for flexible structures include modal superposition method (MSM), finite element method (FEM), lumped mass method (LMM), finite segment method (FSM), and absolute coordinate method (ASM). In [71], Ferhatoglu et al. proposed a new MSM utilizing novel hybrid mode shapes for the dynamic analysis of nonlinear systems. The accuracy and computational efficiency of the proposed MSM are verified by several cases. However, obtaining the mode shape functions always requires complex derivations. FEM and assumed mode method (AMM) were compared to investigate the coupled

Dynamic Modeling and Neural Network-Based Intelligent Control of Flexible Systems,
First Edition. Hejia Gao, Wei He, and Changyin Sun.

vibrations of a flexible-disk rotor system with lacing wires in [366]. Although the precision of both methods can be improved with the increasing number of the elements, the issue of the heavy computational demand will be followed. Sun et al. [263] developed a lumped spring-mass method for a flexible robotic manipulator to analyze the system vibration. The flexible mechanical structures can be successfully and clearly modeled by LMM, but the accuracy of this method is lower than that of FEM with the same degrees of freedom [207]. In [169], a computational-efficiency method, which combines FSM with the discrete time transfer matrix method, was presented for modeling robot manipulators with flexible links and joints. However, FSM is only applicable to slender beam systems with small deformations. The aspect ratio of the bionic flexible flapping wing is generally between 1 and 2, so FSM is not suitable for dynamic analysis of flexible wings. In the literature, various software modeling methods, in particular to flexible systems, have also attracted the attention of a large number of researchers. Tahmasebi and Esmailzadeh proposed an accurate modeling approach based on ADAMS for a large flexible satellite, and co-simulation in ADAMS and MATLAB was carried out in [267]. However, these traditional modeling methods have some distinct disadvantages, such as high complexity, poor scalability, and slow response [233, 317]. It is also noted that the modification of the model parameters is complicated once the model is established in the software.

In addition to the flexible system modeling issues, how to design an intelligent control strategy is vitally important [29, 77, 82, 114, 353]. Kebria et al. [151] proposes a robust adaptive algorithm that effectively copes with time-varying delay and uncertainties in Internet-based teleoperation systems, Haidegger et al. [99] presented a proportional-integral-derivative (PID)-fuzzy control method for long distance telesurgical applications. In [231], Precup et al. proposed a new fuzzy logic controller for a general class of chaotic processes to further enhance the system performance. In addition, Deng et al. [54] investigated a hierarchical control approach which enables an exoskeleton robot to achieve cooperative manipulation with humans. To address unknown actuator hysteresis issues, Wu et al. [313] proposed an adaptive fuzzy quantized controller for a class of nonlinear systems.

At present, the control theory research of the bionic flapping wing robotic aircraft is still at its initial stage [79, 129, 249]. In [232], an adaptive position tracking controller with input constraint was proposed for flapping wing vehicle. In [118], the attitude and position control based on neural networks was designed for the flapping wing micro aerial vehicle (FWMAV) with uncertainties. By neglecting the oscillating signals, the strategies in [232] and [118] guarantee the uniform ultimate boundedness of the closed-loop system even in the presence of various disturbances. However, different from traditional rigid mechanical structures [31, 183, 184, 338], there are obvious elastic deformations and vibrations during the movement of the bionic flexible flapping wings. Vibration of the flexible flapping wing will result in structural damage and lifespan reduction.

As a consequence, vibration control problem during the tracking also needs to be solved urgently in addition to the dynamic modeling of the flexible flapping wing robotic aircraft. In [279], a flutter suppression approach was discussed for highly flexible wings using passive and active piezoelectric effects which are involved in the dynamic nonlinear beam equations. In the presence of dynamic uncertainties, Bialy et al. [15] and Gao et al. [84], respectively, presented an adaptive boundary control and a neural network control based on partial differential equations (PDEs) model of the flexible wing. The control inputs of these three controllers [15, 85, 279] all applied to the wing tip. However, the servo system is generally installed at the fuselage in practice for control flexibility [17]. In [248], Shao et al. proposed two types of control laws for a flexible wing model based on fuzzy control. Kumar et al. designed an adaptive control for a wing-fuselage model based on neural networks in [158]. The fuzzy logic and neural network methods, which have the superficial characteristics of strong representation and self learning, respectively, are of importance in solving control problems of uncertain systems [110, 315, 318].

Additionally, due to complex and changing environment [317, 326, 352, 359], the bionic flexible flapping-wing robotic aircraft is susceptible to certain types of faults [185, 342, 360]. In [225], the design and validation of a fault-tolerant gust load alleviation control system on a flexible wing were investigated based on a detailed linear ordinary differential equation (ODE) aeroelastic model. In [357], an adaptive fault-tolerant control scheme was developed based on a PDEs-ODEs model for a flexible aircraft wing against actuator faults. Fan et al. [66] designed a mixed-norm H_2/H_∞ fault-tolerant control for a flexible aircraft model with control efficiency loss. Moreover, sliding mode control has been widely applied to bionic flapping wing robotic aircrafts [302, 350]. Compared to the conventional sliding mode control design [340], high-order sliding mode control proposed by Chen and Tao [28] not only preserved system robustness, but also avoided chatter phenomenon. Hu et al. [126] presented a fast terminal fuzzy sliding mode attitude control scheme for the flapping-wing micro aerial vehicle to improve the system robustness. Nevertheless, the study of sliding mode method is of great significance to improve the system robustness [76, 170]. The vast majority of the aforementioned studies have shown that an accurate mathematical model of the flexible systems always necessitates complex derivations and heavy computation load [45, 71, 263, 366]. Even if a few studies have employed the software modeling methods, the distinct disadvantages, such as high complexity, poor scalability, and slow response are followed [233, 267, 317]. Another finding of the comprehensive literature review is that the control theory research of the bionic flapping-wing aircraft is at its initial stage. The previous control researches have mostly neglected the vibrations [118, 232], dynamic uncertainties [279], and actuator failures, which are extremely unfavorable in view of the practical application. Even among the few vibration studies, most control inputs can only be applied to the wing tip [15, 85], which are impractical in real flight situations. It is thus desirable to obtain a high-efficiency modeling method and design a feasible intelligent vibration control strategy, while solving the system uncertainties and actuator failures.

Motivated by the above aspects, this chapter develops an improved rigid finite element (IRFE) method and presents a new bionic flexible flapping-wing aircraft model by combining MapleSim multi-domain modeling tool. Based on the presented model, the dynamic analysis and control design as well as performance optimization of the flexible flapping wing system are studied from the view of multi-domains. An adaptive finite-time (AFT) fault-tolerant control strategy is proposed to suppress vibration of the flexible wings. The stability of the closed-loop system is analyzed through Lyapunov's direct method. Finally, co-simulation on MapleSim and MATLAB/Simulink is performed and the performance of the proposed control is compared with partial-state feedback (PSF) control and fuzzy neural network (FNN) control. The studies will contribute to the control theory development of distributed parameter systems.

10.2 Problem Formulation

The flapping wings are different from fixed wings and rotary wings. The flight of the bionic flapping wing robotic aircraft consists of flapping flight mode and gliding flight mode, which is consistent with the basic pattern of "flapping-gliding-flapping-gliding" of birds. Figure 10.1 shows a 3D model of a flexible flapping wing system. It is worth noting that there are obvious elastic vibrations during the movement of the bionic flexible flapping wings.

Remark 10.1 Based on the bionics, due to the extreme complexity of biological flight, the flexible wing system is represented by some simplifications (only flapping motion is considered and twisting motion is ignored).

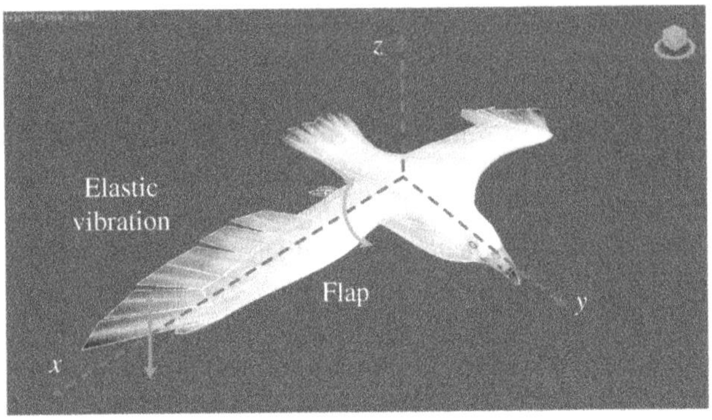

Figure 10.1 3D model of a flexible flapping wing system.

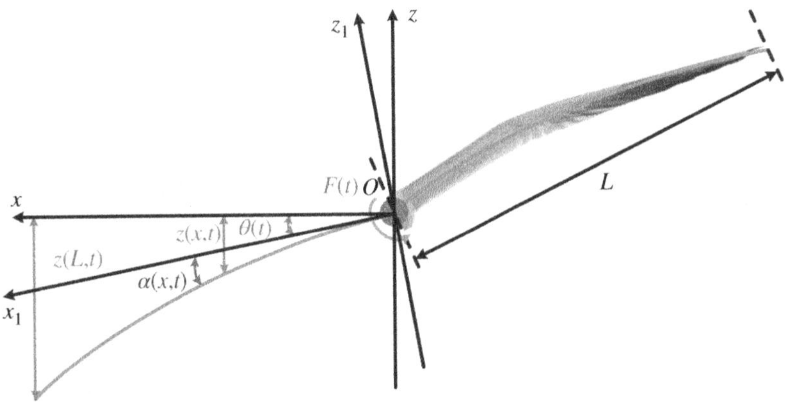

Figure 10.2 System representation and notations.

10.2.1 Dynamic Modeling

As illustrated in Figure 10.2, the axis x and y denote the wing span direction and the wing chord direction, respectively. xOy is the flapping plane that coincides with the horizontal plane, and the axis z is perpendicular to the horizontal plane. O represents the root of the flexible wing mounted with the servo system. xOz and x_1Oz_1 represent the fixed frame and local frame, respectively. $F(t)$ represents the control input, and L is the length of the flexible wing. $\theta(t)$ and $\alpha(x, t)$ represent the flapping angle and the elastic vibration, respectively. $z(x, t)$ is the position of the flexible wing at position x time t.

In order to obtain the dynamic model of the flexible wing system, an IRFE method is presented in this section. The front view and top view of the flexible wing model using the IRFE method are shown in Figures 10.3a and 10.3b, respectively. As illustrated in Figure 10.3a, the bionic flexible flapping wing is divided into flexible parts (spring-damping elements) and rigid parts (rigid finite elements). The rigid finite elements (RFEs) are connected to each other with spring-damping elements (SDEs), where RFEs and SDEs are relatively independent, the number of the RFEs is n which is the same with that of the SDEs. The stiffness and damping parameters of SDEs can be adjusted according to the specific conditions of different flexible wings. As shown in Figure 10.3b, c is the length of the wing chord, and RFEs are small rigid strips with the width i.

To further improve the modeling efficiency and the fidelity of the model, MapleSim multi-domain modeling tool is employed to establish a bionic flexible flapping-wing robotic aircraft model.

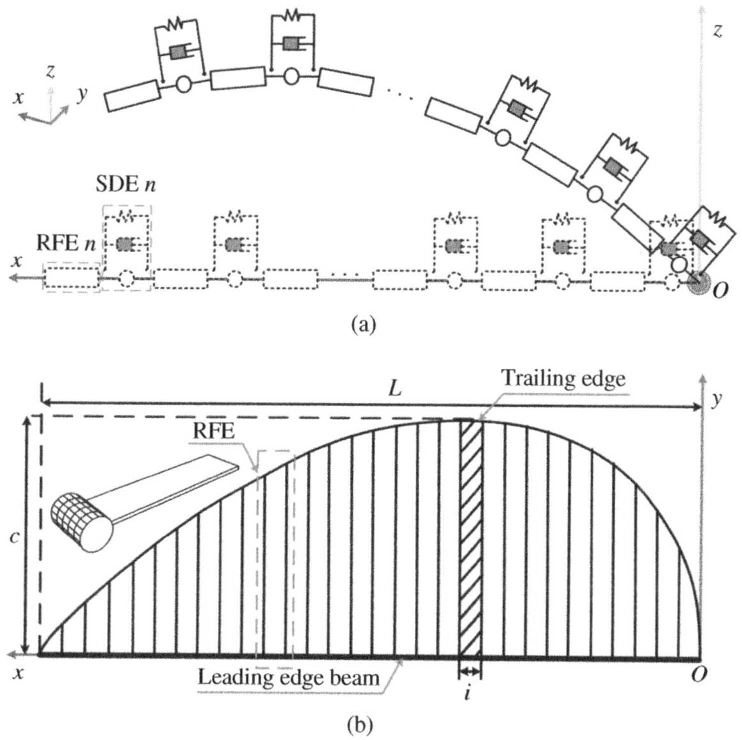

Figure 10.3 Kinematic model of flapping wing using IRFE method. (a) Front view (Frame xOz). (b) Top view (Frame xOy).

This model is composed of RFEs, SDEs, joints, servo system, and sensors. The dynamics of the bionic flexible flapping-wing robotic aircraft has the form as:

$$K(\gamma)\ddot{\gamma} + D(\gamma, \dot{\gamma})\dot{\gamma} + S(\gamma) = F + A(\gamma, \dot{\gamma}, F). \tag{10.1}$$

where $\gamma = [\theta, \alpha, \beta_2, \dots, \beta_n] \in \mathbb{R}^{n+1}$ is the state vector, and β_k represents the joint angle of the kth SDE. $K(\gamma)$, $D(\gamma, \dot{\gamma})$, and $S(\gamma)$ are the mass matrix, damping matrix, and stiffness matrix, respectively. $F \in \mathbb{R}^{n+1}$ denotes the control input, and $A(\gamma, \dot{\gamma}, F)$ represents the actuator failures.

10.2.2 Model Discretization

Based on the bionics, the parameters of the flexible flapping wing model are listed as Table 10.1.

The number of the RFEs or SDEs is set as 10. The stiffness and damping parameters of the kth SDE ($k = 1, 2, \dots, 10$) are chosen as 118, 112, 106, 100, 94, 88, 82, 76, 70, 64 Nm/rad.

Table 10.1 Model parameters of the flexible flapping wing.

Symbol	Description	Value
L	Flexible link length	41.9 cm
m	Flexible link mass	0.065 kg
B_{eq}	Viscous friction coefficient	0.004 Nm
J_{eq}	Moment of inertia	0.002 kgm^2

Table 10.2 Mode frequencies and amplitudes.

Order	1	2	3	4	5
Fre.[Hz]	1.55	3.21	16.34	41.18	72.47
Amp.[m]	0.3	0.6	0.2	0.1	0.08

Order	6	7	8	9	10
Fre.[Hz]	105.75	137.55	164.89	187.28	210.11
Amp.[m]	0.06	0.04	0.02	0.01	0.005

The mode visualization results are given to analyze the undamped modes of the flexible wing system. Mode frequencies and amplitudes are shown in Table 10.2.

The visualization results of the 1st–4th order modes are shown as Figure 10.4.

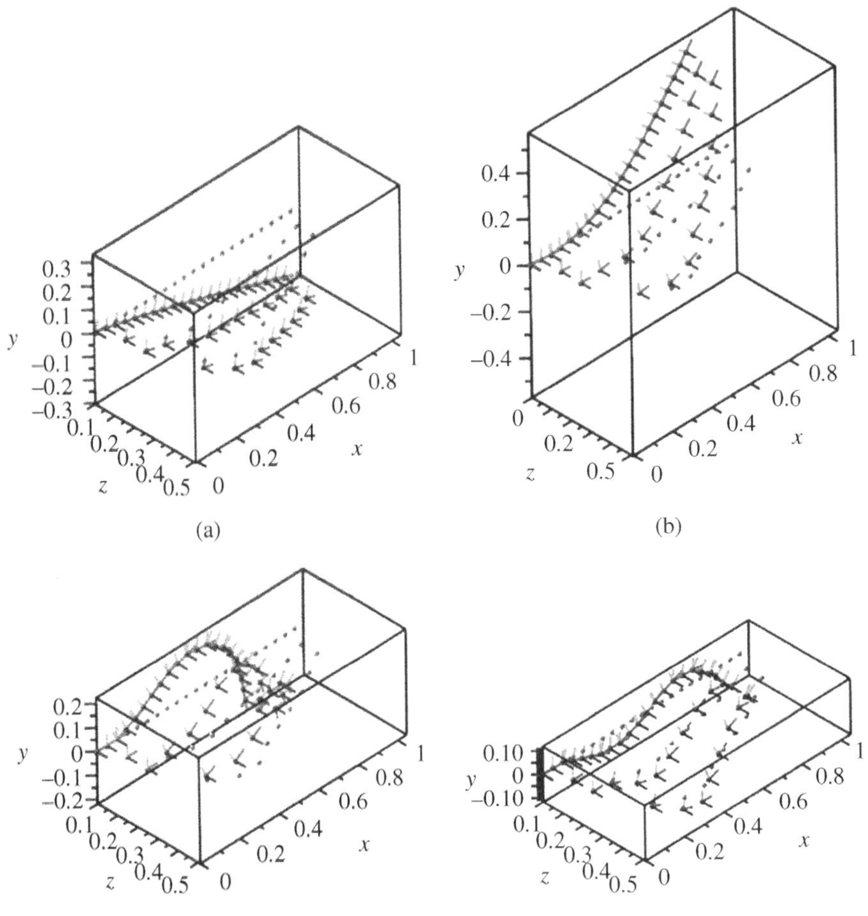

Figure 10.4 Mode visualization results (a) Frist order mode. (b) Second order mode. (c) Third order mode. (d) Fourth order mode.

10.3 Adaptive Finite-Time Control

Based on the proposed model of the bionic flexible flapping-wing aircraft (10.1), the dynamic uncertainties ($K(\gamma)$, $D(\gamma, \dot{\gamma})$, and $S(\gamma)$) and two types of actuator failures ($A(\gamma, \dot{\gamma}, F)$) are approximated by fuzzy neural networks. Besides, NFTSM is presented to improve the system robustness and accelerate the recovery of system stability after the occurrence of actuator failures. Compared with the traditional sliding mode method, the NFTSM method ensures that the system state error on the sliding mode surface can converge to zero in a finite-time and avoids the singularity issue of the FTSM variable. The diagram of the control system design is shown in Figure 10.5.

10.3.1 Control Design

Design a desired trajectory for the flexible wing as γ_d. Subsequently, the state error κ_1 and the speed error κ_2 can be expressed as $\kappa_1 = \gamma - \gamma_d$ and $\kappa_2 = \dot{\gamma} - \varrho$. $\varrho = -\Lambda\kappa_1 + \dot{\gamma}_d$ denotes a virtual variable, and $\Lambda > 0$ is a control gain. Thus, the derivative of κ_1 and the derivative of κ_2 are expressed as:

$$\dot{\kappa}_1 = \dot{\gamma} - \dot{\gamma}_d = \kappa_2 - \Lambda\kappa_1, \tag{10.2}$$

$$\dot{\kappa}_2 = K(\gamma)^{-1}[F + A(\gamma, \dot{\gamma}, F) - D(\gamma, \dot{\gamma})\dot{\gamma} \\ -S(\gamma) - K(\gamma)\dot{\varrho}]. \tag{10.3}$$

Design a Lyapunov function candidate as $\Theta_1 = \frac{1}{2}\kappa_1^T\kappa_1$. Differentiating Θ_1,

$$\dot{\Theta}_1 = -\kappa_1^T\Lambda\kappa_1 + \kappa_1^T\kappa_2. \tag{10.4}$$

To avoid the coupling term $\kappa_1^T\kappa_2$ in (10.4), design a Lyapunov function candidate as:

$$\Theta_2 = \frac{1}{2}\kappa_1^T\kappa_1 + \frac{1}{2}\kappa_2^T K(\gamma)\kappa_2. \tag{10.5}$$

Differentiating Θ_2,

$$\dot{\Theta}_2 = \kappa_1^T\kappa_2 - \kappa_1^T\Lambda\kappa_1 + \kappa_2^T[F + A(\gamma, \dot{\gamma}, F) \\ - D(\gamma, \dot{\gamma})\dot{\gamma} - S(\gamma) - K(\gamma)\dot{\varrho}]. \tag{10.6}$$

Design a control law as:

$$F_0 = -\kappa_1 - \Upsilon\kappa_2 - A(\gamma, \dot{\gamma}, F) \\ + D(\gamma, \dot{\gamma})\dot{\gamma} + S(\gamma) + K(\gamma)\dot{\varrho}. \tag{10.7}$$

where $\Upsilon > 0$ is a control gain. Therefore, $\dot{\Theta}_2 = -\kappa_1^T\Lambda\kappa_1 - \kappa_2^T\Upsilon\kappa_2$ is negative-definite.

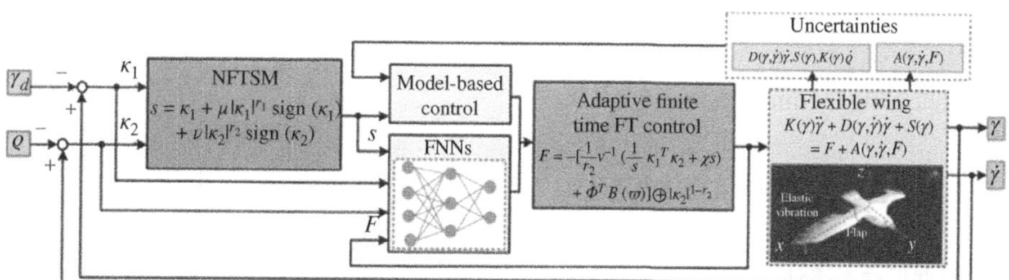

Figure 10.5 Control design diagram.

However, most bionic flexible flapping-wing robotic aircrafts work in extreme environment or under complex tasks, which may result in actuator failures. In addition, the actuator failures and dynamic information usually cannot be obtained accurately. Therefore, the proposed control (10.7) is not feasible for the bionic flexible flapping wing-robotic aircraft. To deal with the dynamic uncertainties ($K(\gamma)$, $D(\gamma, \dot{\gamma})$, $S(\gamma)$) and the unknown actuator failures ($A(\gamma, \dot{\gamma}, F)$), fuzzy logic and neural networks are introduced. Define a fuzzy logic system as:

$$f(x) = \frac{\Sigma_{j=1}^{l} y_j (\Pi_{i=1})^n u_{A_i^j}(x_i)}{\Sigma_{j=1}^{l} (\Pi_{i=1})^n u_{A_i^j}(x_i)}. \tag{10.8}$$

where Gaussian membership function is described as $u_{A_i^j}(x_i) = \exp[-(x_i - c_{ij})^2/\sigma_{ij}^2]$. c and σ are constants. And the inference rules are expressed as:

IF x_i is A_i^j, THEN y is W^j, $i = 1, 2 \ldots, n+1, j = 1, 2 \ldots, l$.

Therefore, the approximation can be expressed as:

$$\begin{aligned} f(\varpi) &= \Phi^{*T} B(\varpi) + \epsilon \\ &= D(\gamma, \dot{\gamma})\dot{\gamma} + S(\gamma) + K(\gamma)\dot{\varrho} - A(\gamma, \dot{\gamma}, F). \end{aligned} \tag{10.9}$$

where Φ^* and B are the ideal value of the fuzzy neural network weight and the fuzzy basis function, respectively. $\varpi = [\gamma, \dot{\gamma}, \varrho, \dot{\varrho}]^T$ denotes the inputs of the FNNs. ϵ is the approximate error and $|\epsilon| \leq \bar{\epsilon}$. Design the FNN control as:

$$F_{FNN} = -\kappa_1 - \Upsilon\kappa_2 + \hat{\Phi}^T B(\varpi), \tag{10.10}$$

where $\hat{\Phi}$ denotes the predictive value of the fuzzy neural network weight. The adaptive law is proposed as:

$$\dot{\hat{\Phi}}_i = \Gamma_i [B_i(\varpi)\kappa_{2,i} - \Psi_i \hat{\Phi}_i], \tag{10.11}$$

where $\Gamma_i > 0$ is the control gain and $\Psi_i > 0$ is a positive constant. Design a Lyapunov function candidate as:

$$\Theta_3 = \frac{1}{2}\kappa_1^T\kappa_1 + \frac{1}{2}\kappa_2^T K(\gamma)\kappa_2 + \frac{1}{2}\sum_{i=1}^{n+1} \tilde{\Phi}_i^T \Gamma_i^{-1} \tilde{\Phi}_i. \tag{10.12}$$

where $\tilde{\Phi} = \hat{\Phi} - \Phi^*$ denotes the error between predicted value and ideal value of the fuzzy neural network weight. Differentiating Θ_3 and substituting (10.10) and (10.11),

$$\begin{aligned} \dot{\Theta}_3 = &-\kappa_1^T \Lambda \kappa_1 - \kappa_2^T \Upsilon \kappa_2 + \kappa_2^T [\tilde{\Phi}^T B(\varpi) - \epsilon] \\ &- \sum_{i=1}^{n+1} \tilde{\Phi}_i^T [B_i(\varpi)\kappa_{2,i} - \Psi_i \hat{\Phi}_i]. \end{aligned} \tag{10.13}$$

Applying the following inequalities:

$$-\sum_{i=1}^{n+1} \Psi_i \tilde{\Phi}_i^T \hat{\Phi}_i \leq -\frac{1}{2}\sum_{i=1}^{n+1} \Psi_i (\|\tilde{\Phi}_i\|^2 - \|\Phi_i^*\|^2), \tag{10.14}$$

$$-2\kappa_2^T \epsilon \leq (\|\bar{\epsilon}\|^2 + \kappa_2^T \kappa_2). \tag{10.15}$$

$\dot{\Theta}_3$ can be obtained as:

$$\begin{aligned}\dot{\Theta}_3 &\le -\kappa_1^T\Lambda\kappa_1 - \kappa_2^T(\Upsilon - \frac{1}{2}I)\kappa_2 + \frac{1}{2}\|\bar{\epsilon}\|^2 \\ &- \frac{1}{2}\sum_{i=1}^{n+1}\Psi_i\|\tilde{\Phi}_i\|^2 + \frac{1}{2}\sum_{i=1}^{n+1}\Psi_i\|\Phi_i^*\|^2 \\ &\le -\lambda\Theta_3 + d,\end{aligned} \tag{10.16}$$

where

$$\lambda = \min\left[2\mathop{\Xi}_{\min}(\Lambda), \frac{\mathop{\Xi}_{\min}(2\Upsilon - I)}{\mathop{\Xi}_{\max}(K(\gamma))}, \min_{i=1,2\ldots,n+1}\left(\frac{\Psi_i}{\mathop{\Xi}_{\max}(\Gamma_i^{-1})}\right)\right],$$

$$d = \frac{1}{2}\|\bar{\epsilon}\|^2 + \frac{1}{2}\sum_{i=1}^{n}\Psi_i\|\Phi_i^*\|^2.$$

To guarantee the system stability, the control parameters need to satisfy the condition $\Xi_{\min}(2\Upsilon - I) > 0$, where $\Xi_{\min}(\diamond)$ and $\Xi_{\max}(\diamond)$ represent the minimum and maximum eigenvalues of matrix $\diamond$, respectively.

Based on (10.16), we can obtain $\dot{\Theta}_3 e^{\lambda t} \le (-\lambda\Theta_3 + d)e^{\lambda t}$. Integrating both sides of this inequality, we have

$$\begin{aligned}\Theta_3 &\le [\Theta_3(0) - d/\lambda]e^{-\lambda t} + d/\lambda \le \Theta_3(0)e^{-\lambda t} + d/\lambda \\ &\le \Theta_3(0) + d/\lambda.\end{aligned} \tag{10.17}$$

Theorem 10.1 For the bionic flexible flapping-wing robotic aircraft (10.1), under the proposed fuzzy neural network control law (10.10) and the adaptive law (10.11) together with the bounded initial conditions, the semi-globally uniformly ultimately boundedness (SGUUB) of the closed-loop system is guaranteed through Lyapunov's stability theory. κ_1, κ_2, and $\tilde{\Phi}$ will remain within the compact sets Ω_{κ_1}, Ω_{κ_2}, and $\Omega_{\tilde{\Phi}}$, respectively.

$$\Omega_{\kappa_1} = \left\{\kappa_1 \in \mathbb{R}^{n+1} \mid \|\kappa_1\|^2 \le \sqrt{2\Theta_3(0) + 2d/\lambda}\right\},$$

$$\Omega_{\kappa_2} = \left\{\kappa_2 \in \mathbb{R}^{n+1} \mid \|\kappa_2\|^2 \le \sqrt{\frac{2\Theta_3(0) + 2d/\lambda}{\Xi_{\max}(K(\gamma))}}\right\},$$

$$\Omega_{\tilde{\Phi}} = \left\{\tilde{\Phi} \in \mathbb{R}^{l\times(n+1)} \mid \|\tilde{\Phi}_i\|^2 \le \sqrt{\frac{2\Theta_3(0) + 2d/\lambda}{\Xi_{\max}(\Gamma_i^{-1})}}\right\}.$$

10.3.2 Stability Analysis

To further improve the system robustness and accelerate the recovery of system stability after failure, construct a nonsingular fast terminal sliding mode surface as follows:

$$s = \kappa_1 + \mu|\kappa_1|^{r_1}\mathrm{sign}(\kappa_1) + \nu|\kappa_2|^{r_2}\mathrm{sign}(\kappa_2), \tag{10.18}$$

where $\mu > 0, \nu > 0, 1 < r_2 < 2, r_1 > r_2, r_1 = e/f, r_2 = g/h$. $e, f, g, h \in \mathbb{N}$ are odd numbers. Then, the derivative of s has the following form:

$$\dot{s} = \dot{\kappa}_1 + \mu r_1\dot{\kappa}_1 \odot |\kappa_1|^{r_1-1} + \nu r_2\dot{\kappa}_2 \odot |\kappa_2|^{r_2-1}, \tag{10.19}$$

where $a \odot b = [a_1b_1, a_2b_2, \ldots, a_nb_n]^T$.

Design a Lyapunov function candidate as:

$$\Theta_4 = \frac{1}{2}\kappa_1^T\kappa_1 + \frac{1}{2}s^T K(\gamma)s. \tag{10.20}$$

Differentiating Θ_4, due to the property that $\dot{K}(\gamma) - 2D(\gamma,\dot{\gamma})$ is a skew-symmetric matrix [81], $\dot{\Theta}_4$ has the form as:

$$\begin{aligned}\dot{\Theta}_4 = &-\kappa_1^T\Lambda\kappa_1 + \kappa_1^T\kappa_2 + s^T[D(\gamma,\dot{\gamma})s \\ &+K(\gamma)\dot{\kappa}_1 + K(\gamma)\mu r_1\dot{\kappa}_1 \odot |\kappa_1|^{r_1-1}] \\ &+ s^T\nu r_2[F + A(\gamma,\dot{\gamma},F) - D(\gamma,\dot{\gamma})\dot{\gamma} \\ &- S(\gamma) - K(\gamma)\dot{\varrho}] \odot |\kappa_2|^{r_2-1}.\end{aligned} \tag{10.21}$$

A model-based finite-time fault-tolerant control can be expressed as:

$$F_{MB} = F_1 + F_2 + F_3, \tag{10.22}$$

where

$$F_1 = -\frac{1}{r_2}\nu^{-1}\left(\frac{1}{s}\kappa_1^T\kappa_2 + \chi s\right) \odot |\kappa_2|^{1-r_2},$$

$$\begin{aligned}F_2 = &-\frac{1}{r_2}\nu^{-1}[D(\gamma,\dot{\gamma})s + K(\gamma)\dot{\kappa}_1 + K(\gamma)\mu r_1\dot{\kappa}_1 \odot \\ &|\kappa_1|^{r_1-1}] \odot |\kappa_2|^{1-r_2},\end{aligned}$$

$$F_3 = D(\gamma,\dot{\gamma})\dot{\gamma} + S(\gamma) + K(\gamma)\dot{\varrho} - A(\gamma,\dot{\gamma},F),$$

and χ is a constant. Substituting (10.22) into (10.21),

$$\dot{\Theta}_4 = -\kappa_1^T\Lambda\kappa_1 - s^T\chi s. \tag{10.23}$$

Remark 10.2 The finite-time convergence is achieved with the NFTSM method (10.18) [331]. When the system state is close to the equilibrium point ($0 < \kappa_1 \ll 1$), the low-order term of κ_1 dominates. The change rate of the system state on the NFTSM surface is similar to that on the NTSM surface ($\dot{\kappa}_1 = -(\frac{1}{\nu})^{1/r_2}\mathrm{sign}(\kappa_1)|\kappa_1|^{1/r_2}, 0 < 1/r_2 < 1$). When the system state is far from the equilibrium point ($\kappa_1 \gg 1$), the higher order term of κ_1 dominates. The change rate of the system state on the NFTSM surface is faster than that on the NTSM surface ($\dot{\kappa}_1 = -(\frac{\mu}{\nu})^{1/r_2}\mathrm{sign}(\kappa_1)|\kappa_1|^{r_1/r_2}$, $r_1/r_2 > 1$). Thus, it shows that NFTSM has a higher convergence rate than NTSM in the sliding phase. Compared with FTSM, $r_2 > 1$ can avoid the negative index term caused by the derivative of s and ensure the nonsingularity.

Similar to the Subsection 10.3.1, the proposed model-based control (10.22) is not feasible for the flexible flapping-wing aircraft due to the dynamic uncertainties and unknown actuator failures. Redefine a fuzzy neural network approximation as:

$$\begin{aligned}f(\varpi) &= \Phi^{*T}B(\varpi) + \epsilon \\ &= [A(\gamma,\dot{\gamma},F) - D(\gamma,\dot{\gamma})\dot{\gamma} - S(\gamma) - K(\gamma)\dot{\varrho}] \\ &\odot |\kappa_2|^{r_2-1} + \frac{1}{r_2}\nu^{-1}[D(\gamma,\dot{\gamma})s + K(\gamma)\dot{\kappa}_1 \\ &+K(\gamma)\mu r_1\dot{\kappa}_1 \odot |\kappa_1|^{r_1-1}],\end{aligned} \tag{10.24}$$

where $\varpi = [\kappa_1^T, \kappa_2^T, s^T, F^T]^T$ denotes the inputs of the FNNs. Design the adaptive finite-time (AFT) fault-tolerant control as:

$$F_{AFT} = -\left[\frac{1}{r_2}\nu^{-1}\left(\frac{1}{s}\kappa_1^T\kappa_2 + \chi s\right) + \hat{\Phi}^T B(\varpi)\right] \odot |\kappa_2|^{1-r_2}, \tag{10.25}$$

where $\hat{\Phi}$ denotes the estimation weight of FNNs. The adaptive law is proposed as:

$$\dot{\hat{\Phi}}_i = \Gamma_i[s_i^T \nu_i r_2 B_i(\varpi) - \Psi_i \hat{\Phi}_i], \tag{10.26}$$

where $\Gamma_i > 0$ and $\Psi_i > 0$ are control gains.

Construct a new Lyapunov function candidate as:

$$\Theta_5 = \frac{1}{2}\kappa_1^T\kappa_1 + \frac{1}{2}s^T K(\gamma)s + \frac{1}{2}\sum_{i=1}^{n+1}\tilde{\Phi}_i^T\Gamma_i^{-1}\tilde{\Phi}_i. \tag{10.27}$$

Applying (10.25) and (10.26), derive Θ_5 with respect to the time t,

$$\begin{aligned}\dot{\Theta}_5 &= -\kappa_1^T\Lambda\kappa_1 - s^T\chi s + s^T\nu r_2\epsilon \\ &\quad - s^T\nu r_2\sum_{i=1}^{n+1}\tilde{\Phi}_i B(\varpi) + \sum_{i=1}^{n+1}\tilde{\Phi}_i^T\Gamma_i^{-1}\dot{\hat{\Phi}}_i \\ &= -\kappa_1^T\Lambda\kappa_1 - s^T\chi s + s^T\nu r_2\epsilon - \sum_{i=1}^{n+1}\Psi_i\tilde{\Phi}_i^T\hat{\Phi}_i.\end{aligned} \tag{10.28}$$

Applying (10.14) and $s^T\epsilon \leq \frac{1}{2}(\|\bar{\epsilon}\|^2 + s^T s)$, $\dot{\Theta}_5$ has the form as:

$$\begin{aligned}\dot{\Theta}_5 &\leq -\kappa_1^T\Lambda\kappa_1 - s^T\chi s + \frac{1}{2}\nu r_2(\|\bar{\epsilon}\|^2 + s^T s) \\ &\quad - \frac{1}{2}\sum_{i=1}^{n+1}\Psi_i(\|\tilde{\Phi}_i\|^2 - \|\Phi_i^*\|^2), \\ &\leq -\kappa_1^T\Lambda\kappa_1 - s^T\left(\chi I - \frac{1}{2}\nu_i r_2\right)s + \frac{1}{2}\nu_i r_2\|\bar{\epsilon}\|^2 \\ &\quad - \frac{1}{2}\sum_{i=1}^{n+1}\Psi_i\|\tilde{\Phi}_i\|^2 + \frac{1}{2}\sum_{i=1}^{n+1}\Psi_i\|\Phi_i^*\|^2, \\ &\leq -\ell\Theta_5 + b.\end{aligned} \tag{10.29}$$

where

$$\ell = \min\left[2\underset{\min}{\Xi}(\Lambda), \frac{\underset{\min}{\Xi}(2\chi I - \nu_i r_2)}{\underset{\max}{\Xi}(K(\gamma))}, \min_{i=1,2\ldots,n}\left(\frac{\Psi_i}{\underset{\max}{\Xi}(\Gamma_i)}\right)\right],$$

$$b = \frac{1}{2}\nu_i r_2\|\bar{\epsilon}\|^2 + \frac{1}{2}\sum_{i=1}^{n}\Psi_i\|\Phi_i^*\|^2.$$

To guarantee the system stability, the control parameters need to satisfy the condition $\Xi_{\min}(2\chi I - \nu_i r_2) > 0$.

Based on (10.29), we can obtain $\dot{\Theta}_5 e^{\ell t} \leq (-\ell\Theta_5 + b)e^{\ell t}$. Integrating both sides of this inequality, we have

$$\begin{aligned}\Theta_5 &\leq [\Theta_5(0) - b/\ell]e^{-\ell t} + b/\ell \leq \Theta_5(0)e^{-\ell t} + b/\ell \\ &\leq \Theta_5(0) + b/\ell.\end{aligned} \tag{10.30}$$

Theorem 10.2 For the bionic flexible flapping-wing robotic aircraft (10.1), with the bounded initial conditions, the proposed adaptive finite-time fault-tolerant control law (10.25) with the adaptive law (10.26) can guarantee SGUUB stability of the closed-loop system. κ_1, s, and $\tilde{\Phi}$ will remain within the compact sets Ω_{κ_1}, Ω_s, and $\Omega_{\tilde{\Phi}}$, respectively.

$$\Omega_{\kappa_1} = \left\{ \kappa_1 \in \mathbb{R}^{n+1} \mid \|\kappa_1\|^2 \le \sqrt{2\Theta_5(0) + 2b/\ell} \right\},$$

$$\Omega_s = \left\{ s \in \mathbb{R}^{n+1} \mid \|s\|^2 \le \sqrt{\frac{2\Theta_5(0) + 2b/\ell}{\Xi_{\max}(K(\gamma))}} \right\},$$

$$\Omega_{\tilde{\Phi}} = \left\{ \tilde{\Phi} \in \mathbb{R}^{l\times(n+1)} \mid \|\tilde{\Phi}_i\|^2 \le \sqrt{\frac{2\Theta_5(0) + 2b/\ell}{\Xi_{\max}(\Gamma_i^{-1})}} \right\}.$$

Remark 10.3 The theoretical results show that the convergence speed of the system is related to the variable ℓ. Appropriate control parameters can improve the convergent performance of the closed-loop system. Both the increase of the control gains Λ, χ, and the decrease of the adaptive gain Γ_i can speed up the convergence of the system.

10.4 Numerical Simulations

In order to verify the feasibility of the proposed adaptive finite-time (AFT) fault-tolerant control algorithm, the co-simulation in the MapleSim and MATLAB/Simulink environment is investigated for the flexible flapping wing system against two types of actuator failures (efficiency loss and bias-type fault). Two different flight modes (flapping/gliding flight) are also simulated. It should be noted that the flexible flapping wing aircraft model is composed of flexible wings, leading edge beam, joint, servo system, and sensors. The detailed dynamic model information of the flexible wing system has been given in Section II. The tracking error, vibration suppression, and robustness are all analyzed and discussed. In the simulation process ($0 < t \le 15$ s), the desired trajectory of the flexible wing is set as:

$$\theta_d = \begin{cases} 30\sin(t), & 0 < t \le 10 \text{ s} \\ 30, & 10 < t \le 15 \text{ s}, \end{cases} \tag{10.31}$$

where the aircraft is in flapping flight mode when $0 < t \le 10$ s and in gliding flight mode when $10 < t \le 15$ s. In addition, the actuator failure problem is described as:

$$A(\gamma, \dot{\gamma}, F) = \begin{cases} -50\%F, & 5 < t \le 7 \text{ s} \\ 3\sin(t) + 10, & 11 < t \le 13 \text{ s}, \end{cases} \tag{10.32}$$

where the efficiency loss and the bias-type fault occur when $5 < t \le 7$ s and $11 < t \le 13$ s, respectively.

To further illustrate the effectiveness of the proposed control algorithm, the final performance is compared with the partial-state feedback (PSF) control and the fuzzy neural network (FNN) control.

The PSF control law has the form as:

$$F_{PSF} = -K_1(\theta - \theta_d) - K_2(\dot{\theta} - \dot{\theta}_d), \tag{10.33}$$

where K_1 and K_2 are the control gain parameters. Simulations with three sets of gain parameters (Set 1: $K_1 = 230, K_2 = 1.7$; Set 2: $K_1 = 230, K_2 = 2.0$; Set 3: $K_1 = 250, K_2 = 2.0$) have been performed. For convenience, the tracking error's oscillation time after the occurrence of actuator failures is represented as TEO time. The effects of the gain parameters on the control performance can be seen from Figures 10.6–10.9. The tracking error, the vibration and the TEO time are all reduced as K_2 increases. However, a longer TEO time is required with increase of K_1, though both the tracking error and the vibration are reduced. Based on this, Set 3 is selected with combined consideration of tracking error, vibration suppression, and robustness. A detailed result analysis under Set 3 is provided in the following section.

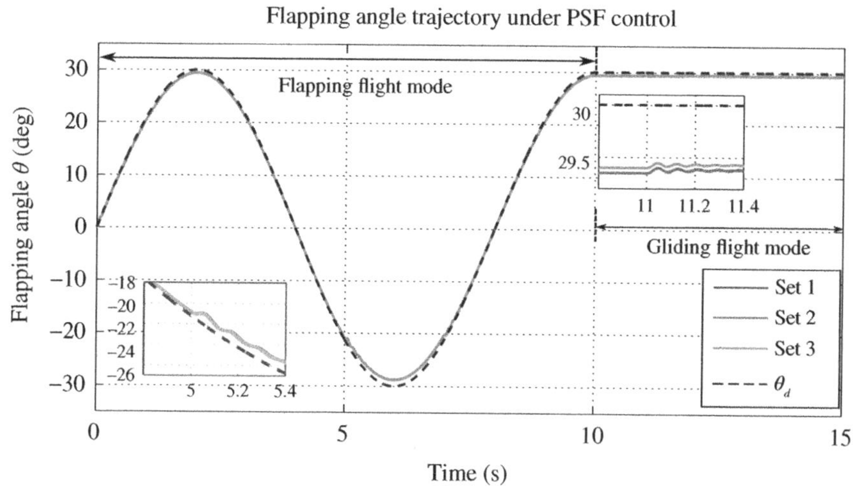

Figure 10.6 Flapping angle trajectory under PSF control.

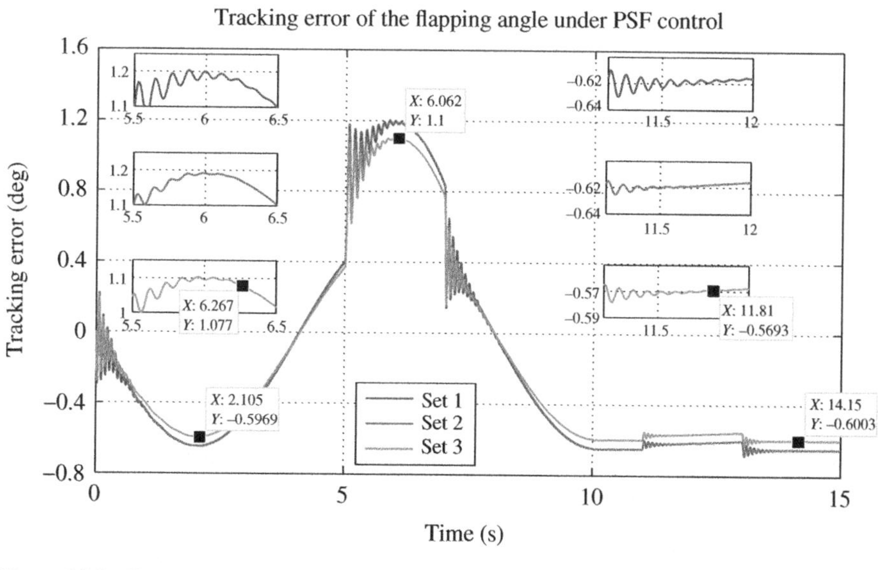

Figure 10.7 Tracking error of the flapping angle under PSF control.

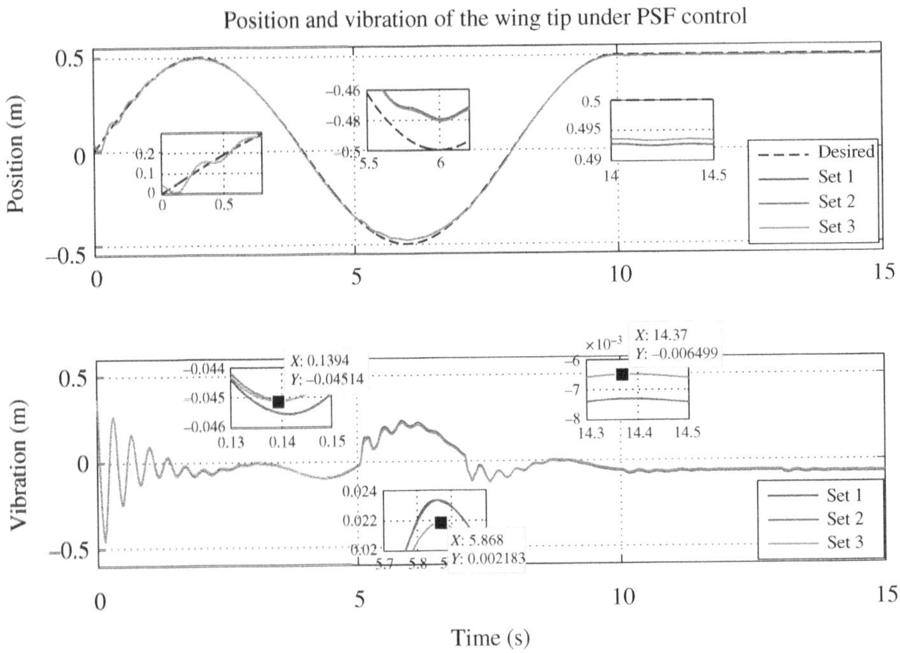

Figure 10.8 Position and vibration of the wing tip under PS control.

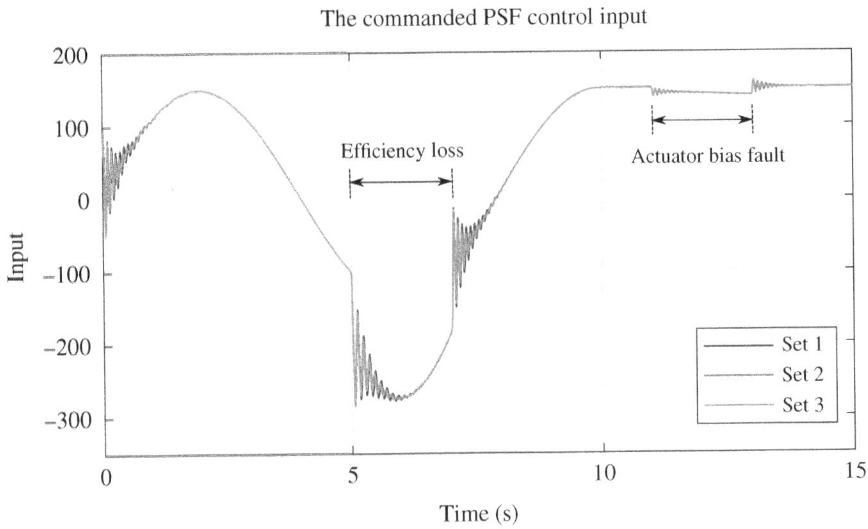

Figure 10.9 PSF control input of the flexible flapping wing.

The flapping angle trajectory and tracking error of the flapping angle under PSF control are shown in Figures 10.6 and 10.7, respectively. Figure 10.6 shows that the flexible flapping wing system can achieve tracking control performance under PSF control. However, there are slight oscillations in tracking the actual trajectory when $t = 5$ and $t = 11$ s. These are caused by the sudden occurrence of actuator failures. As can be seen from Figure 10.7, the peak-to-peak value of the tracking error is 1.697 deg, and the tracking error of the flexible flapping wing reaches the maximum of 1.100 deg at 6.1 s due to the actuator failure. When the flexible flapping wing is in gliding flight mode, the steady-state tracking error is 0.600 deg. In addition, when the flexible flapping wing system is subjected to the sudden actuator failures, the tracking performance will be affected, resulting in the oscillation of the tracking error. As illustrated in Figure 10.7, the TEO times under PSF control when the system encounters an efficiency loss or bias-type actuator failure are 1.27 and 0.81 s, respectively. Besides, the position and the vibration of the flexible flapping wing tip under PSF control are shown in Figure 10.8. As shown in Figure 10.8, the maximum vibration of the flexible wing tip is 4.51 cm. During the actuator failures, the wing tip vibration reaches up to 2.18 cm. The vibration of the flexible wing tip is eventually stable at 0.65 cm. Furthermore, the commanded PSF control input is provided in Figure 10.9.

The PSF control results of the bionic flexible flapping-wing robotic aircraft are illustrated in Figures 10.6–10.9. Although the tracking control performance and vibration suppression are achieved without actuator failures, they are deteriorated when unexpected actuator failures suddenly occur. That is, the adaptive capacity of the system is very limited. However, most of the flexible flapping-wing systems should be able to work in the extreme complex environment. If the unknown actuator failures of the flexible wing system cannot be solved timely and effectively, it will greatly affect the system control performance, and even damage the system structure, leading to unpredictable serious consequences.

Motivated by the above results, the proposed FNN control solves the problems of the uncertain dynamic information and unknown actuator failures for the flexible wing system. The FNN control has the form as (10.10). The fuzzy neural network system parameters are set as: the number of fuzzy inference rules l is 8, the parameters of Gaussian membership function c_{ij} and σ_{ij} are ± 1 and 1, respectively. The node number for each $B(\omega)$ is 256 and variance of centers is 2. The initial weights of fuzzy neural networks are chosen as $\hat{\Phi}_i = 0$ $(i = 1, 2, \dots, 256)$. The adaptive gains are set as $\Gamma = 10$ and $\Psi = 0.2$. Simulations with three sets of gain parameters (Set 1: $\Lambda = 100$, $\Upsilon = 4$; Set 2: $\Lambda = 90$, $\Upsilon = 5$; Set 3: $\Lambda = 100$, $\Upsilon = 5$) have been conducted and the results are shown in Figures 10.10–10.13. With the increase of Υ, the tracking performance, vibration suppression, and robustness are all strengthened. However, when Λ increases, the TEO time gets longer, though

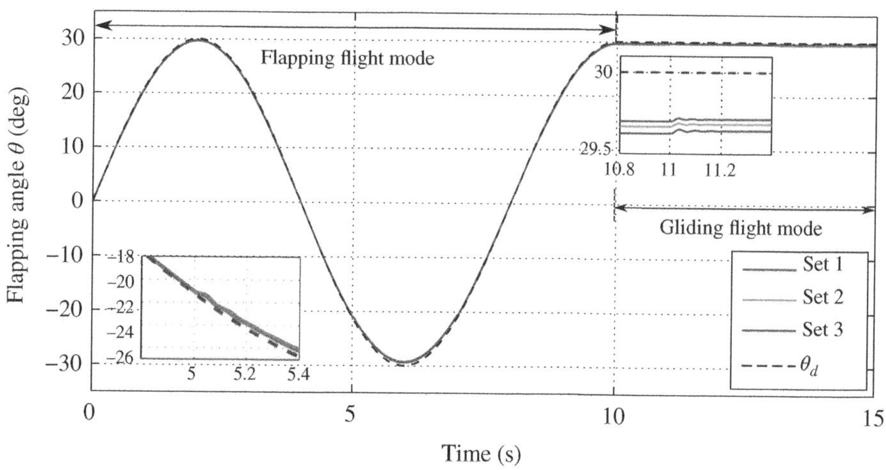

Figure 10.10 Flapping angle trajectory under FNN control.

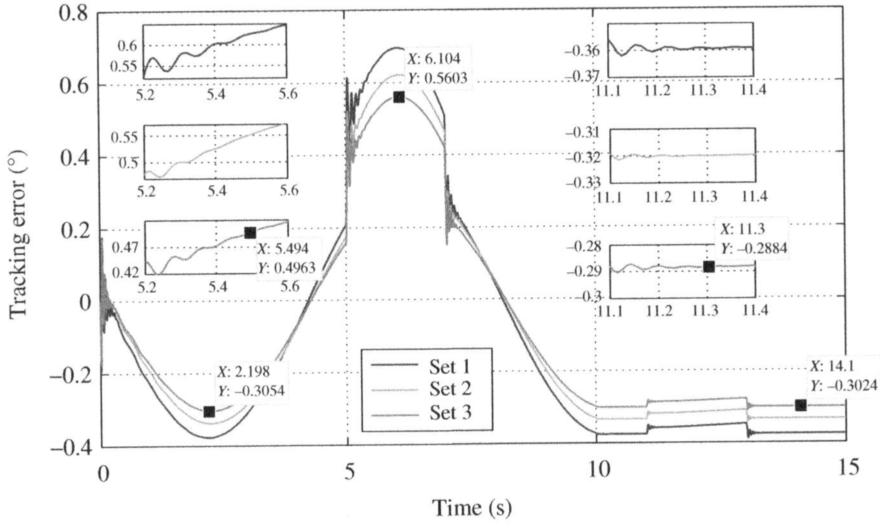

Figure 10.11 Tracking error of the flapping angle under FNN control.

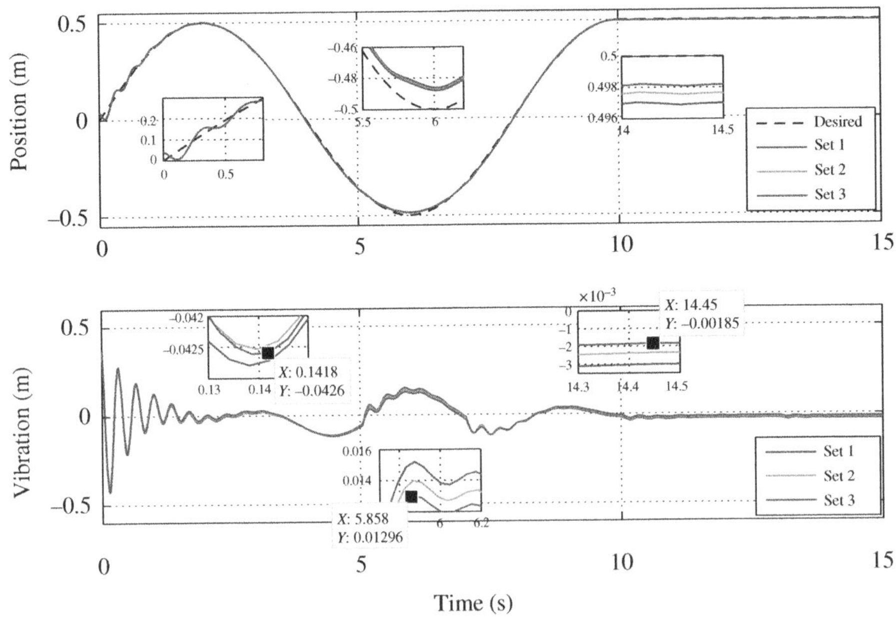

Figure 10.12 Position and vibration of the wing tip under FNN control.

tracking performance and vibration suppression are better. Thus, Set 3 is chosen and detailed discussion is provided as follows.

The flapping angle trajectory and tracking error of the flapping angle under FNN control are shown in Figures 10.10 and 10.11, respectively. Compared with PSF control, when the actuator failures suddenly occur ($t = 5$ and $t = 11$ s), the FNN control system not only has higher accuracy in tracking, but also reduces the oscillations in tracking the actual trajectory in Figure 10.6. Additionally, as can be seen from Figure 10.11, the peak-to-peak value of the tracking error is

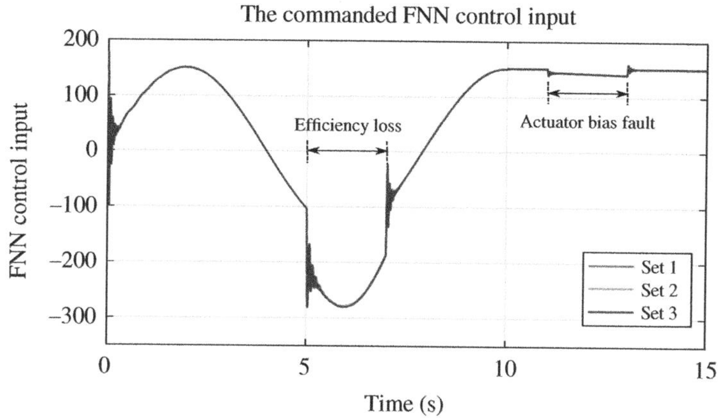

Figure 10.13 FNN control input of the flexible flapping wing.

0.866°. The maximum tracking error of the flexible flapping wing is 0.560° at $t = 6.1$ s due to the actuator failure. When the flexible wing is in gliding flight mode, the steady-state tracking error is 0.302°. The TEO times against the efficiency loss and bias-type actuator failures are 0.50 and 0.29 s, respectively. In addition, the position and vibration of the flexible wing tip under FNN control are shown in Figure 10.12. As illustrated in Figure 10.12, the maximum vibration of the flexible wing tip is 4.26 cm. During the unexpected actuator failures, the wing tip vibration reaches the maximum of 1.30 cm. The vibration of the steady-state is 0.19 cm when the flexible wing is in gliding flight mode. Compared with PSF control, FNN control achieves better vibration suppression performance. Furthermore, the commanded FNN control input is provided in Figure 10.13.

As shown in Figures 10.10–10.13, compared with PSF control, FNN control system has higher accuracy in tracking, and stronger ability in vibration suppression. Even when the actuator failures suddenly occur, the system can still achieve the corresponding control performance within a shorter time. However, it is still essential to further improve accuracy and ensure timeliness.

In order to further improve the control performance of the flexible-wing system, the proposed AFT fault-tolerant control addresses the issue of the finite-time control based on the NFTSM method. The AFT control has the form as (10.25). The control gain is chosen as $\Lambda = 100$. The parameters of the NFTSM surface are set as: the gains of the index terms are chosen as $\mu = 300$ and $\nu = 350$. Index parameters are set as $e = 5, f = 3, g = 7$, and $h = 5$. The fuzzy neural network system parameters are set as: the number of fuzzy inference rules l is 8, the parameters of Gaussian membership function c_{ij} and σ_{ij} are ± 1 and 1, respectively. The node number for each $B(\varpi)$ is 256 and variance of centers is 2. The initial weights of fuzzy neural networks are chosen as $\hat{\Phi}_i = 0$, $(i = 1, 2, \ldots, 256)$. The adaptive gains are set as $\Gamma = 200$ and $\Psi = 0.2$. Simulations with three sets of gain parameter (Set 1: $\chi = 12$; Set 2: $\chi = 18$; Set 3: $\chi = 24$) are presented to observe the effect of χ on control performance. As illustrated in Figures 10.14–10.17, with the increasing of χ, tracking performance and robustness are enhanced, though vibration suppression performance is deteriorated. So we select Set 3 and analyze the corresponding result in this section.

The flapping angle trajectory and the tracking error under AFT control are shown in Figures 10.14 and 10.15, respectively. Compared with PSF and FNN control, when the actuator failures occur suddenly ($t = 5$ and $t = 11$ s), the control system can not only remove the oscillations in tracking the actual trajectory in Figures 10.6 and 10.10 successfully, but also realize more accurate tracking and better robustness. As can be seen from Figure 10.15, the

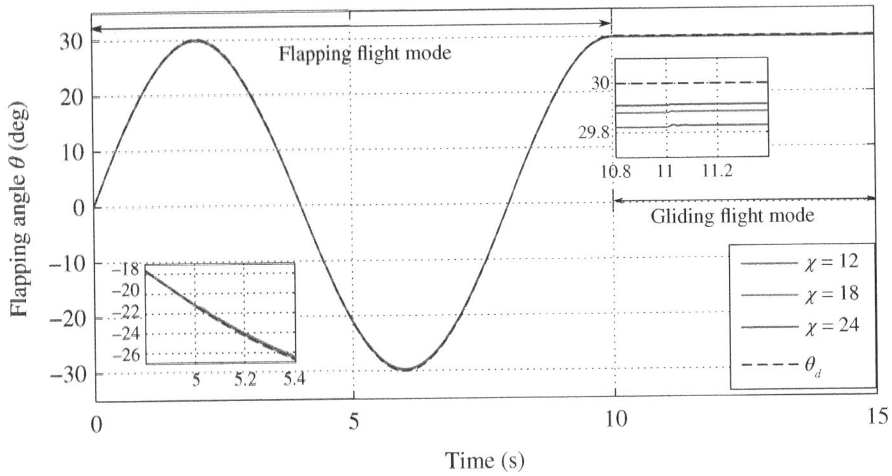

Figure 10.14 Flapping angle trajectory under AFT control.

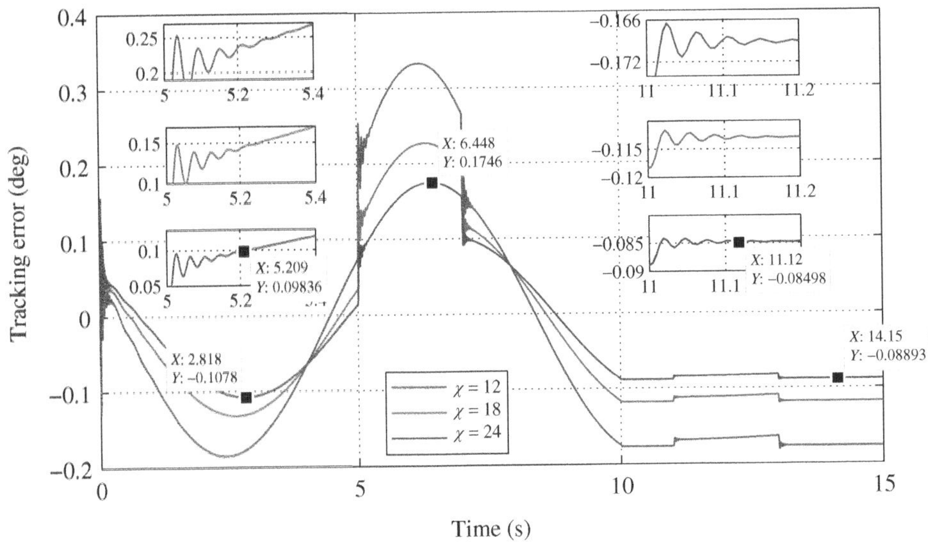

Figure 10.15 Tracking error of the flapping angle under AFT control.

peak-to-peak value of tracking error is 0.282°, and the maximum tracking error is 0.175° during the actuator failure. When the flexible wing is in gliding flight mode, the steady-state tracking error is 0.089°. The TEO times when the system encounters an efficiency loss and bias-type actuator failure are 0.21 and 0.12 s, respectively. Additionally, the position and the vibration of the flexible flapping wing tip under the AFT control are shown in Figure 10.16. As illustrated, the maximum vibration of the wing tip is 4.17 cm. When unexpected actuator failures suddenly occur, the wing tip vibration reaches up to 0.68 cm. The vibration of the flexible wing tip eventually stabilizes at 0.15 cm. Compared with PSF and FNN control, AFT control achieves better vibration suppression performance. Furthermore, the commanded AFT control input is provided in Figure 10.17.

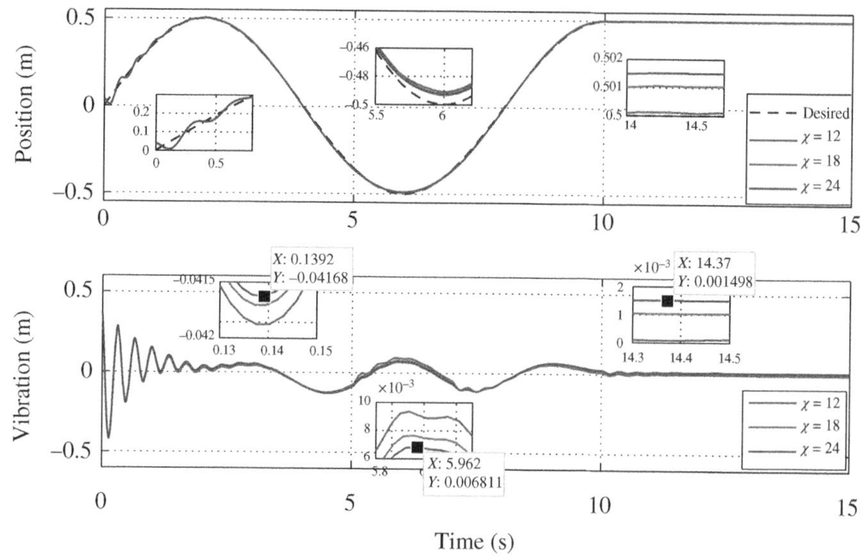

Figure 10.16 Position and vibration of the wing tip under AFT control.

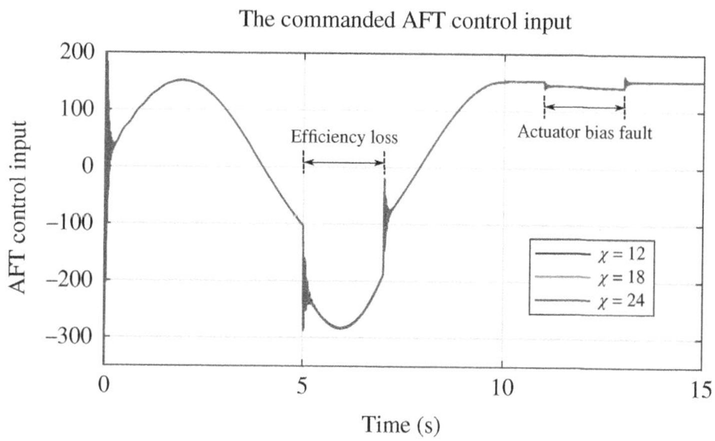

Figure 10.17 AFT control input of the flexible flapping wing.

The tracking performance, vibration suppression ability, and the robustness under PSF, FNN, and AFT control are summarized in Table 10.3. The corresponding indexes under the three controllers are compared in detail. In regards to tracking performance, the peak-to-peak value of the tracking error and the steady-state tracking error under the AFT control only account 16.6% and 14.8%, for the respective values under the PSF control, respectively. In addition, the peak-to-peak value of the tracking error and the steady-state tracking error under the AFT control are 32.6% and 29.5% of those values under the FNN control, respectively. In terms of the vibration suppression ability, the maximum vibration of the flexible wing tip is reduced by 7.5% and 2.1% using AFT control compared to PSF control and FNN control, respectively. When the system is in gliding flight mode, the steady-state vibration under AFT control is reduced by 76.9% and 21.1% compared to PSF control and FNN control, respectively. As four main indexes reflecting the robustness, the maximum tracking error and the maximum vibration together with the

Table 10.3 Comparison of PSF, FNN and AFT control.

Methods		Tracking performance			Vibration suppression			Robustness	
		Peak-to-peak error value	Max. tracking error	Steady-state error	Max. vibration (cm)	Max. vibration during failures (cm)	Steady-state vibration (cm)	Oscillation time (Efficiency loss) (s)	Oscillation time (Bias fault) (s)
PSF control	Set 1	1.851°	1.203°	0.651°	4.56	2.34	0.74	1.46	0.99
	Set 2	1.839°	1.191°	0.650°	4.52	2.33	0.73	1.10	0.69
	Set 3	**1.697°**	**1.100°**	**0.600°**	**4.51**	**2.18**	**0.65**	**1.27**	**0.81**
FNN control	Set 1	1.074°	0.696°	0.377°	4.28	1.52	0.30	0.56	0.36
	Set 2	0.960°	0.621°	0.336°	4.26	1.40	0.24	0.41	0.21
	Set 3	**0.866°**	**0.560°**	**0.302°**	**4.26**	**1.30**	**0.19**	**0.50**	**0.29**
AFT control	Set 1	0.518°	0.333°	0.177°	4.20	0.93	0.01	0.33	0.19
	Set 2	0.360°	0.227°	0.119°	4.18	0.76	0.10	0.27	0.15
	Set 3	**0.282°**	**0.175°**	**0.089°**	**4.17**	**0.68**	**0.15**	**0.21**	**0.12**

TEO time due to actuator failures (efficiency loss and bias-type fault) have been analyzed and discussed. The maximum tracking error due to failures under AFT control only account 15.9% for the respective value under PSF control, and 31.3% for the respective value under FNN control. Besides, the maximum wing-tip vibration during failures under AFT control is reduced by 68.8% and 47.7% compared to PSF control and FNN control, respectively. Moreover, compared with PSF control, the TEO times under AFT control during efficiency loss and bias-type actuator failures are shortened by 83.5% and 85.2%, respectively. Compared with FNN control, the TEO times under AFT control during efficiency loss and bias-type actuator failures are shortened by 58.0% and 58.6%, respectively. Therefore, it can be concluded that the proposed AFT controller performs better in terms of effectiveness and robustness compared to both PSF and FNN controllers.

Moreover, in order to discuss the computational complexity of the proposed approach, repeated co-simulations for PSF, FNN, and AFT control have been performed to show the code efficiency. The computer is configured with Intel(R) Core(TM) i3-9100 CPU @ 3.60 GHz process and 8.00 GB RAM. We test the running time of each algorithm 30 times and record their average value. The average running time of PSF control algorithm is 0.044 s, FNN control algorithm is 0.043 s, and AFT control algorithm is 0.044 s. The simulation result shows the running time of the proposed AFT control algorithm is comparable to those of PSF and FNN control algorithms.

10.5 Summary

Modeling and control problems of a bionic flexible flapping-wing robotic aircraft with system uncertainties are investigated in this study. The flexible flapping wing model has been presented by combining the IRFE method with MapleSim multi-domain modeling tool. An adaptive finite-time fault-tolerant control approach based on FNNs and NFTSM has been proposed to achieve tracking control and vibration suppression. By comparing with the PSF and FNN control algorithms, the co-simulation results have shown the effectiveness and the robustness of the proposed AFT control algorithm.

As future works, we will design and construct a full-scale experimental platform to simulate the bionic flexible flapping-wing robotic aircraft. Implementing the designed algorithm on the real experimental platform is also expected to further verify its effectiveness on vibration reduction and fault immunity.

11

Adaptive Vibration Control for Two-Stage Bionic Flapping Wings Based on Neural Network Algorithm

After reading this chapter, you should be able to:

- Vibration control is crucial for bionic flapping-wing robotic aircraft and autonomous ornithopter applications
- A rigid-flexible coupled bionic flapping-wing model is created using MapleSim software
- An adaptive vibration controller based on a neural network algorithm is proposed to suppress wing vibration and track the desired trajectory accurately

11.1 Introduction

The bionic flapping aircraft are inspired by the flight attitude of natural creatures such as seagulls, bats, and insects [229]. Considering that the flapping-wing aircraft has significant advantages over rotor and fixed-wing, the bionic flapping-wing aircraft has broad application prospects in recent years [335, 367]. The vast majority of the existing works on flapping-wing aircraft only focus on the single-segment wing [84], the research on the rigid-flexible coupled bionic flapping wing is thus in urgent need.

The infinite-dimensional dynamics of the rigid-flexible coupled flapping wing lead to the complexity of system modeling [26, 205]. In [120], He et al. derived a series of partial differential equations (PDEs) and ordinary differential equations (ODEs) through the Hamilton's principle to describe the rigid-flexible wing system. Although this modeling method has high accuracy, the complex derivations and heavy computational demand will be followed. To improve the modeling efficiency, Sun et al. [263] presented a lumped spring-mass method for a class of flexible systems, and this method realized the conversion from infinite dimension to finite dimension [81]. However, this method still does not overcome the heavy computation and the model accuracy is also reduced. Regarding the above issues, many new software modeling methods emerge as the required. Tahmasebi and Esmailzadeh [267] designed a large flexible satellite model based on ADAMS software. Considering these traditional software modeling methods have poor scalability, research on high-efficient and accuracy modeling methods of the rigid-flexible coupled flapping-wing system is thus vitally important.

Dynamic Modeling and Neural Network-Based Intelligent Control of Flexible Systems,
First Edition. Hejia Gao, Wei He, and Changyin Sun.

As the key technology in aircraft application, the research on the control theory of the bionic flapping wing has attracted wide attention [79, 129, 249]. Focused on the flapping wing micro aerial vehicles (FWMAV), Qian et. al [232] and He et.al [118], respectively proposed an adaptive hybrid attitude tracking control algorithm, which successfully addressed the issues of the noises, and an adaptive position controller based on neural networks, which dealed with the unknown disturbance. It should be noted that the strategies in [232] and [118] both neglected the vibration signals. Different from traditional rigid structures [183], the stability of the system cannot be guaranteed unless the vibration problem of rigid-flexible coupled flapping wing is successfully solved. In [279], Tsushima et al. proposed a linear quadratic Gaussian control strategy for highly flexible wings, which is capable of suppressing the flexible wings' vibration. To further deal with the issues of the dynamic uncertainties, Bialy et al. [15] and Gao et al. [84] designed two boundary vibration control approaches for the flexible robotic aircraft. However, all of the three methods [15, 85, 279] referred to installing an actuator at the end of the flapping wing, which resulted in far too much complexity of the mechanical design.

Motivated by the above aspects, this chapter establishes the visualization model of the rigid-flexible coupled bionic flapping wing by the advanced system-level modeling software MapleSim. It is the first time to combine the mathematical method with MapleSim multi-domain modeling tool to achieve higher accuracy and reduce difficulties of modeling. Based on the above model, a novel adaptive vibration controller (AVC) based on neural network (NN) algorithm is proposed to compensate the system uncertainties. The proposed method can successfully suppress the vibration of the flapping wing while accurately track the desired trajectory. Furthermore, the semi-global uniformly ultimately boundness (SGUUB) of the closed-loop system is proved through the Lyapunov's direct method. Finally, simulation results from the advanced system-level modeling software MapleSim and Matlab/Simulink validate the effectiveness of the proposed method. The research results will provide new ideas for the control design of bionic robot system and further contribute to the creative application of such system by intelligent control theory.

11.2 Problem Formulation

11.2.1 Dynamic Modeling

Inspired by the flying way of seagulls, a rigid-flexible coupled type flapping-wing model, ignoring the the twist of the flapping wing, is considered and shown in Figure 11.1. As illustrated in Figure 11.1a, the bionic flapping wing is modeled with two-stage structure consisting of a rigid part (with the weight M_1) close to the fuselage and a flexible part (with the weight M_2) away from the fuselage. According to the strip coating method, the rigid-flexible coupled type flapping wing is divided into n equal strips and each strip has length of $c_i(r_i)$ and width of $d(r_i)$. The simplified model of the multi-element airfoil is shown in Figure 11.1b, where the axis x, y, and z denote the spanwise, chord-wise, and vertical direction, respectively. $c_i(r_i)$ represents the wing chord length. L_1 and L_2 are the length of the rigid part and flexible part, respectively. As shown in Figure 11.1c, $\theta_i(t)$ represents the flapping angles and $\beta(x_2, t)$ is the elastic vibration of the flexible part.

To construct a high-efficient and accuracy modeling approach, the advanced system-level modeling software MapleSim is employed to establish the visualization model of the rigid-flexible coupled bionic flapping wing. The bionics and statistics-based parameters of the rigid-flexible coupled type flapping-wing model are listed in Table 11.1.

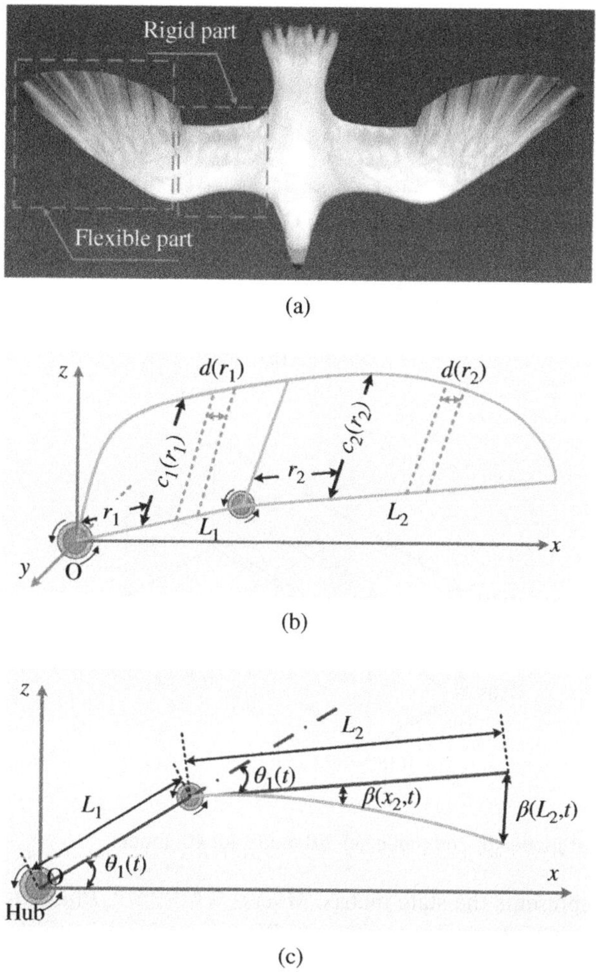

Figure 11.1 System representation and notations. (a) The bionic rigid-flexible coupled flapping wing. (b) Dynamic model of the multi-element airfoil. (c) Dynamic model of the flapping wing.

Table 11.1 The bionics and statistics-based parameters of the flapping-wing model.

Symbol	L_1	L_2	M_1	M_2	c_1	c_2	n
Value	18	32	288	171	16	$16 - \frac{(r_2 - 18)^2}{64}$	10
Unit	cm	cm	g	g	cm	cm	–

11.2.2 Model Discretization

The visualization model of the bionic rigid-flexible coupled flapping wing is built by various multibody components. The modes affecting the multibody components are identified and plotted, as shown in Figure 11.2. The dynamics of the two-stage bionic flapping wings has the form as:

$$M(\alpha)\ddot{\alpha} + D(\alpha, \dot{\alpha})\dot{\alpha} + S(\alpha) = F(t). \tag{11.1}$$

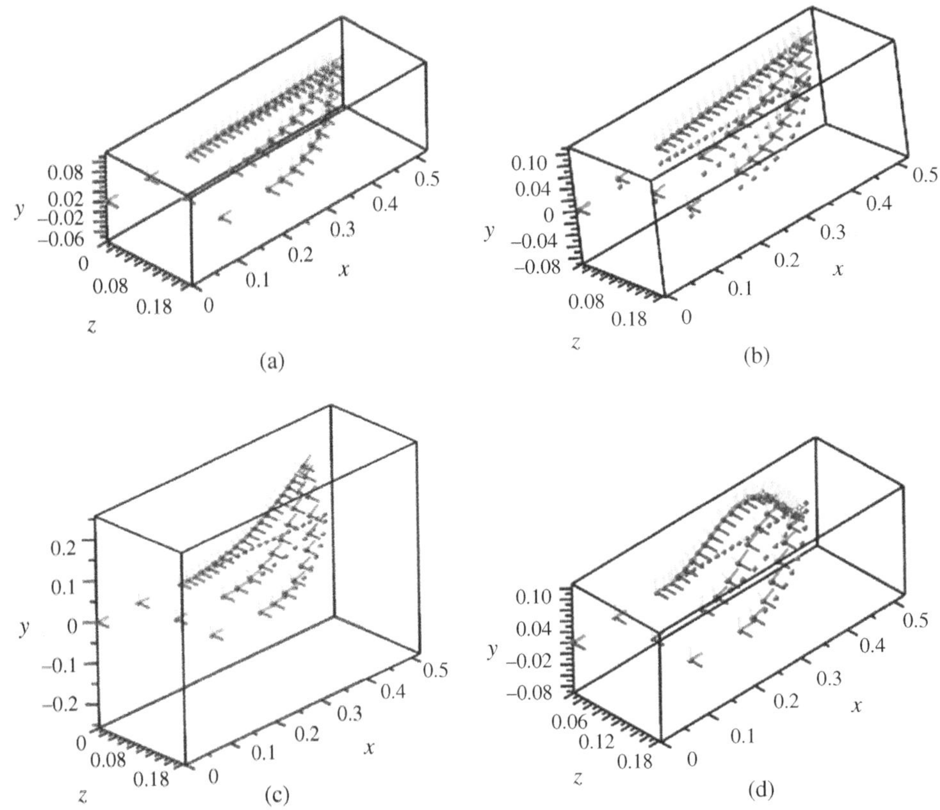

Figure 11.2 Mode visualization results. (a) 1st mode. (b) 2nd mode. (c) 3rd mode. (d) 4th mode.

where $\alpha = [\theta_1, \theta_2, \beta_1, \beta_2, \dots, \beta_n] \in \mathbb{R}^{n+2}$ represents the state matrix. $M(\alpha) \in \mathbb{R}^{(n+2)\times(n+2)}$, $D(\alpha, \dot{\alpha}) \in \mathbb{R}^{(n+2)\times(n+2)}$ and $S(\alpha) \in \mathbb{R}^{n+2}$ are the mass matrix, damping matrix, and stiffness matrix, respectively. $F(t) \in \mathbb{R}^{n+1}$ denotes the control input of the bionic flapping wing.

11.3 Adaptive Vibration Control

11.3.1 Control Design

Considering that the two-stage bionic flapping-wing aircraft usually works in complex environments, the system's dynamic information is difficult to obtain accurately. In this section, a novel adaptive vibration controller (AVC) based on neural network (NN) algorithm is proposed to compensate the system uncertainties. The diagram of the adaptive vibration control is displayed as Figure 11.3. The proposed method can successfully suppress the vibration of the flapping wing while accurately track the desired trajectory.

Based on the dynamics of the two-stage bionic flapping wings (11.1), the transformation form can be expressed as:

$$\ddot{\alpha} = M^{-1}(\alpha)[F(t) - D(\alpha, \dot{\alpha})\dot{\alpha} - S(\alpha)]. \tag{11.2}$$

Define two track error of the system as:

$$\vartheta_1(t) = \alpha(t) - \alpha_d(t), \ \vartheta_2(t) = \dot{\alpha}(t) - \lambda_1(t), \tag{11.3}$$

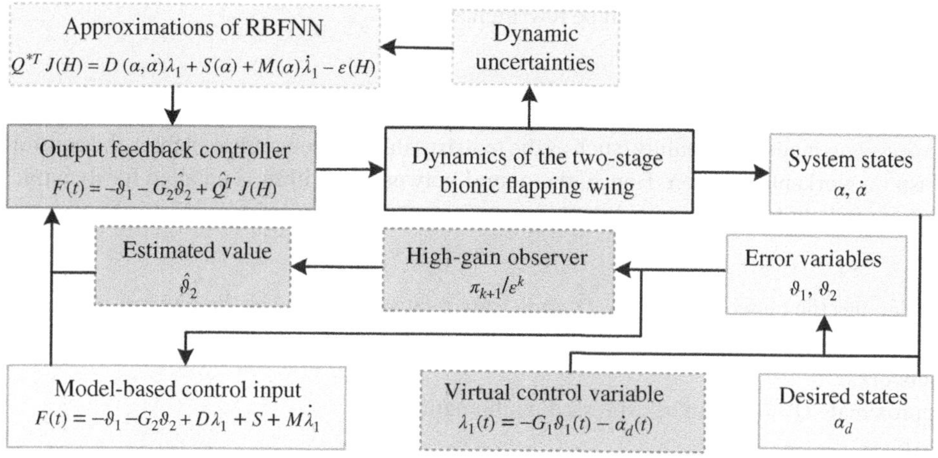

Figure 11.3 The diagram of the adaptive vibration control.

where $\lambda_1 = -G_1\vartheta_1 + \dot{\alpha}_d$ is a virtual variable, and $G_1 = G_1^T > 0$ represents the control gain. The derivatives of ϑ_1 and ϑ_2 are:

$$\dot{\vartheta}_1 = \vartheta_2 - G_1\vartheta_1, \tag{11.4}$$

$$\dot{\vartheta}_2 = M^{-1}(\alpha)[F(t) - D(\alpha,\dot{\alpha})\dot{\alpha} - S(\alpha)] - \dot{\lambda}_1(t). \tag{11.5}$$

Construct a Lyapunov function $V_1 = \frac{1}{2}\vartheta_1^T\vartheta_1$ and its first order derivation is:

$$\dot{V}_1 = \vartheta_1^T\vartheta_2 - \vartheta_1^T G_1\vartheta_1. \tag{11.6}$$

A new constructed Lyapunov function can deal with the coupling term $\vartheta_1^T\vartheta_2$ in the above equation.

$$V_2 = \frac{1}{2}\vartheta_1^T\vartheta_1 + \frac{1}{2}\vartheta_2^T M(q)\vartheta_2. \tag{11.7}$$

Lemma 11.1 If X is a skew-symmetric determinant, there thus is $X = -X^T$.

Considering that the matrix $\dot{M}(\alpha) - 2D(\alpha,\dot{\alpha})$ is a skew-symmetric determinant, so there is

$$\begin{aligned} &-\vartheta_2^T[\dot{M}(\alpha) - 2D(\alpha,\dot{\alpha})]\vartheta_2 \\ &= \vartheta_2^T[\dot{M}(\alpha) - 2D(\alpha,\dot{\alpha})]^T\vartheta_2 \\ &= \vartheta_2^T[\dot{M}(\alpha) - 2D(\alpha,\dot{\alpha})]\vartheta_2. \end{aligned} \tag{11.8}$$

Consequently, we have $\frac{1}{2}\vartheta_2^T[\dot{M}(\alpha) - 2D(\alpha,\dot{\alpha})]\vartheta_2 = 0$.

In order to find the control input that makes the system stable, we take the time derivative of (11.7) as:

$$\begin{aligned} \dot{V}_2 = {} & \vartheta_1^T\vartheta_2 - \vartheta_1^T G_1\vartheta_1 + \vartheta_2^T[F(t) \\ & - S(\alpha) - D(\alpha,\dot{\alpha})\dot{\alpha} - M(\alpha)\dot{\lambda}_1(t)]. \end{aligned} \tag{11.9}$$

According to Lyapunov stability theory, the control input which makes the system satisfy $\dot{V}_2 < 0$, has the form as:

$$\begin{aligned} F(t) = {} & -\vartheta_1 - G_2\vartheta_2 + D(\alpha,\dot{\alpha})\dot{\alpha} \\ & + S(\alpha) + M(\alpha)\dot{\lambda}_1(t). \end{aligned} \tag{11.10}$$

where the gain matrix $G_2 = G_2^T > 0$.

Substituting (11.10) into (11.9), $\dot{V}_2$ can be rewritten as:

$$\begin{aligned} \dot{V}_2 &= -\vartheta_1^T G_1 \vartheta_1 - \vartheta_2^T G_2 \vartheta_2 \\ &\leq 0. \end{aligned} \tag{11.11}$$

In case of system model uncertainty (such as the matrixes information is unavailable), the control law $F(t)$ isn't a workable solution. Hence, the neural network algorithm is adopted for dynamic uncertainties prediction.

$$F(t) = -\vartheta_1 - G_2\vartheta_2 + \hat{Q}^T J(H). \tag{11.12}$$

where Q_i^* denotes the networks weights, $\hat{Q}_i$ and $\tilde{Q}_i$ are relatively defined as the estimate and error of networks weights. $J(H)$ is the basis function. $H = [\alpha^T, \dot{\alpha}^T, \lambda_1^T, \dot{\lambda}_1^T]$ represents the inputs of the neural networks.

The approximate $Q^{*T}J(H)$ can be expressed mathematically as:

$$\begin{aligned} &Q^{*T}J(H) = \\ &D(\alpha,\dot{\alpha})\dot{\alpha} + S(\alpha) + M(\alpha)\dot{\lambda}_1(t) - \epsilon(H). \end{aligned} \tag{11.13}$$

where $\epsilon(H)$ denotes the approximation error of the neural networks algorithm.

The adaptive law of the neural networks algorithm could be written as follows:

$$\dot{\hat{Q}}_i = -\Gamma_i[J_i(H)_{\vartheta_2,i} + \sigma_i\hat{Q}_i]. \tag{11.14}$$

where Γ_i denotes the gain, and $\sigma_i > 0$ is a small constant.

Remark 11.1 If there are other unknown items in the system dynamics, the proposed control strategy will also work for this case.

11.3.2 Stability Analysis

To prove the feasibility of the designed control system, this chapter designs a new Lyapunov function as:

$$V_3 = \frac{1}{2}\vartheta_1^T\vartheta_1 + \frac{1}{2}\vartheta_2^T M\vartheta_2 + \frac{1}{2}\sum_{i=1}^{n+2}\tilde{Q}_i^T\Gamma_i^{-1}\tilde{Q}_i. \tag{11.15}$$

Take the derivative versus time of this Lyapunov equation,

$$\begin{aligned} \dot{V}_3 &\leq \vartheta_1^T\vartheta_2 - \vartheta_1^T G_1\vartheta_1 + \vartheta_2^T[F(t) - D(\alpha,\dot{\alpha})\dot{\alpha} - S(\alpha) - M(\alpha)\dot{\lambda}_1(t)] + \sum_{i=1}^{n+2}\tilde{Q}_i^T\Gamma_i^{-1}\dot{\hat{Q}}_i, \\ &\leq \vartheta_1^T\vartheta_2 - \vartheta_1^T G_1\vartheta_1 + \vartheta_2^T[F(t) - Q^{*T}J(H) - \epsilon(H)] \\ &\quad + \sum_{i=1}^{n+2}\tilde{Q}_i^T\Gamma_i^{-1}\dot{\hat{Q}}_i. \end{aligned} \tag{11.16}$$

By means of these two inequalities respectively,

$$\begin{aligned} -\sum_{i=1}^{n}\sigma_i\tilde{Q}_i^T\hat{Q}_i &= -\sum_{i=1}^{n}\sigma_i\tilde{Q}_i^T(\tilde{Q}_i + Q_i^*) \\ &\leq -\sum_{i=1}^{n}\frac{\sigma_i}{2}(\|\tilde{Q}_i\|^2 - \|Q_i^*\|^2), \end{aligned} \tag{11.17}$$

$$-\vartheta_2^T\epsilon(H) \leq \frac{1}{2}(\|\overline{\epsilon}(H)\|^2 + \vartheta_2^T\vartheta_2), \tag{11.18}$$

the derivative $\dot{V}_3$ is acquired by substituting (11.12) and (11.14) into (11.16),

$$\begin{aligned}\dot{V}_3 &\leq -\vartheta_1^T G_1 \vartheta_1 - \vartheta_2^T (G_2 - \frac{1}{2} I_{(n+2)\times(n+2)}) \vartheta_2 \\ &+ \frac{1}{2} \|\bar{\epsilon}(H)\|^2 - \sum_{i=1}^{n+2} \frac{\sigma_i}{2} (\|\tilde{Q}_i\|^2 - \|Q_i^*\|^2) \\ &\leq -\mu_1 V_3 + \kappa_1. \end{aligned} \tag{11.19}$$

where:

$$\mu_1 = \min\left(2\Lambda_{\min}(G_1), \frac{2\Lambda_{\min}(G_2 - \frac{1}{2} I_{(n+2)\times(n+2)})}{\Lambda_{\max}(M)}, \min_{i=1,\ldots n+2}\left(\frac{\sigma_i}{\Gamma_i^{-1}}\right)\right), \tag{11.20}$$

$$\kappa_1 = \sum_{i=1}^{n+2} \frac{\sigma_i}{2} \|Q_i^*\|^2 + \frac{1}{2} \|\bar{\epsilon}(H)\|^2. \tag{11.21}$$

where $\Lambda_{\max}(X)$ and $\Lambda_{\min}(X)$ donate the maximum and minimum eigenvalues of matrix X, respectively. $\Lambda_{\min}(G_2 - \frac{1}{2}I) > 0$.

Remark 11.2 The above control inputs are only feasible when the system states $\alpha, \dot{\alpha}$ are known. The Lyapunov stability theory is used to obtain the neural network control law. According to this, it is proved in this chapter that the tracking errors will finally converge to a small neighborhood of zero by appropriate gains.

When only α is available, a high-gain state observer is required to estimate $\dot{\alpha}$. Design the state observer as:

$$\alpha = \pi_1,\ \dot{\alpha} = \frac{\pi_2}{\varepsilon},\ \hat{\vartheta}_2 = \frac{\pi_2}{\varepsilon} - \lambda_1, \tag{11.22}$$

and thereby the control input (11.12) and the adaptive law (11.14) can be rewritten as:

$$F(t) = -\vartheta_1 - G_2\hat{\vartheta}_2 + \hat{Q}^T J(\hat{H}), \tag{11.23}$$

$$\dot{\hat{W}}_i = -\Gamma_i (J_i(\hat{H})_{\hat{\vartheta}_{2,i}} + \sigma_i \hat{Q}_i), \tag{11.24}$$

The obtaining process is given with stability proving applied Lyapunov Function, which has the form as:

$$V_4 = \frac{1}{2}\vartheta_1^T\vartheta_1 + \frac{1}{2}\vartheta_2^T M \vartheta_2 + \frac{1}{2}\sum_{i=1}^{n+2} \tilde{Q}_i^T \Gamma_i^{-1} \tilde{Q}_i. \tag{11.25}$$

The derivation of (11.25) can be obtained with term-by-term substitution of (11.23) and (11.24):

$$\begin{aligned}\dot{V}_4 &\leq -\vartheta_1^T G_1 \vartheta_1 - \vartheta_2^T \left(G_2 - \frac{1}{2} I\right) \vartheta_2 - \vartheta_2^T G_2 \tilde{\vartheta}_2 \\ &+ \sum_{i=1}^{n+2} \vartheta_{2,i} [\hat{Q}_i^T J_i(\hat{H}) - Q_i^{*T} J_i(H)] + \frac{1}{2}\|\bar{\epsilon}\|^2 \\ &- \sum_{i=1}^{n+2} [\tilde{Q}_i^T J_i(\hat{H})_{\hat{\vartheta}_{2,i}} + \sigma_i \hat{Q}_i^T \hat{Q}_i]. \end{aligned} \tag{11.26}$$

By means of the following inequalities or equations:

$$-\sigma_i \tilde{Q}_i^T \hat{Q}_i \leq -\frac{\sigma_i}{2}(||\tilde{Q}_i||^2 - ||Q_i^*||^2), \tag{11.27}$$

$$||J_i(\hat{H})||^2 \leq l_i, \tag{11.28}$$

$$\hat{Q}_i^T J_i(\hat{H}) = Q_i^{*T} \epsilon J_{ti} + Q_i^{*T} J_i(H) + \tilde{Q}_i^T J_i(H). \tag{11.29}$$

where $l_i > 0$, $\epsilon > 0$, and R_t is a bounded vector function. The new inequality on $\dot{V}_4$ can be obtained:

$$\begin{aligned}\dot{V}_4 \leq & -\vartheta_1^T G_1 \vartheta_1 - \vartheta_2^T (G_2 - \frac{1}{2} I)\vartheta_2 - \vartheta_2^T G_2 \tilde{\vartheta}_2 \\ & + \sum_{i=1}^{n+2} \vartheta_{2,i} [Q_i^{*T} \epsilon J_{ti} + \tilde{Q}_i^T J_i(H)] + \frac{1}{2}||\bar{\epsilon}||^2 \\ & - \sum_{i=1}^{n+2} \tilde{Q}_i^T J_i(\hat{H})_{\hat{\vartheta}_{2,i}} - \frac{\sigma_i}{2}(||\tilde{Q}_i||^2 - ||Q_i^*||^2).\end{aligned} \tag{11.30}$$

Further, if we consider the following inequalities:

$$\sum_{i=1}^{n} \vartheta_{2,i} Q_i^{*T} \epsilon J_{ti} \leq \frac{1}{2}\vartheta_2^T \vartheta_2 + \sum_{i=1}^{n} \frac{||Q_i^*||^2 \epsilon^2 ||J_{ti}||^2}{2},$$

$$\begin{aligned}-\sum_{i=1}^{n} \tilde{Q}_i^T J_i(\hat{H})_{\hat{\vartheta}_{2,i}} \leq & \sum_{i=1}^{n} \frac{\sigma_i ||\tilde{Q}_i||^2}{4} \\ & + \sum_{i=1}^{n} \frac{2||J_i(\hat{H})||^2}{\sigma_i} \frac{1}{2} \tilde{\vartheta}_2^T \tilde{\vartheta}_2,\end{aligned}$$

$\dot{V}_4$ can be updated to:

$$\begin{aligned}\dot{V}_4 \leq & -\vartheta_1^T G_1 \vartheta_1 - \vartheta_2^T (G_2 - I)\vartheta_2 - \vartheta_2^T G_2 \tilde{\vartheta}_2 + \frac{1}{2}||\bar{\epsilon}||^2 \\ & - \sum_{i=1}^{n+2} \frac{\sigma_i}{4} ||\tilde{Q}_i||^2 + \sum_{i=1}^{n+2} \frac{2||J_i(\hat{H})||^2}{\sigma_i} \frac{1}{2} \tilde{\vartheta}_2^T \tilde{\vartheta}_2 \\ & + \sum_{i=1}^{n+2} \left(\frac{\sigma_i}{2} + \frac{\epsilon^2 ||J_{ti}||^2}{2} \right) ||Q_i^*||^2.\end{aligned} \tag{11.31}$$

Considering $\tilde{\vartheta}_2 = \zeta$, $||\zeta|| \leq \varepsilon h_2$ and $\vartheta_2^T G_2 \tilde{\vartheta}_2 \leq \frac{1}{2}\vartheta_2^T \vartheta_2 + \frac{1}{2}(G_2\tilde{\vartheta}_2)^T (G_2 \tilde{\vartheta}_2)$, $\dot{V}_4$ can be further updated to:

$$\begin{aligned}\dot{V}_4 \leq & -\vartheta_1^T G_1 \vartheta_1 - \vartheta_2^T \left(G_2 - \frac{3}{2} I \right) \vartheta_2 + \frac{1}{2}||\bar{\epsilon}||^2 \\ & - \sum_{i=1}^{n+2} \frac{\sigma_i}{4} ||\tilde{Q}_i||^2 + \Lambda_{\max} \left(G_2^T G_2 + \text{diag}\left[\frac{2l_i}{\sigma_i} \right] \right) \frac{1}{2} \zeta_2^T \zeta_2 \\ & + \frac{1}{2} \sum_{i=1}^{n+2} (\sigma_i + \epsilon^2 ||J_{ti}||^2) ||Q_i^*||^2, \\ \leq & -\vartheta_1^T G_1 \vartheta_1 - \vartheta_2^T \left(G_2 - \frac{3}{2} I \right) \vartheta_2 + \frac{1}{2}||\bar{\epsilon}||^2 \\ & - \sum_{i=1}^{n+2} \frac{\sigma_i}{4} ||\tilde{Q}_i||^2 + \Lambda_{\max} \left(G_2^T G_2 + \text{diag}\left[\frac{2l_i}{\sigma_i} \right] \right) \frac{1}{2} \varepsilon^2 h_2^2\end{aligned}$$

$$+\frac{1}{2}\sum_{i=1}^{n+2}(\sigma_i + \epsilon^2\|J_{ti}\|^2)\|Q_i^*\|^2$$

$$\leq -\mu_2 V_4 + \kappa_2, \tag{11.32}$$

where

$$\mu_2 = \min\left(2\Lambda_{\min}(G_1), \frac{2\Lambda_{\min}(K_2 - \frac{3}{2}I)}{\Lambda_{\max}(M)}, \min_{i=1,2\ldots n+2}\left(\frac{\sigma_i}{2\Lambda_{\max}(\Gamma_i^{-1})}\right)\right), \tag{11.33}$$

$$\kappa_2 = \frac{1}{2}\sum_{i=1}^{n+2}(\epsilon^2\|J_{ti}\|^2 + \sigma_i)\|W_i^*\|^2 + \frac{1}{2}\|\overline{\epsilon}\|^2$$

$$+\Lambda_{\max}\left(G_2^T G_2 + \operatorname{diag}\left[\frac{2l_i}{\sigma_i}\right]\right)\frac{1}{2}\varepsilon^2 h_2^2. \tag{11.34}$$

where $\Lambda_{\min}(G_2 - \frac{3}{2}I) > 0$.

Remark 11.3 The above controller is feasible even if the unknown state exists.

11.4 Numerical Simulations

In order to verify the effectiveness of the proposed algorithm, the simulative analysis of the two-stage bionic flapping-wings system adoptes the Maple/MapleSim and MATLAB joint simulation method. To simulate the flying of seagulls, the desired flapping angles of the two-stage flapping wings are set as:

$$\theta_{1d} = \begin{cases} 30\sin(t), & 0 < t \leq 8\text{ s} \\ 0, & 8 < t \leq 15\text{ s} \end{cases} \tag{11.35}$$

$$\theta_{2d} = \begin{cases} 15\sin(t), & 0 < t \leq 8\text{ s} \\ 0, & 8 < t \leq 15\text{ s} \end{cases} \tag{11.36}$$

where the flapping-wing flying and glide flying are simulated corresponding to $0 < t \leq 10$ s and $10 < t \leq 15$ s, respectively.

The partial state feedback (PSF) control, which has the form as $F_{PSF} = -G_p(\theta - \theta_d) - G_d(\dot{\theta} - \dot{\theta}_d)$, and the proposed AVC are applied to visual bionic flapping-wing model. The control gains of the PSF controller are set as $G_p = [250, 250]^T$ and $G_d = [2, 2]^T$. Accordingly, the control gains of the AVC are set as $G_1 = [20, 20]^T$ and $G_2 = [60, 60]^T$.

The flapping angle trajectories of the rigid wing and the flexible wing are shown in Figures 11.4 and 11.5, respectively. These trajectories show that the tracking performance under PSF control is worse than AVC. As can be seen from the tracking errors (Figures 11.6 and 11.7), the tracking errors of the rigid wing and flexible wing under PSF control are up to 0.58° and 0.28°, respectively, while those under AVC are only 0.37° and 0.18°. The wing-tip vibration of the bionic flapping-wing aircraft is revealed as Figure 11.8. When the flapping-wing aircraft is switched from flapping-wing mode to gliding mode, the vibration under the PSF control converges to a neighborhood of zero within 1.2 s, while that under AVC converges to a neighborhood of zero is only 0.3 s. In addition, the control inputs of the bionic flapping-wing aircraft are displayed in Figures 11.9 and 11.10.

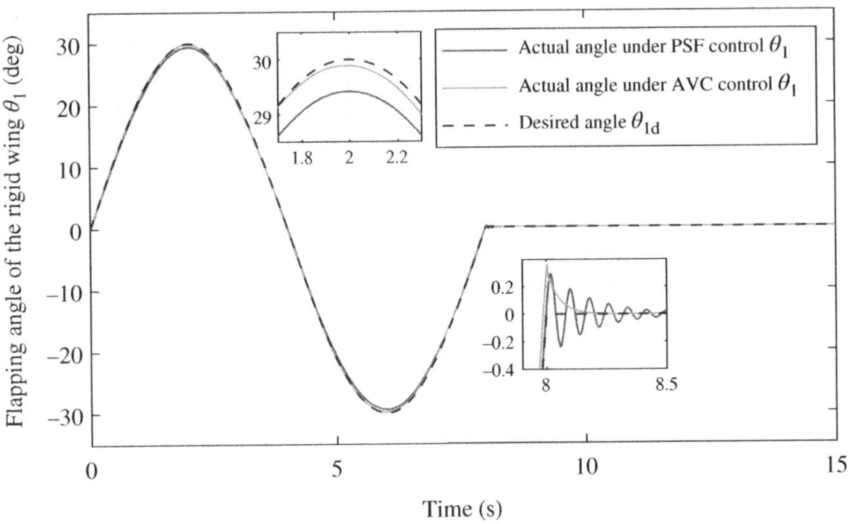

Figure 11.4 Flapping angle trajectory of the rigid wing.

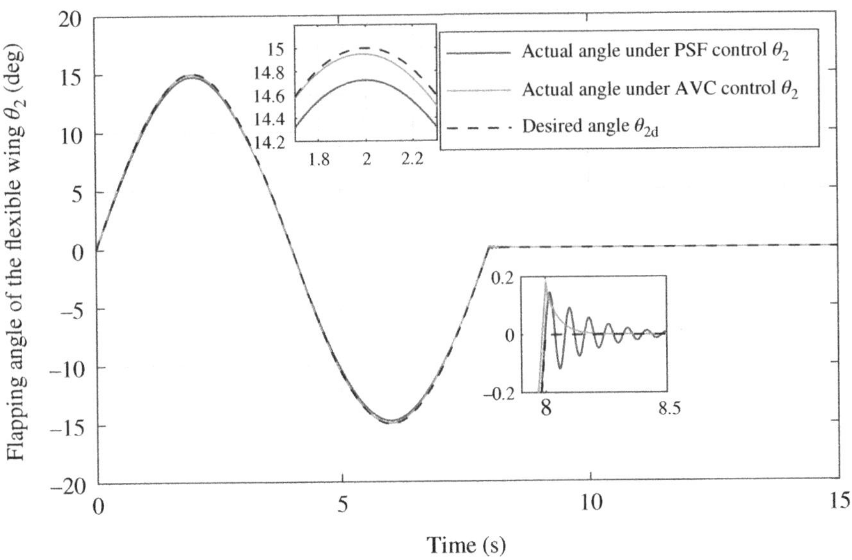

Figure 11.5 Flapping angle trajectory of the flexible wing.

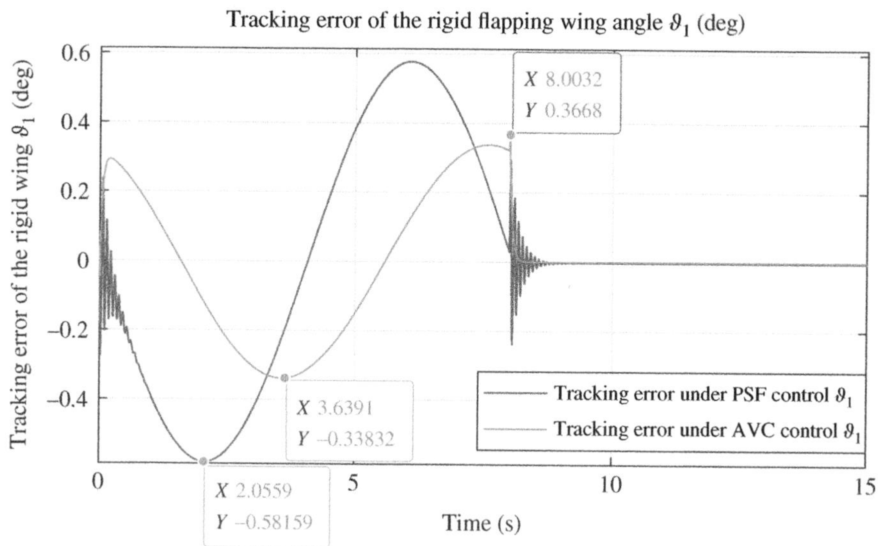

Figure 11.6 Tracking error of the rigid flapping-wing angle.

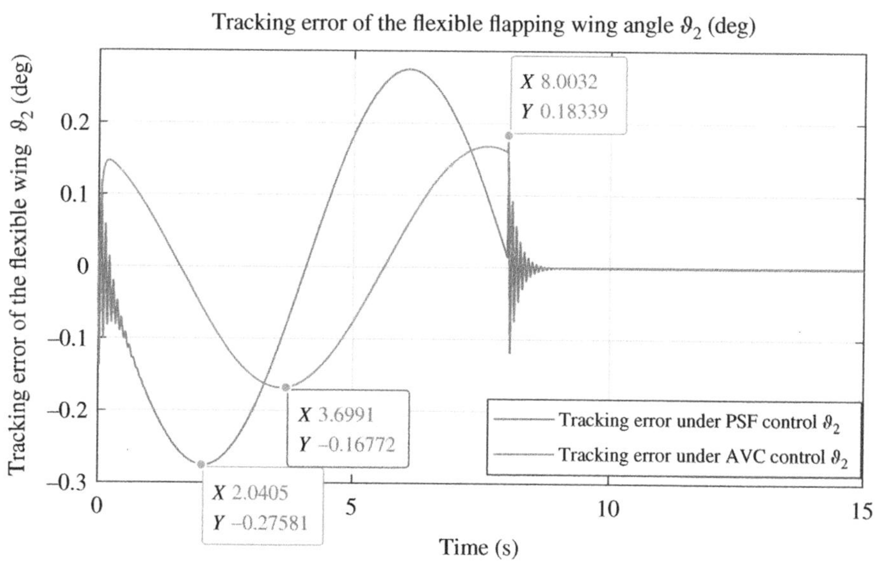

Figure 11.7 Tracking error of the flexible flapping-wing angle.

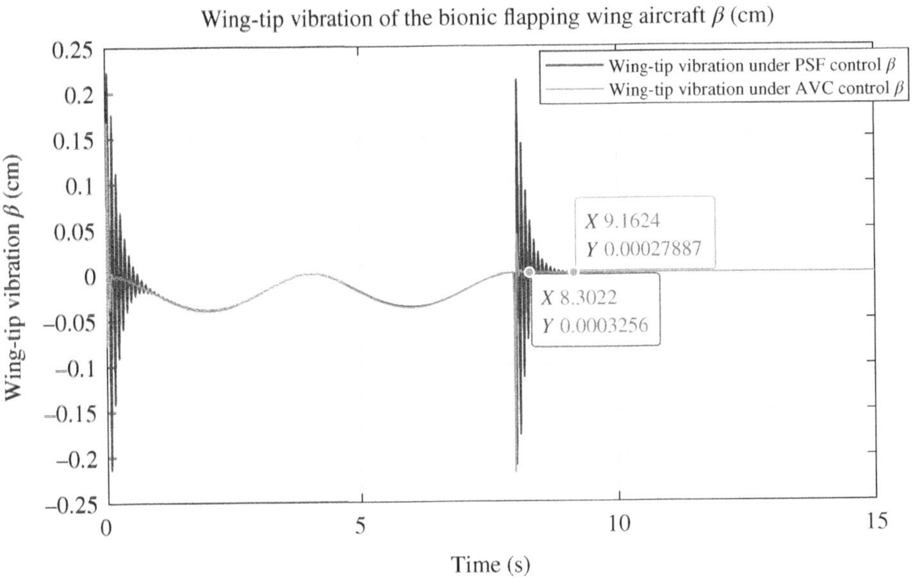

Figure 11.8 Wing-tip vibration of the bionic flapping-wing aircraft.

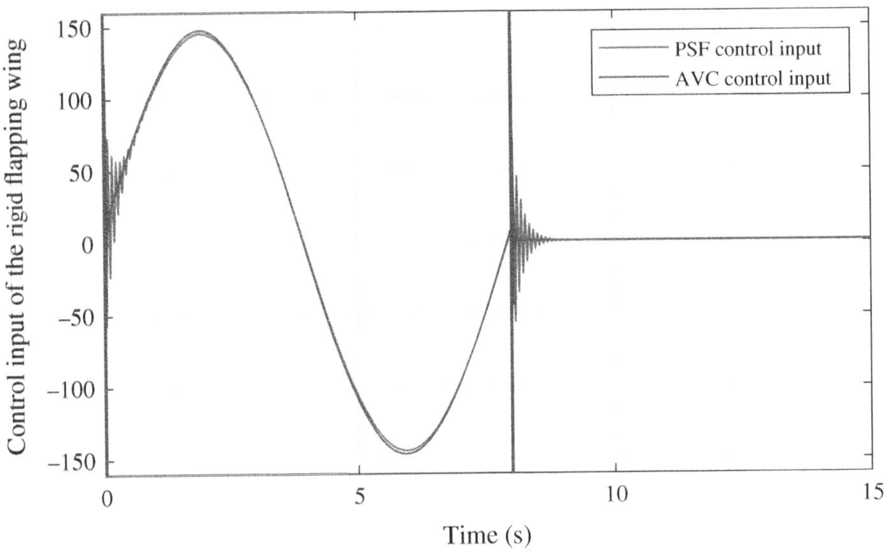

Figure 11.9 Control input of the rigid flapping wing.

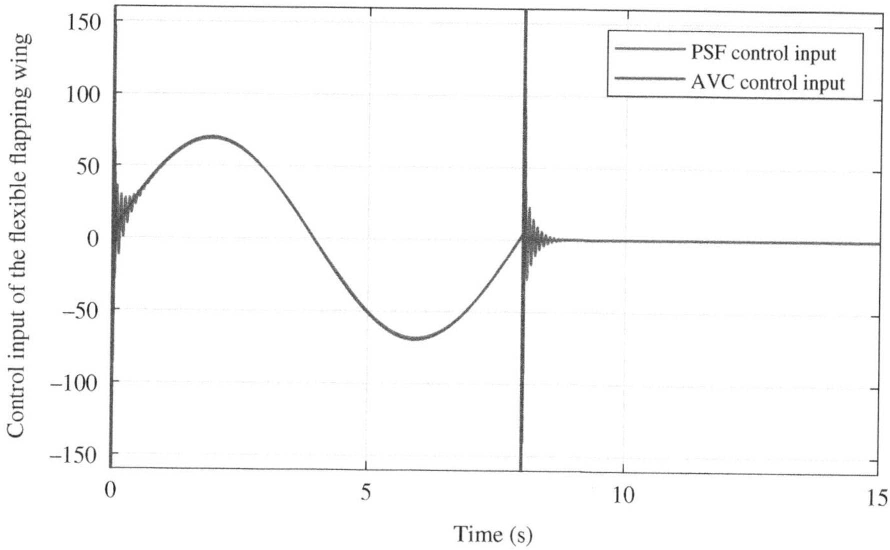

Figure 11.10 Control input of the flexible flapping wing.

11.5 Summary

This chapter has established the visualization model of the two-stage bionic flapping wing by the advanced system-level modeling software MapleSim. A novel adaptive vibration controller based on NN algorithm has been proposed to compensate the system uncertainties. The proposed method can successfully suppress the vibration of the flapping wing while accurately track the desired trajectory. The co-simulation results from the advanced system-level modeling software MapleSim and Matlab/Simulink have validated the effectiveness of the proposed method.

As future works, we will design a set of control algorithm to realize autonomous tracking flight of the bionic flapping wing aircraft. In addition, considering the unknown environment, the effective approach to deal with the wing collisions will be studied for potential applications in flying robots. Furthermore, a bionic flying robot experimental platform will be set-up to verify the effectiveness of the proposed strategies.

12

Boundary Vibration Control of a Floating Wind Turbine System with Mooring Lines

After reading this chapter, you should be able to:

- The study involves dynamic modeling, active boundary control design, and stability analysis for a coupled floating wind turbine (FWT) system
- Two boundary controllers are designed to restrain the vibrations of this flexible system caused by external disturbances
- The performance of the designed boundary controllers and systems stability are analyzed theoretically, and simulation results validate the efficacy of the proposed approach

12.1 Introduction

The offshore area has abundant wind energy resource, especially in deep water area (>100 m), and floating platforms are widely used in energy exploitation in offshore engineering. In offshore area, floating wind turbine (FWT) system, Figure 12.1, is a large wind energy conversion equipment including wind generator tower, floating platform, and mooring lines. It is suitable for a harsh deep-sea environment where the bottom-mounted tower is infeasible [162]. However, the complex marine environment has a great impact on FWT system, and turbine tower and mooring lines will generate wave-induced oscillation, which may cause irreversible damage and premature fatigue. Thus, for the conversion efficiency and safety concern, structural vibrations of FWT system need to be studied and addressed.

In recent years, many works focus on vibration control design and dynamic analysis for the FWT systems. In previous works, many researchers use passive control methods to address the vibration reduction of the system, In general, passive control is a control without external energy, and the control force is generated passively by vibration deformation of the control device along the structure or the movement of the device itself. Common passive control includes tuned mass damper [51, 260, 278], tuned liquid damper [75, 358], friction damper, etc. In [254], the OC3-Hywind FWT system is discussed with a passive structural controller. Another vibration is active control which is to use external energy to exert the control force and the control device can use external energy to tune its own parameters for adjusting control force. Therefore, active control may have a better performance than passive control. In [10], the authors analyze dynamic behavior of an offshore wind turbine system by considering linear varying parameter and constrained information, and establish a dynamic model. In [9], an output-feedback control method for an offshore FWT system

Dynamic Modeling and Neural Network-Based Intelligent Control of Flexible Systems,
First Edition. Hejia Gao, Wei He, and Changyin Sun.

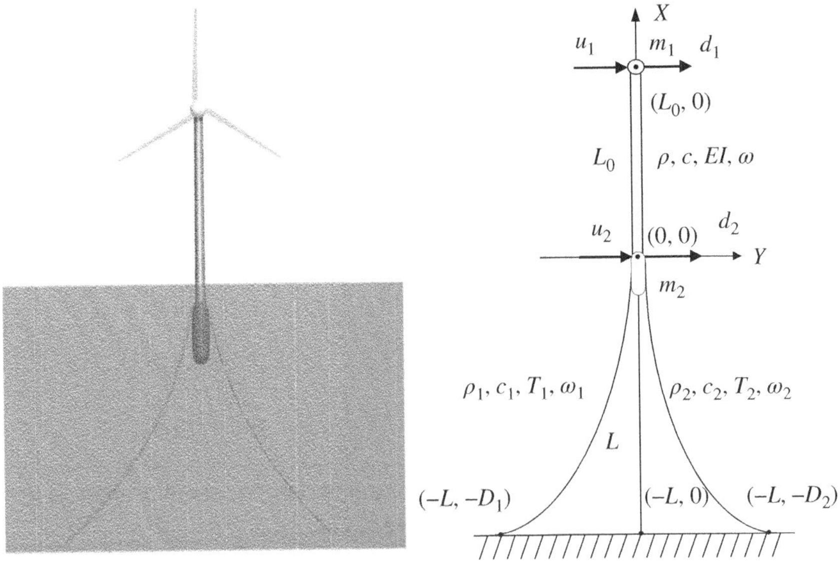

Figure 12.1 The floating wind turbine system.

with constraint is proposed to reduce wave-oscillations. In [8], many kinds of linear models are introduced to represent a wind turbine system. In addition, a robust output-feedback controller integration using H_∞ control with pole setting constraints is investigated for a wind turbine system. In [178], the authors propose a novel H_∞ control method based on a linear model and apply it to an FWT system. These works make great achievements in the modeling and control of the FWT system. However, they analyze dynamics of the FWT and address the control problem based on the linear or finite-dimensional model because of regarding the wind turbine tower as a rigid body system. From the structure of FWT, it is a coupled beam-strings system which is characterized with infinite modes [94, 240, 310], namely, using PDEs to describe the system's dynamics. In this work, we consider the FWT system as a flexible coupled structure and address the vibration suppression problem of the turbine tower and mooring lines.

For vibration reduction of flexible structures, many vibration control methods are developed [14, 29, 114, 122, 146, 201], e.g., adaptive control method [39, 124, 169, 180, 241, 317], optimal control method [262, 312], and sliding mode control method [290], force control method [63, 64], etc., are also investigated. Compared with rigid structure [176, 212], flexible structures has the advantages of light weight and strong flexibility. In [14], the vibration control problem of a flexible beam with collocated sensor/actuator pairs is investigated. Notice that some flexible systems mentioned above are discretized as a truncated model, namely considering finite modes of the system. Furthermore, the control design for flexible systems is conducted based on the discretized model. However, these unmodeled modes, including high-frequency modes, may have a great impact on the system performance, and it is valuable to consider the influences of these modes.

Boundary control is a control method which considers the dynamics of entire modes and has been applied to various infinite-dimensional systems widely, e.g., heat equations [132, 288], moving string system [104, 219, 220], etc. For the wind turbine tower system, there exist some literature to address the vibration reduction problem of the tower based on boundary control strategy. A nonlinear vibration isolator is proposed to reduce the structural impact of turbine under seismal loading and wind in [281], which is a novelty work used in the wind turbine. In [101], a vibration controller is investigated for a nonuniform wind turbine tower with random wind loads. However, these works consider the dynamics of a single flexible structure and focus on the vibration suppression of wind turbine tower. In this work, we consider the vibration control of FWT system, which is a coupled beam-strings system. The main challenges are to consider the strong coupling between turbine tower and mooring lines, and to develop boundary controllers to reduce vibrations of FWT system. For mooring systems, motions of the mooring lines are usually regarded as an external force acted to floating platforms in the previous works. In [37, 270], the authors study dynamics of mooring lines by ignoring the connection of the mooring lines and the ship. Nevertheless, the coupling relationship has a great influence on the performance of mooring system. Notice that we will consider the strong coupling between mooring lines, floating platform, and flexible turbine tower, and establish a nonlinear dynamic model. Furthermore, we will propose two boundary controllers simultaneously to reduce the vibration of flexible turbine tower and mooring lines and to compensate for the performance degradation.

In brief, main works of this chapter are summarized as

1) A coupled beam-strings model is used to illustrate the dynamical behavior of FWT system, and the coupled relationship among turbine tower, floating platform, and mooring lines is considered in dynamic analysis. A coupled hybrid PDE-ODEs model is established to describe FWT system by applying Hamilton's principle.
2) Two boundary controllers are designed to ensure the stability of the system simultaneously based on nonlinear coupled model under unknown environment loads. Theoretical analysis and simulation results are presented to demonstrate the effectiveness of proposed controllers.

The rest of this chapter are summarized as follows. In Section 12.2, we establish the system model described by PDEs using Hamilton's principle, to derive the governing equations (described by PDEs) and boundary conditions (described by ODEs) of FWT system. In Section 12.3, the control objectives and uniform boundedness of the closed-loop system are derived by Lyapunov's direct method. In Section 12.4, simulation results demonstrate the efficacy of the proposed approach. In Section 12.5, we give conclusions of this chapter and future works.

12.2 System Modeling and Preliminaries

As shown in Figure 12.1, FWT system is described by a coupled beam-strings structure, where, the turbine tower is regarded as a flexible beam with a lumped-mass-like payload, and mooring lines are regarded as two coupled strings [349]. The control force $u_1(t)$ is mounted on the top of tower and the control force $u_2(t)$ is mounted on the floating platform. L_0 and L are the tower's and mooring lines' length, m_1 and m_2 are the mass of top of tower and floating platform, respectively. The mass per unit length of the tower is ρ. D_1 and D_2 are original locations of mooring lines. ρ_1 and ρ_2 are the mass per unit length of mooring lines, respectively. c, c_1, and c_2 are the damping coefficients of flexible tower and mooring lines. w_1 represents the displacement of left line, and w_2 represents the displacement of right line. w represents the displacement of wind turbine tower. Moreover, EI represents the bending stiffness of beam-like tower, and tensions of two mooring lines are expressed as T_1 and T_2, respectively. $X - Y$ coordinate system is a plane coordinate system, where, X is used to determine the position and Y is the displacement of mooring lines and flexible wind tower. Meanwhile, floating platform is fixed on the sea level by mooring lines, its position will not follow the FWT system.

Remark 12.1 For conciseness, the notations $w'' = \partial^2 w/\partial x^2, \ddot{w} = \partial^2 w/\partial t^2, w' = \partial w/\partial x, \dot{w} = \partial w/\partial t, w = w(x,t), w_{L_0} = w(L_0, t)$, and $w_0 = w(0,t)$ are used.

12.2.1 Dynamical Model of Floating Wind Turbine Vibrations

The FWT system's kinetic energy $E_k(t)$ can be obtained as

$$E_k(t) = \frac{\rho_1}{2}\int_{-L}^{0}\left(\frac{\partial\left[w_1 + D_1\right]}{\partial t}\right)^2 dx + \frac{\rho_2}{2}\int_{-L}^{0}\left(\frac{\partial\left[w_2 - D_2\right]}{\partial t}\right)^2 dx$$
$$+\frac{\rho}{2}\int_{0}^{L_0}\left(\frac{\partial w}{\partial t}\right)^2 dx + \frac{m_1}{2}\left(\frac{\partial w(L_0,t)}{\partial t}\right)^2 + \frac{m_2}{2}\left(\frac{\partial w(0,t)}{\partial t}\right)^2, \tag{12.1}$$

and the potential energy $E_p(t)$ can be written as

$$E_p(t) = \frac{EI}{2}\int_{0}^{L_0}\left(\frac{\partial^2 w}{\partial x^2}\right)^2 dx + \frac{T_1}{2}\int_{-L}^{0}\left(\frac{\partial\left[w_1 + D_1\right]}{\partial x}\right)^2 dx$$
$$+\frac{T_2}{2}\int_{-L}^{0}\left(\frac{\partial\left[w_2 - D_2\right]}{\partial x}\right)^2 dx, \tag{12.2}$$

where, t and x denote the time variable and the position variable, respectively.

Remark 12.2 From Figure 12.1, because the strong coupling of the two flexible structures, we have $w = w_1 = w_2$ at $x = 0$ and $\partial w/\partial t = \partial w_1/\partial t = \partial w_1/\partial t$ at $x = 0$.

The virtual work done by the model is

$$\delta W_c(t) = -c\int_{0}^{L}\frac{\partial w}{\partial t}\delta w dx - c_1\int_{0}^{L}\frac{\partial\left[w_1 + D_1\right]}{\partial t}\delta w_1 dx$$
$$-c_2\int_{0}^{L}\frac{\partial\left[w_2 - D_2\right]}{\partial t}\delta w_2 dx. \tag{12.3}$$

where the FWT system damping coefficients are represented as c, c_1, and c_2.

The virtual work done by external disturbances $d_1(t)$ and $d_2(t)$ is obtained by

$$\delta W_d(t) = d_1(t)\delta w(L_0, t) + d_2(t)\delta w(0,t). \tag{12.4}$$

Remark 12.3 The lumped mass at the top boundary of the tower suffers from the boundary disturbance $d_1(t)$. $d_1(t)$ is the synthesis of all disturbances to the components at the tip boundary of tower. $d_2(t)$ is the wave disturbance mainly about waves with different height, speed etc., for example, the wave with significant wave height and the theoretical wave height [3, 127] that can result in large excitation forces. All these excitation forces acting on the FWT system as external disturbances can cause vibrations of tower.

The virtual work $W_u(t)$ done by the controllers $u_1(t)$ and $u_2(t)$ is denoted as

$$\delta W_u(t) = u_1(t)\delta w(L_0, t) + u_2(t)\delta w(0,t). \tag{12.5}$$

Then, the entire virtual work $W(t)$ done on the system can be represented as

$$\delta W(t) = \delta W_c(t) + \delta W_d(t) + \delta W_u(t). \tag{12.6}$$

According to the Hamilton's principle [91], governing equations of the FWT system including the turbine tower and mooring lines are obtained as

$$\rho \ddot{w} + EIw'''' + c\dot{w} = 0, \quad (x,t) \in (0, L_0) \times [0, \infty). \tag{12.7}$$

$$\rho_i \ddot{w}_i - T_i w_i'' + c_i \dot{w}_i = 0, i = (1,2) \text{ and } (x,t) \in (-L, 0) \times [0, \infty). \tag{12.8}$$

with boundary conditions

$$w_0'' = w_{L_0}'' = 0, \tag{12.9}$$

$$m_1 \ddot{w}_{L_0} - EIw_{L_0}''' = u_1(t) + d_1(t) \tag{12.10}$$

$$w_1(-L, t) = -D_1, \tag{12.11}$$

$$w_2(-L, t) = D_2, \tag{12.12}$$

$$m_2 \ddot{w}_0 + EIw_0''' + T_1 w_1'(0,t) + T_2 w_2'(0,t) = u_2(t) + d_2(t). \tag{12.13}$$

$\forall t \in [0, \infty)$.

Remark 12.4 The governing equations (12.7) and (12.8) describe the dynamics of flexible structures including two mooring lines and a flexible tower. Meanwhile, the boundary conditions (12.9), (12.11), and (12.12) describe the dynamics of the boundary position of flexible structures and they satisfy the Newton's second law from the form of these equations, $\ddot{w}_0$ and $\ddot{w}_{L_0}$ are the acceleration of floating platform and top of tower. A part of boundary conditions is nonconservative force which includes that $u_2(t)$ and $u_1(t)$ are boundary control forces acting on platform and top of tower, and $d_2(t)$ and $d_1(t)$ are external boundary disturbances. EIw_{L_0}''', EIw_0''', and $T_1 w_1'(0,t)$, $T_2 w_2'(0,t)$ are strain forces of two boundary of flexible tower and tensions of two mooring lines. From the equation (12.12), it is easy known that the motion of platform is related to the dynamics of mooring lines and flexible tower, and the coupling relation has an effect on the dynamic modeling of FWT system.

Remark 12.5 For the FWT system, the distributed disturbances have less influence on the dynamics of the flexible system comparing with the boundary disturbances. In this chapter, we ignore the effect of disturbances for its less influence and simplifying the theoretical analysis.

12.2.2 Preliminaries

For stability analysis, we present lemmas, remark, and assumption for the subsequent development as follows.

Lemma 12.1 [103] For any continuously differentiable ϕ on $[L_1, L_2]$, we have

$$[\phi - \phi(L_1, t)]^2 \leq (L_2 - L_1) \int_{L_1}^{L_2} [\phi']^2 dx. \tag{12.14}$$

Lemma 12.2 [103] For any continuously differentiable ϕ on $[L_1, L_2]$, we have

$$[\phi]^2 \leq 2(L_2 - L_1) \int_{L_1}^{L_2} [\phi']^2 dx + 2[\phi(L_1, t)]^2. \tag{12.15}$$

Lemma 12.3 Poincaré Inequality [256]: For any ϕ, continuously differentiable on $[L_1, L_2]$, we have

$$\int_{L_1}^{L_2} [\phi]^2 dx \leq 2(L_2 - L_1)\phi^2(L_2, t) + 4(L_2 - L_1)^2 \int_{L_1}^{L_2} [\phi']^2 dx, \tag{12.16}$$

$$\int_{L_1}^{L_2} [\phi]^2 dx \leq 2(L_2 - L_1)\phi^2(L_1, t) + 4(L_2 - L_1)^2 \int_{L_1}^{L_2} [\phi']^2 dx. \tag{12.17}$$

Remark 12.6 From Poincaré inequality (12.16), we further have

$$[\phi]^2 \leq 4(L_2 - L_1)^2 \phi'^2(L_1) + 8(L_2 - L_1)^3 \int_{L_1}^{L_2} [\phi'']^2 dx + 2[\phi(L_1)]^2. \tag{12.18}$$

Assumption 12.1 According to the unknown external boundary disturbances $d_1(t)$ and $d_2(t)$, we assume that the two positive values $\overline{d}_1 \in \mathbb{R}^+$ and $\overline{d}_2 \in \mathbb{R}^+$ exist, to make $|d_1(t)| \leq \overline{d}_1$ and $|d_2(t)| \leq \overline{d}_2$ hold. This assumption is appropriate because the disturbances have finite energy.

12.3 Controller Design

In this part, we design controllers $u_1(t)$ and $u_2(t)$ for vibrations suppression of the FWT system under external disturbances.

Theorem 12.1 Based on the dynamic model (12.7) and (12.8), and (12.9)–(12.13), the boundary controllers are proposed as follow

$$u_1(t) = -k_1 \left[\dot{w}_{L_0} + \frac{\alpha}{\beta} w_{L_0}\right] - \text{sgn}\left[\dot{w}_{L_0} + \frac{\alpha}{\beta} w_{L_0}\right] \overline{d}_1 - \frac{\alpha}{\beta} m_1 \dot{w}_{L_0}, \tag{12.19}$$

$$u_2(t) = -k_2 \left[\dot{w}_0 + \frac{\alpha}{\beta} w_0\right] - k_p w_0 - \text{sgn}\left[\dot{w}_0 + \frac{\alpha}{\beta} w_0\right] \overline{d}_2 - \frac{\alpha}{\beta} m_2 \dot{w}_0, \tag{12.20}$$

where k_1, k_2, and k_p are three positive control gains, α and β are the positive constants, and $\text{sgn}(\cdot)$ represents the signum function. The closed-loop system is uniformly bounded as the initial conditions are bounded.

Proof: Consider the Lyapunov candidate function as

$$\Theta(t) = \Theta_1(t) + \Theta_2(t) + \Theta_3(t) + \Delta(t), \tag{12.21}$$

where $\Theta_1(t)$, $\Theta_2(t)$, $\Theta_3(t)$, and $\Delta(t)$ are defined as

$$\Theta_1(t) = \frac{\beta}{2} \int_0^{L_0} \rho[\dot{w}]^2 dx + \frac{\beta}{2} EI \int_0^{L_0} [w'']^2 dx + \frac{\beta}{2} \int_{-L}^{0} \left(\rho_1[\dot{w}_1]^2 + \rho_2[\dot{w}_2]^2\right) dx$$
$$+ \frac{\beta}{2} \int_{-L}^{0} \left(T_1[w_1']^2 + T_2[w_2']^2\right) dx, \tag{12.22}$$

$$\Theta_2(t) = \frac{\beta}{2} m_1 \left[\dot{w}_{L_0} + \frac{\alpha}{\beta} w_{L_0}\right]^2 + \frac{\beta}{2} m_2 \left[\dot{w}_0 + \frac{\alpha}{\beta} w_0\right]^2 + \frac{\beta}{2} k_p [w_0]^2, \tag{12.23}$$

$$\Theta_3(t) = \frac{\alpha}{2} \int_0^{L_0} c[w]^2 dx + \frac{\alpha}{2} \int_{-L}^{0} c_1[w_1]^2 dx + \frac{\alpha}{2} \int_{-L}^{0} c_2[w_2]^2 dx, \tag{12.24}$$

$$\Delta(t) = \alpha \int_0^{L_0} \rho \dot{w} w dx + \alpha \int_{-L}^{0} \left[\rho_1 \dot{w}_1 w_1 + \rho_2 \dot{w}_2 w_2\right] dx$$
$$- \gamma \int_{-L}^{0} x \left[\rho_1 \dot{w}_1 w_1' + \rho_2 \dot{w}_2 w_2'\right] dx, \tag{12.25}$$

where γ is a positive weighting constant. ■

Consider a positive function $\kappa(t)$ as

$$\kappa(t) = \int_0^{L_0} \left\{[\dot{w}]^2 + [w]^2 + [w'']^2\right\} dx + \int_{-L}^{0} \left\{[\dot{w}_1]^2 + [w_1]^2 + [w_1']^2\right.$$
$$\left. + [\dot{w}_2]^2 + [w_2]^2 + [w_2']^2\right\} dx. \tag{12.26}$$

From the expression of $\Theta_1(t)$ and $\Theta_3(t)$, we have

$$\mu_1 \kappa(t) \leq \Theta_1(t) + \Theta_3(t) \leq \mu_2 \kappa(t), \tag{12.27}$$

where

$$\mu_1 = \min\left(\frac{\beta\rho}{2}, \frac{\beta EI}{2}, \frac{\alpha c}{2}, \frac{\beta\rho_1}{2}, \frac{\beta T_1}{2}, \frac{\alpha c_1}{2}, \frac{\beta\rho_2}{2}, \frac{\beta T_2}{2}, \frac{\alpha c_2}{2}\right), \tag{12.28}$$

$$\mu_2 = \max\left(\frac{\beta\rho}{2}, \frac{\beta EI}{2}, \frac{\alpha c}{2}, \frac{\beta\rho_1}{2}, \frac{\beta T_1}{2}, \frac{\alpha c_1}{2}, \frac{\beta\rho_2}{2}, \frac{\beta T_2}{2}, \frac{\alpha c_2}{2}\right). \tag{12.29}$$

From the expression of $\Delta(t)$, we know that $\Delta(t)$ is bounded as $|\Delta(t)| \leq \mu_3 \kappa(t)$, where $\mu_3 = \max\left(\alpha\rho, \alpha\rho_1 + \gamma L\rho_1, \alpha\rho_2 + \gamma L\rho_2\right)$.

Considering α, β, and γ satisfying $\mu_1 - \mu_3 > 0$, we have

$$0 < \mu_4 \left[\Theta_1(t) + \Theta_3(t)\right] \leq \Theta_1(t) + \Theta_3(t) + \Delta(t) \leq \mu_5 \left[\Theta_1(t) + \Theta_3(t)\right], \tag{12.30}$$

where $\mu_4 = \frac{\mu_1 - \mu_3}{\mu_2}$ and $\mu_5 = \frac{\mu_2 + \mu_3}{\mu_1}$. Then the Lyapunov candidate function is bounded as

$$\lambda_1 \left[\Theta_1(t) + \Theta_2(t) + \Theta_3(t)\right] \leq \Theta(t) \leq \lambda_2 \left[\Theta_1(t) + \Theta_2(t) + \Theta_3(t)\right], \tag{12.31}$$

where $\lambda_1 = \min\left(\mu_4, 1\right)$, $\lambda_2 = \max\left(\mu_5, 1\right)$.

Lemma 12.4 With above assumptions and defines, the time derivative of Lyapunov candidate function (12.21) is bounded with

$$\dot{\Theta}(t) \leq -\lambda\Theta(t) + \varepsilon, \tag{12.32}$$

where λ and ε are two positive constants.

Proof: Differentiation $\Theta(t)$ leads to

$$\dot{\Theta}(t) = \dot{\Theta}_1(t) + \dot{\Theta}_2(t) + \dot{\Theta}_3(t) + \dot{\Delta}(t), \tag{12.33}$$

From Remark 12.2 and boundary condition (12.10), using (12.17), and substituting the proposed control (12.19) and (12.20), and let $\gamma L - \alpha \geq 0$, we further have

$$\dot{\Theta}(t) \leq -\beta k_1 \left[\dot{w}_{L_0} + \frac{\alpha}{\beta} w_{L_0}\right]^2 - \beta k_2 \left[\dot{w}_0 + \frac{\alpha}{\beta} w_0\right]^2 - \left(\alpha k_p - \sigma_1 - \sigma_2 - \sigma_3\right) [w_0]^2$$
$$- (\beta c - \alpha\rho) \int_0^{L_0} [\dot{w}]^2 dx - \left(\alpha EI - 8\sigma_1 L_0^3\right) \int_0^{L_0} [w'']^2 dx$$

$$
\begin{aligned}
&-\left(\beta c_1-\alpha\rho_1-\frac{\gamma c_1 L}{2}-\frac{\gamma}{2}\rho_1\right)\int_{-L}^{0}[\dot{w}_1]^2dx\\
&-\left(\beta c_2-\alpha\rho_2-\frac{\gamma c_2 L}{2}-\frac{\gamma}{2}\rho_2\right)\int_{-L}^{0}[\dot{w}_2]^2dx\\
&-\left(\alpha T_1-\frac{\gamma}{2}T_1-\frac{\gamma c_1 L}{2}-2\sigma_2 L\right)\int_{-L}^{0}[w_1']^2dx\\
&-\left(\alpha T_2-\frac{\gamma}{2}T_2-\frac{\gamma c_2 L}{2}-2\sigma_3 L\right)\int_{-L}^{0}[w_2']^2dx-\frac{\sigma_1}{2L_0}\int_0^{L_0}w^2dx\\
&-\frac{\sigma_2}{2L}\int_{-L}^{0}w_1^2dx-\frac{\sigma_3}{2L}\int_{-L}^{0}w_2^2dx+\frac{\alpha T_1}{2}D_1^2+\frac{\alpha T_2}{2}D_2^2.
\end{aligned}
\tag{12.34}
$$

where σ_1, σ_2, and σ_3 are positive constants. Then, we have

$$
\dot{\Theta}(t)\le-\lambda_3\left[\Theta_1(t)+\Theta_2(t)+\Theta_3(t)\right]+\varepsilon,
\tag{12.35}
$$

where $\varepsilon=\frac{\alpha T_1}{2}D_1^2+\frac{\alpha T_2}{2}D_2^2$, and

$$
\begin{aligned}
&\epsilon_1=\beta c-\alpha\rho>0, \qquad \epsilon_2=\alpha EI-8\sigma_1 L_0^3>0,\\
&\epsilon_3=\beta c_1-\alpha\rho_1-\frac{\gamma c_1 L}{2}-\frac{\gamma}{2}\rho_1>0,\\
&\epsilon_4=\beta c_2-\alpha\rho_2-\frac{\gamma c_2 L}{2}-\frac{\gamma}{2}\rho_2>0,\\
&\epsilon_5=\alpha T_1-\frac{\gamma}{2}T_1-\frac{\gamma c_1 L}{2}-2\sigma_2 L>0,\\
&\epsilon_6=\alpha T_2-\frac{\gamma}{2}T_2-\frac{\gamma c_2 L}{2}-2\sigma_3 L>0,\\
&\epsilon_7=\beta k_1>0, \qquad \epsilon_8=\beta k_2>0,\\
&\epsilon_9=\frac{\sigma_1}{2L_0}>0, \qquad \epsilon_{10}=\frac{\sigma_2}{2L}>0,\\
&\epsilon_{11}=\frac{\sigma_3}{2L}>0, \qquad \epsilon_{12}=\alpha k_p-\sigma_1-\sigma_2-\sigma_3>0,\\
&\lambda_3=\min\left(\frac{2\epsilon_1}{\beta\rho},\frac{2\epsilon_2}{\beta EI},\frac{2\epsilon_3}{\beta\rho_1},\frac{2\epsilon_4}{\beta\rho_2},\frac{2\epsilon_5}{\beta T_1},\frac{2\epsilon_6}{\beta T_2},\frac{2\epsilon_7}{\beta m_1},\frac{2\epsilon_8}{\beta m_2},\frac{2\epsilon_9}{\alpha c},\frac{2\epsilon_{10}}{\alpha c_1},\frac{2\epsilon_{11}}{\alpha c_2},\frac{2\epsilon_{12}}{\beta k_p}\right)>0.
\end{aligned}
$$

Combining Eqs. (12.31) and (12.35), we have

$$
\dot{\Theta}(t)\le-\lambda\Theta(t)+\varepsilon,
\tag{12.36}
$$

where $\lambda=\lambda_3/\lambda_2$. Subsequently, we have the following main result of this chapter as follows.

Integrating of Eq. (12.36), we obtain

$$
\Theta(t)\le\left(\Theta(0)-\frac{\varepsilon}{\lambda}\right)e^{-\lambda t}+\frac{\varepsilon}{\lambda}\le\Theta(0)e^{-\lambda t}+\frac{\varepsilon}{\lambda}\in\mathcal{L}_\infty.
\tag{12.37}
$$

Utilizing inequality (12.15) and the Poincaré inequality (12.18), we have

$$
\begin{aligned}
\frac{\beta}{16L_0^3}EI[w]^2\le{}&\frac{\beta}{2}EI\int_0^{L_0}[w'']^2dx+\frac{\beta}{2}k_p[w(0,t)]^2\\
&-\left(\frac{\beta}{2}k_p-\frac{\beta}{8L_0^3}EI\right)[w(0,t)]^2.
\end{aligned}
\tag{12.38}
$$

We design k_p satisfying $k_p - \frac{EI}{4L_0^3} \geq 0$, and we further have

$$\frac{\beta}{16L_0^3}EI[w]^2 \leq \frac{\beta}{2}EI\int_0^{L_0}[w'']^2dx + \frac{\beta}{2}k_p[w(0,t)]^2$$
$$\leq \Theta_1(t) + \Theta_2(t) \leq \frac{1}{\lambda_1}\Theta(t) \in \mathcal{L}_\infty. \tag{12.39}$$

Suitably rewritten the above inequality, uniformly bounded of w can be obtained as

$$|w| \leq \sqrt{\frac{16L_0^3}{\beta\lambda_1 EI}\left(\Theta(0)e^{-\lambda t} + \frac{\varepsilon}{\lambda}\right)} \leq D_0, \tag{12.40}$$

where the constant $D_0 = \sqrt{\frac{16L_0^3}{\beta\lambda_1 EI}(\Theta(0) + \frac{\varepsilon}{\lambda})}$, $\forall x \in [0, L_0]$. In addition, as t trends to infinity, we have

$$\lim_{t\to\infty}|w| \leq \sqrt{\frac{16L_0^3\varepsilon}{\beta\lambda_1\lambda EI}}, \quad \forall x \in [0, L_0]. \tag{12.41}$$

In a similar manner, from inequality (12.14), we have

$$\frac{\beta}{2L}T_1[w_1 + D_1]^2 \leq \frac{\beta}{2}T_1\int_{-L}^{0}[w_1']^2dx \leq \Theta_1(t) \leq \frac{1}{\lambda_1}\Theta(t) \in \mathcal{L}_\infty, \tag{12.42}$$

$$\frac{\beta}{2L}T_2[w_2 - D_2]^2 \leq \frac{\beta}{2}T_2\int_{-L}^{0}[w_2']^2dx \leq \Theta_1(t) \leq \frac{1}{\lambda_1}\Theta(t) \in \mathcal{L}_\infty. \tag{12.43}$$

Then, we obtain $w_1 + D_1$ and $w_2 - D_2$ are uniformly bounded as follows

$$|w_1 + D_1| \leq \sqrt{\frac{2L}{\beta\lambda_1 T_1}\left(\Theta(0)e^{-\lambda t} + \frac{\varepsilon}{\lambda}\right)} \leq D, \tag{12.44}$$

$$|w_2 - D_2| \leq \sqrt{\frac{2L}{\beta\lambda_1 T_2}\left(\Theta(0)e^{-\lambda t} + \frac{\varepsilon}{\lambda}\right)} \leq D, \tag{12.45}$$

where the constant $D = \sqrt{\frac{2L}{\beta\lambda_1 \min(T_1,T_2)}(\Theta(0) + \frac{\varepsilon}{\lambda})}$, $\forall x \in [-L, 0]$. Moreover, as t trends to infinity, we have

$$\lim_{t\to\infty}|w_1 + D_1| \leq \sqrt{\frac{2L\varepsilon}{\beta\lambda_1\lambda T_1}}, \tag{12.46}$$

$$\lim_{t\to\infty}|w_2 - D_2| \leq \sqrt{\frac{2L\varepsilon}{\beta\lambda_1\lambda T_2}}, \tag{12.47}$$

$\forall x \in [-L, 0]$. ■

Remark 12.7 For this FWT system, the value of system parameters EI, T_1, T_2, L, and L_0 are relative larger than positive constants α, β, γ, and σ_1 -σ_3. According to inequalities $\epsilon_1 - \epsilon_{12}$, when the system parameters EI, T_1, T_2, L, L_0, ρ, ρ_1, ρ_2, c, c_1, and c_2 are determined, α, β, γ, and σ_1-σ_3 can be chosen to be small. Futher, the value of control gains k_1, k_2, and k_p can be determined to make ϵ_7, ϵ_8, and ϵ_{12} hold.

Remark 12.8 From equations (12.41), (12.46), and (12.47), it is easy to know that the system's displacement w, w_1, and w_2 can converge to bounded range. Therefore, we can obtain that proposed control laws (12.19) and (12.20) are effective for suppressing vibrations of the FWT system.

Remark 12.9 All of the required information in the two controllers can be received by a backward difference method or obtained through a series of sensors. Using the laser displacement one can get w_{L_0} and w_0, which are the tip boundary of the tower and the bottom base for the floating tower respectively [235]. $\dot{w}_{L_0}$ and $\dot{w}_0$ with only one time derivative, which can be obtained by applying the backward difference algorithm.

12.4 Numerical Simulations

In this part, we make numerical simulations to validate the effectiveness of proposed vibration control method for the FWT system. The dynamical behavior of the FWT system is analyzed in the set $\Omega = \{(x_0, x_1, t) : 0 \le x_0 \le L_0, 0 \le x \le L, 0 \le t \le T\}$. The temporal step size k and spacial step sizes h_0 and h as below

$$h_0 = \Delta x_0 = \frac{L_0}{M_0},$$

$$h = \Delta x = \frac{L}{M}$$

$$k = \Delta t = \frac{T}{N}$$

where, M_0, M, and N are the spacial subdivisions of the length of tower and mooring lines, and the time subdivisions, respectively. Referring for previous works [101, 204, 216, 243, 297], system parameters of the FWT system is given in Table 12.1, and disturbances are given as $d_1(t) = [10 + \sin(5\pi t)] \times 10^3$ N and $d_2(t) = [2 + 3\cos(10\pi t)] \times 10^4$ N.

Initial conditions of the FWT system are given as $w(x, 0) = \frac{x}{L_0}$ m, $w_1(x, 0) = \sqrt{\frac{D_1^2 x}{L}} - D_1$ m, $w_2(x, 0) = -\sqrt{\frac{D_2^2 x}{L}} + D_2$ m, and $\dot{w}(x, 0) = \dot{w}_1(x, 0) = \dot{w}_2(x, 0) = 0$m/s, the time duration $t_f = 25$ s.

In the numerical simulation, three control cases will be taken into consideration, namely, without control, with PD control, and with proposed vibration control.

The simulation results are presented in Figures 12.2–12.15. Specifically, Figures 12.2–12.4 illustrate the displacement of the FWT system under the conditions of without control, PD control, and active boundary control, respectively. Figures 12.5–12.7 illustrate the three-dimensional vibrations of the FWT system under the three control methods. Figures 12.8–12.10 illustrate the three-dimensional vibrations of the left mooring line of the FWT system under the three control methods, while Figures 12.11–12.13 correspond to the three-dimensional vibrations of the right mooring line under the three control methods. Figure 12.14 shows the control input for the PD control method, and Figure 12.15 shows the control input for the active boundary control method.

Table 12.1 Parameters of the FWT system.

Parameter	Value	Parameter	Value
L_0	20 m	ρ	10 kg/m
EI	2000 Nm^2	c	8 Ns/m^2
L	100 m	ρ_1, ρ_2	1.5 kg/m
T_1, T_2	10^4 N	c_1, c_2	0.8 Ns/m
m_1	150 kg	m_2	100 kg
D_1	5 m	D_2	5 m

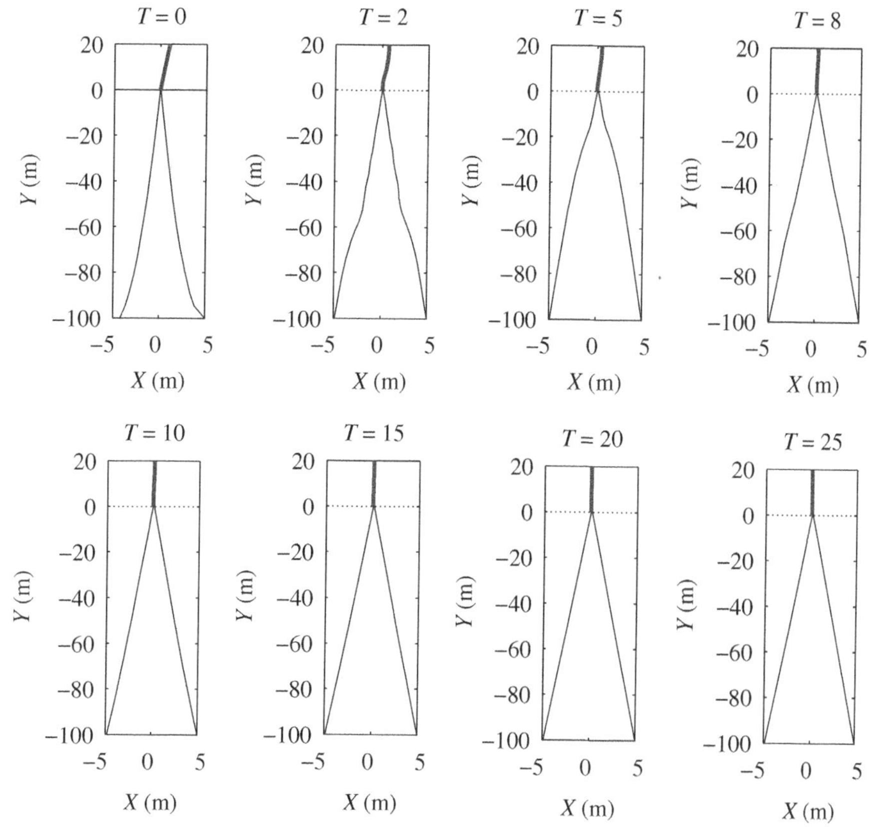

Figure 12.2 Displacement of the FWT system: without control.

Without control: In this case, there is no control input acting on the FWT system, namely, $u_1(t) = 0, u_2(t) = 0$. Figure 12.2 illustrates the movements of the FWT system, where, the bold thick line represents the motion of the tower, and the two bold lines describe the motion of two mooring lines. From Figure 12.2, it is obvious that there exist vibrations in the flexible wind turbine tower and mooring lines under the effect of external disturbances, and the vibrations of FWT system are increasing with increasing time from 0 to 25 s. Moreover, three-dimensional Figures 12.5, 12.8, and 12.11 illustrate displacements of the turbine tower and mooring lines. It is also seen that increasing structural vibrations of FWT system have a great influence on system performance, even will damage the flexible structures. Therefore, it is necessary to propose vibration control to stabilize the FWT system.

PD control: In Figure 12.3, the control performance for suppressing the vibration of FWT system of PD control is shown, where, PD controllers are given as $u_1(t) = 5000\dot{w}_{L_0} + 1000w_{L_0}$, $u_2(t) = 5000\dot{w}_0 + 1000w_0$. From these figures, it can be seen that vibrations of the FWT system are reduced, and the vibration reduction for each mooring lines and tower are shown in Figures 12.6, 12.9, and 12.12, respectively. According to these figures, we can see that there still exist small vibrations in turbine tower and mooring lines under PD control, which implies that PD control has a positive effect on the vibration suppression, while the control performance is not enough. Therefore, it is necessary to develop a better control scheme to reduce the vibration of FWT

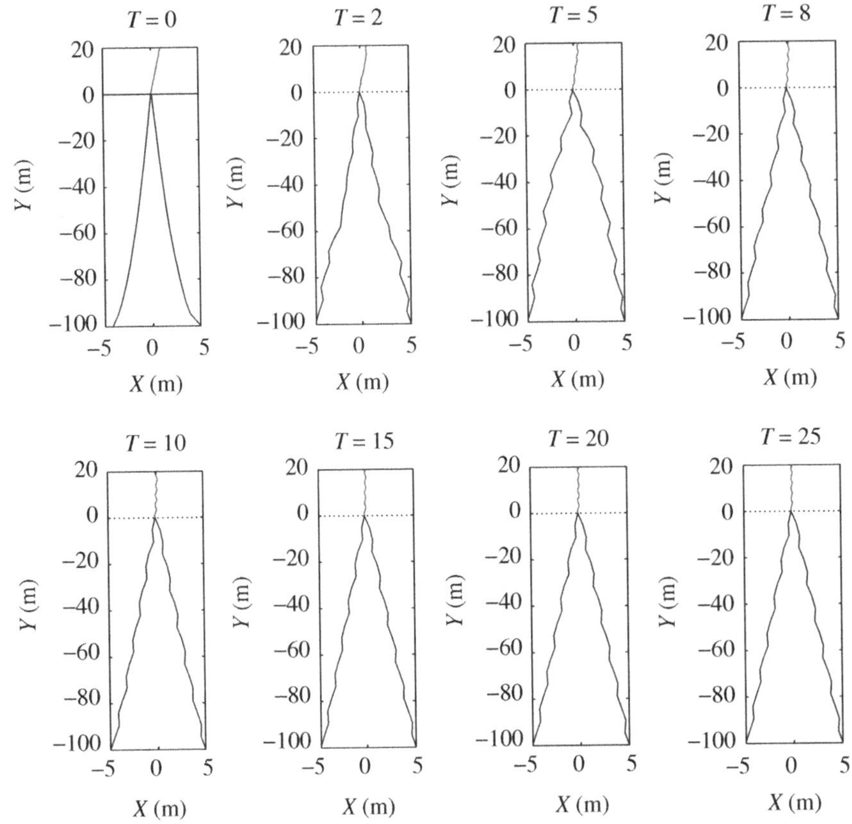

Figure 12.3 Displacement of the FWT system movements: with the PD control.

system. In Figure 12.14, control inputs of the PD control are given. The input of controller $u_1(t)$ is in the range of −1300 to 1000 N from 0 to 25 s, and the input of controller $u_2(t)$ is in the range of -2×10^4 to 2×10^4 N from 0 to 25 s.

Active boundary control: In this case, active boundary controllers $u_1(t)$ (12.19) and $u_2(t)$ (12.20) are implemented in the FWT system with a set of control parameters $k_1 = 5000$, $k_2 = 3000$, $k_p = 800$, $\alpha = 20$, $\beta = 100$, and $\gamma = 0.5$. Similar to the above cases, Figure 12.4 illustrates the whole dynamics of FWT system, and from Figure 12.4, we can see that vibrations of the FWT system are suppressed well. Moreover, the dynamics of each mooring lines and turbine tower are shown in Figures 12.7, 12.10, and 12.13. Figure 12.15 shows the input control signals of two proposed controllers: the input of controller $u_1(t)$ is in the range of −1000N to 1000N from 0 to 25 s, and the input of controller $u_2(t)$ is in the range of -3×10^4 to 3×10^4 N from 0 to 25 s.

According to the Figures 12.3–12.13, it can be seen that the proposed active boundary controllers has a better control performance than PD control in suppressing the vibration of FWT system, where, vibrations of turbine tower and mooring lines have better convergence. Moreover, it is seen

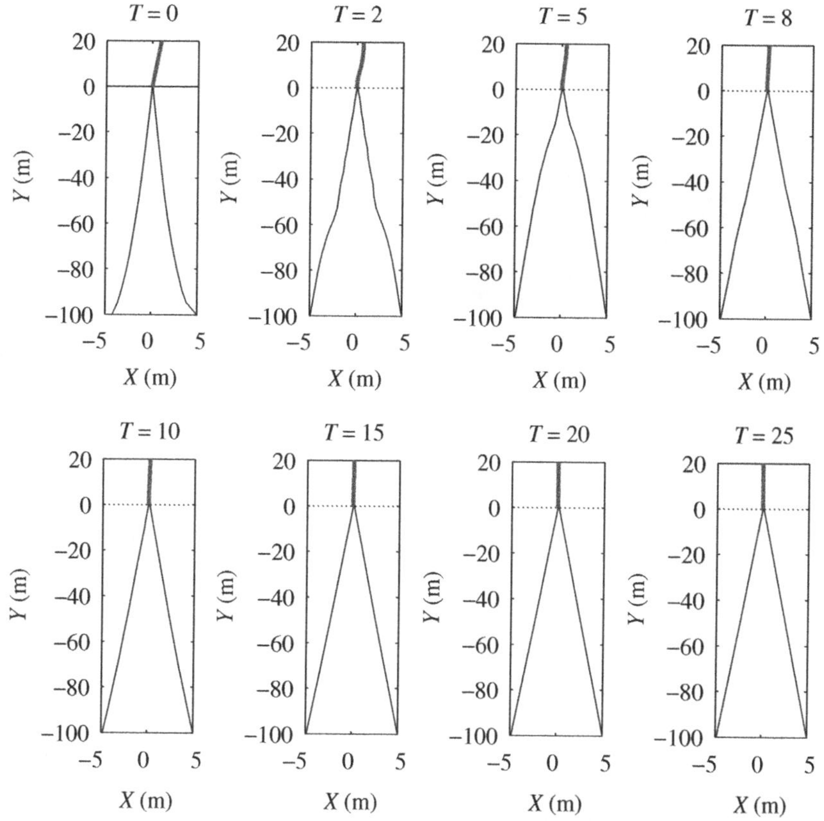

Figure 12.4 Displacement of the FWT system movements: with the boundary control.

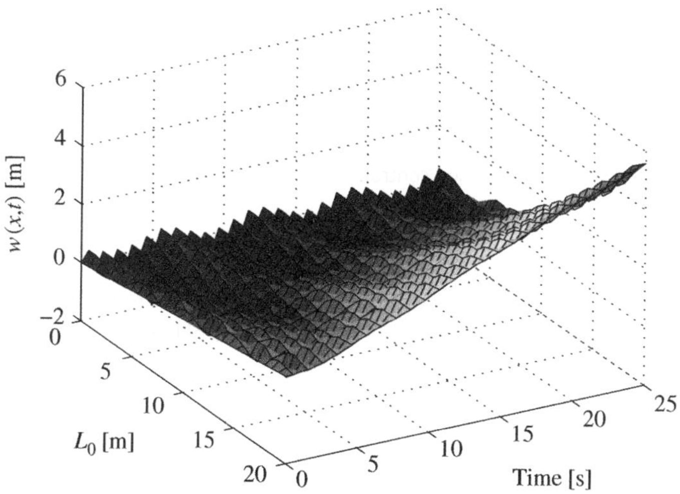

Figure 12.5 $w(x,t)$, vibration of the tower without control.

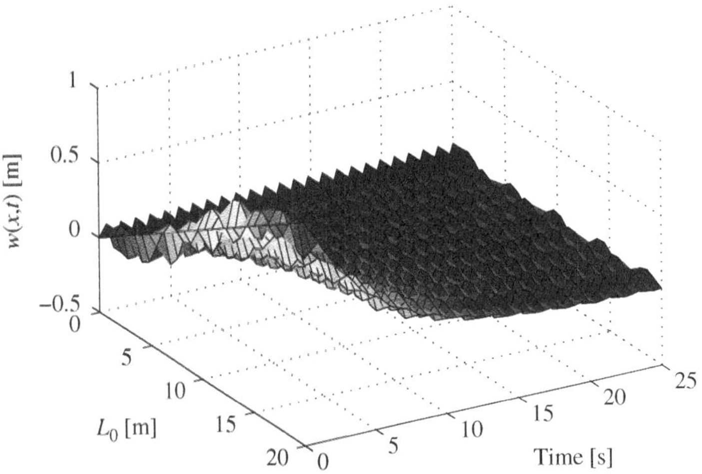

Figure 12.6 $w(x, t)$, vibration of the tower with the PD control.

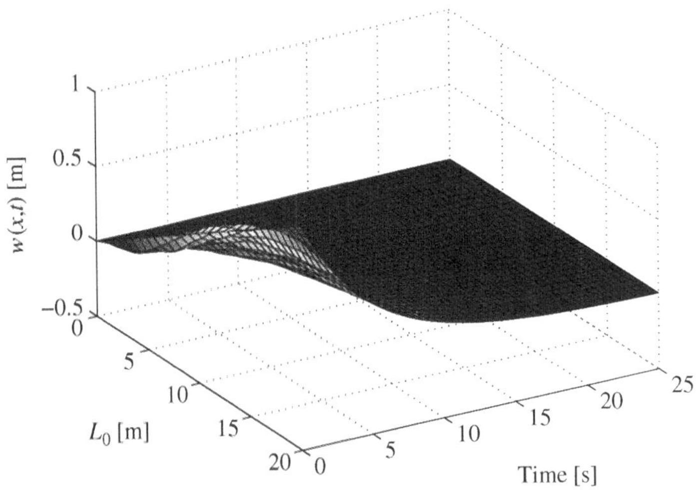

Figure 12.7 $w(x, t)$, vibration of the tower with the proposed control.

that the designed controllers are still able to reject the effect of the external disturbances and to ensure the FWT system stable. These simulations results are consensus with the theoretical results derived in (12.41), (12.46), and (12.47) and demonstrate that the proposed control laws are effective in vibration suppression for the FWT system.

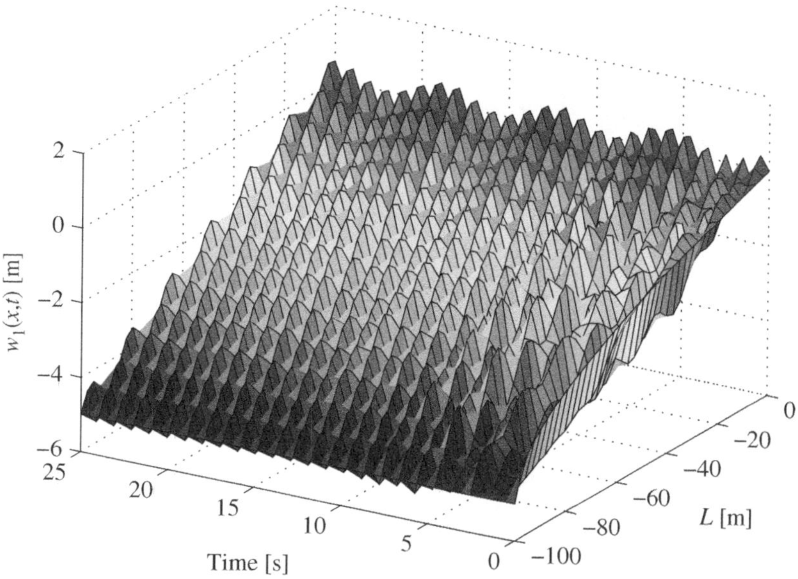

Figure 12.8 $w_1(x,t)$, vibration of the left mooring line without control.

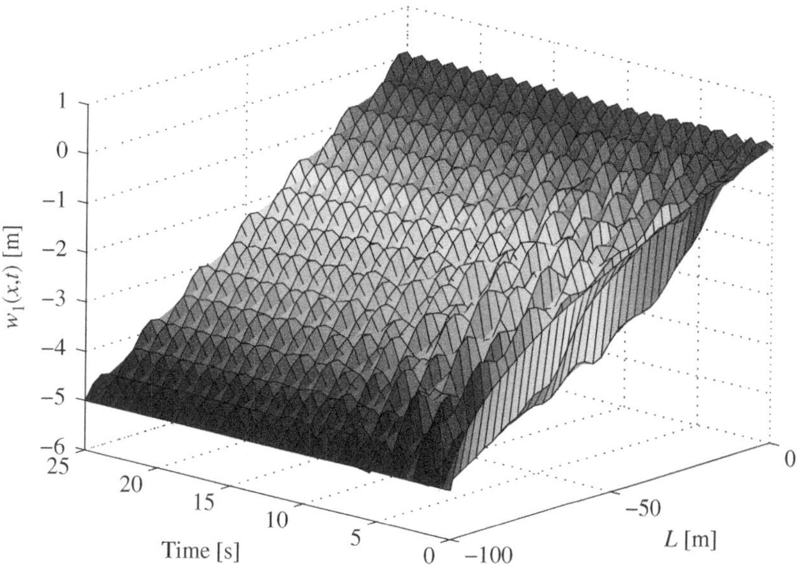

Figure 12.9 $w_1(x,t)$, vibration of the left mooring line with PD control.

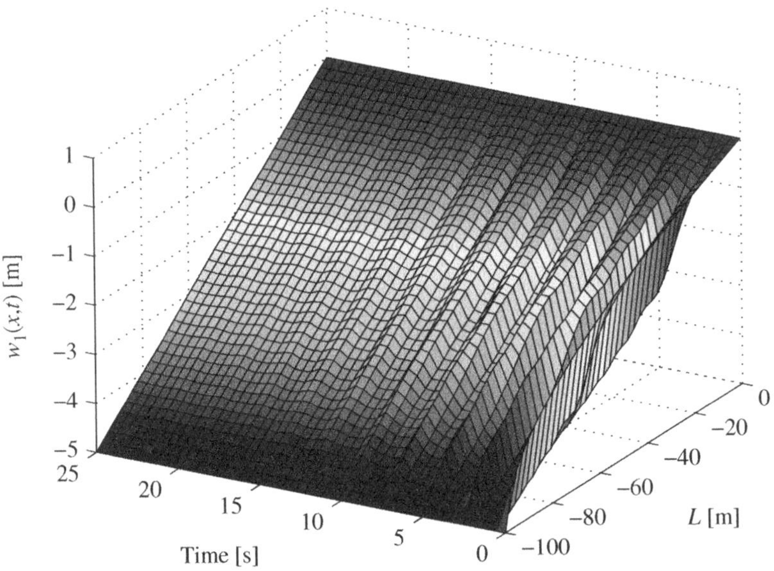

Figure 12.10 $w_1(x, t)$, vibration of the left mooring line with the proposed control.

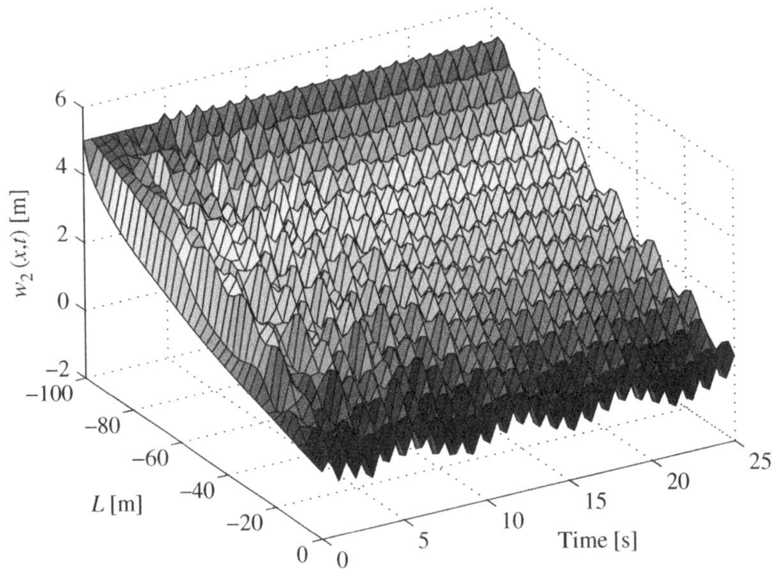

Figure 12.11 $w_2(x, t)$, vibration of the right mooring line without control.

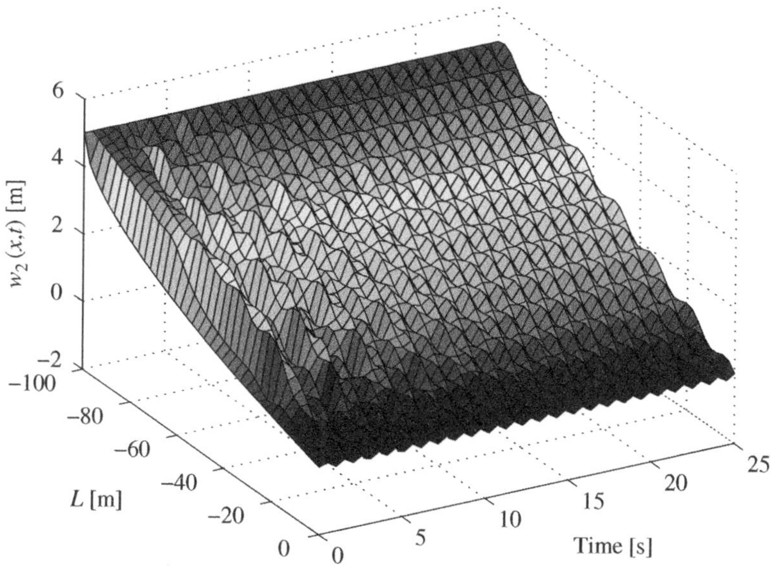

Figure 12.12 $w_2(x, t)$, vibration of the right mooring line with PD control.

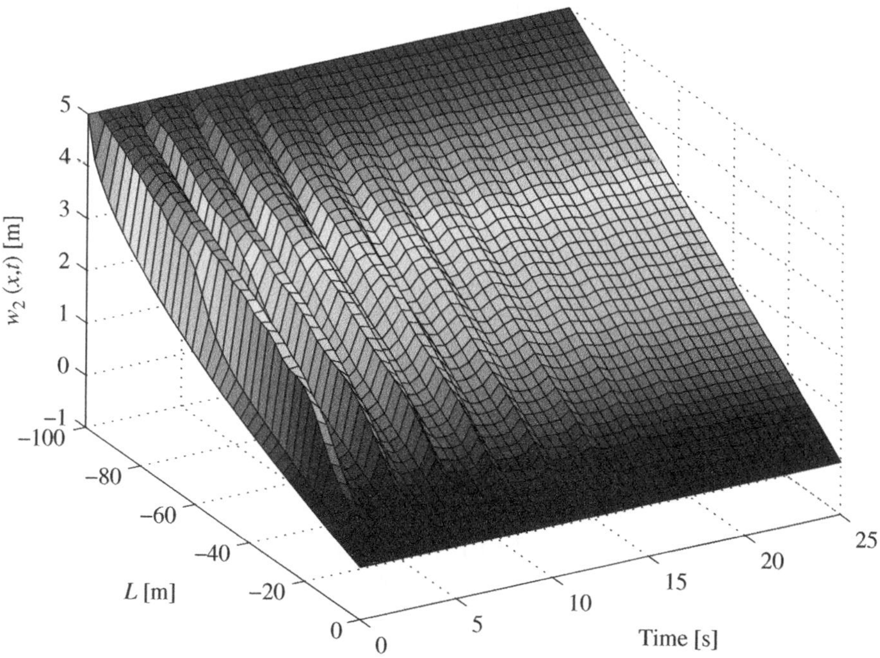

Figure 12.13 $w_2(x, t)$, vibration of the right mooring line with the proposed control.

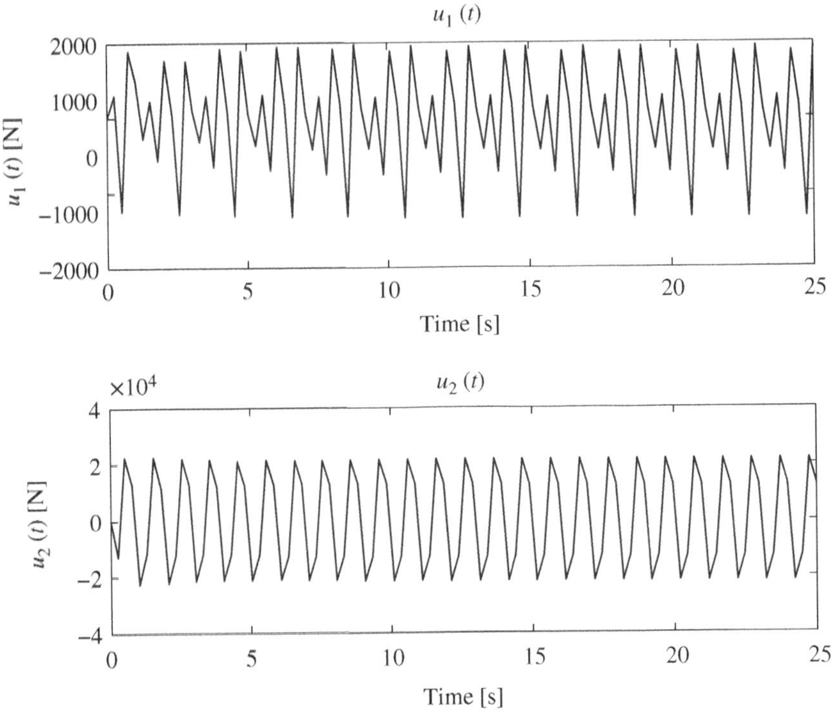

Figure 12.14 PD control inputs.

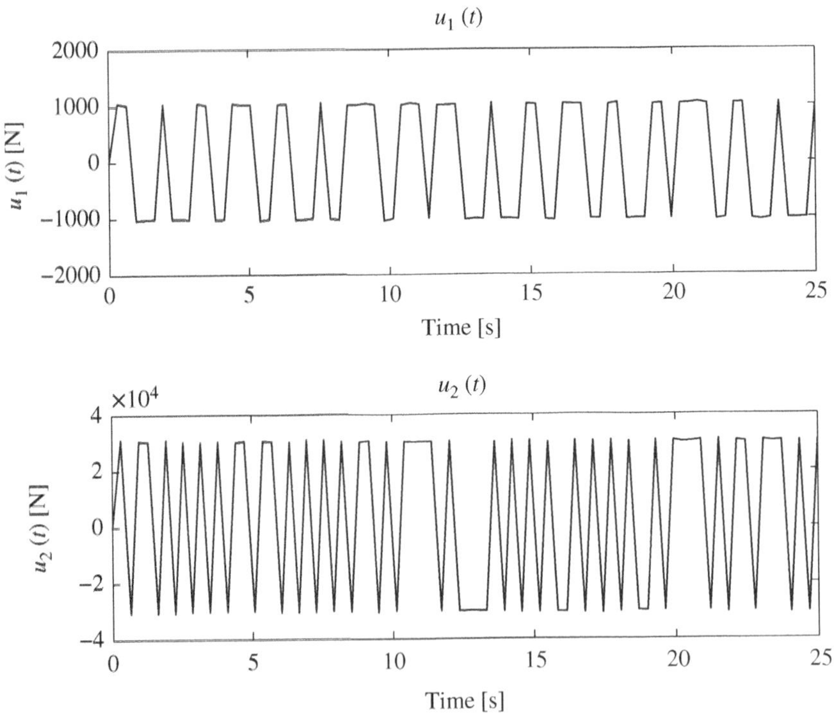

Figure 12.15 Control inputs.

12.5 Summary

In this chapter, we address the stabilization problem of the FWT system, which is a beam-string coupled structure. Through proposing active vibration control method, vibrations of turbine tower and mooring lines can be suppressed obviously. Moreover, the stability of FWT system is also guaranteed by choosing a suitable Lyapunov function. Besides, the effectiveness of proposed controllers is validated by numerical simulation results. In future work, we will focus on the optimal control and finite-time vibration control to improve the control performances, e.g., rate of convergence, transient response, etc.

13

Conclusions

After reading this chapter, you should be able to:

- Summarizes the practical significance in the application of neural network-based intelligent control and proposes some future research directions in this field

This book focuses on the study of flexible systems. High-efficiency modeling methods and novel intelligent control strategies for representative flexible systems are developed using the technology on the basis of neural networks. The book discusses the tracking control of both the single- and multi-link flexible manipulators, the vibration control of string and flexible buildings under natural disasters, the fault-tolerant control of bionic flexible flapping-wing aircraft, and the active boundary control of the floating wind turbine. Besides the theoretical deliberations, the book also demonstrates the effectiveness of the proposed approaches in practice through many case studies. The main highlights of the book are listed as follows.

The developed modeling methods in Chapters 3 and 4 successfully solve the problem of infinite-dimensional system dynamics for the single- and two-link flexible robotic manipulator, respectively. Neural network-based control strategies are designed to deal with system uncertainties and immeasurable states.

Chapter 5 creatively uses a partial differential equation and two ordinary differential equations to describe the dynamics of the string. Based on that, adaptive control is designed to handle the system uncertainties while also giving the stability and solvability of the inequality equations.

Considering disturbances caused by natural disasters such as earthquakes or strong winds, Chapters 6–8 present three intelligent control methods of vibration suppression of stand-along tall-like structure, and single-floor building-like structures. It is noted that the output constraints, unknowing magnitude of disturbance and system uncertainties are all integrated with these intelligent control methods.

The disturbance observer is integrated within the neural network to compensate the system uncertainties of the rigid-flexible coupled bionic flapping wing in Chapter 9. By also taking into account actuator failure, Chapter 10 designs a novel adaptive finite-time controller based on the fuzzy neural network and sliding mode scheme for tracking control and vibration suppression of the flexible wings. Focusing on the same type of platform, Chapter 11 demonstrates another neural network-based vibration controller for bionic flapping-wing robotic aircraft and autonomous ornithopter applications.

Dynamic Modeling and Neural Network-Based Intelligent Control of Flexible Systems,
First Edition. Hejia Gao, Wei He, and Changyin Sun.

Chapter 12 presents full study on dynamic modeling, active boundary control, and stability analysis for the coupled floating wind turbine. Two boundary controllers are nicely designed for restraining the vibrations of such flexible system under external disturbances.

In summary, this book will be useful for researchers and engineers who wish to avoid excessive time in searching high-efficiency modeling methods and neural-network-based control solutions for flexible systems. It is written for industry engineers and researchers who are interested in control theory and the applications. This book can also serve as a complementary reading on modeling and control of the flexible systems for the students and teachers at university who are engaged in self-study of adaptive control for flexible systems.

References

1 Aya H Abdulghany. Generalization of parallel axis theorem for rotational inertia. *American Journal of Physics*, 85(10):791–795, 2017.

2 Satoko Abiko, Yoshikazu Satake, Xin Jiang, Teppei Tsujita, and Masaru Uchiyama. Delay time compensation based on coefficient of restitution for collision hybrid motion simulator. *Advanced Robotics*, 28(17):1177–1188, 2014.

3 Tom Ainsworth and N W S Juneau. *Significant wave height': A closer look at wave forecasts.* NOAA's National Weather Service, 2010.

4 Osman Akyurek. *Lateral and torsional seismic vibration control for torsionally irregular buildings.* PhD thesis, Florida Institute of Technology, 2019.

5 Esmail Ali Alandoli, Tian Soon Lee, Yow Jon Lin, and Vijaya Vijayakumar. Dynamic model and intelligent optimal controller of flexible link manipulator system with payload uncertainty. *Arabian Journal for Science and Engineering*, 46:7423–7433, 2021.

6 Mohammed Aldawod, Bijan Samali, Fazel Naghdy, and Kenny CS Kwok. Active control of along wind response of tall building using a fuzzy controller. *Engineering Structures*, 23(11):1512–1522, 2001.

7 Assaad Alsahlani and Ranjan Mukherjee. Vibration control of a string using a scabbard-like actuator. *Journal of Sound and Vibration*, 330(12):2721–2732, 2011.

8 Tore Bakka and Hamid Reza Karimi. Robust dynamic output feedback control synthesis with pole placement constraints for offshore wind turbine systems. *Mathematical Problems in Engineering*, 2:2012, 2012.

9 Tore Bakka and Hamid Reza Karimi. H∞ static output-feedback control design with constrained information for offshore wind turbine system. *Journal of the Franklin Institute*, 350(8):2244–2260, 2013.

10 Tore Bakka, Hamid-Reza Karimi, and Søren Christiansen. Linear parameter-varying modelling and control of an offshore wind turbine with constrained information. *IET Control Theory & Applications*, 8(1):22–29, 2014.

11 Khaldoon A Bani-Hani. Vibration control of wind-induced response of tall buildings with an active tuned mass damper using neural networks. *Structural Control and Health Monitoring*, 14(1):83–108, 2010.

12 Gang Bao, Yide Zhang, and Zhigang Zeng. Memory analysis for memristors and memristive recurrent neural networks. *IEEE/CAA Journal of Automatica Sinica*, 7(1):96–105, 2020.

13 Saman Behtash. Robust output tracking for non-linear systems. *International Journal of Control*, 51(6):1381–1407, 1990.

Dynamic Modeling and Neural Network-Based Intelligent Control of Flexible Systems,
First Edition. Hejia Gao, Wei He, and Changyin Sun.

14 Bharath Bhikkaji, Reza Moheimani, and Ian Petersen. A negative imaginary approach to modeling and control of a collocated structure. *IEEE/ASME Transactions on Mechatronics*, 17(4):717–727, 2012.

15 Brendan J Bialy, Indrasis Chakraborty, Sadettin C Cekic, and Warren E Dixon. Adaptive boundary control of store induced oscillations in a flexible aircraft wing. *Automatica*, 70:230–238, 2016.

16 Shiba Biswal, Marc Mignolet, and Armando A Rodriguez. Modeling and control of flapping wing micro aerial vehicles. *Bioinspiration & Biomimetics*, 14(2):026004, 2019.

17 Marco Bortolini, Maurizio Faccio, Francesco Gabriele Galizia, Mauro Gamberi, and Francesco Pilati. Design, engineering and testing of an innovative adaptive automation assembly system. *Assembly Automation*, 40(3):531–540, 2020.

18 Delphine Bresch-Pietri and Miroslav Krstic. Output-feedback adaptive control of a wave PDE with boundary anti-damping. *Automatica*, 50(5):1407–1415, 2014.

19 Giacomo Canciello and Alberto Cavallo. Selective modal control for vibration reduction in flexible structures. *Automatica*, 75(1):282–287, 2017.

20 Fangfei Cao and Jinkun Liu. Three-dimensional modeling and input saturation control for a two-link flexible manipulator based on infinite dimensional model. *Journal of the Franklin Institute*, 357(2):1026–1042, 2020.

21 Laura Celentano. Modeling of flexible robots with varying cross section and large link deformations. *Journal of Dynamic Systems, Measurement, and Control*, 138(2):021010, 2016.

22 Laura Celentano and Alessandro Coppola. A computationally efficient method for modeling flexible robots based on the assumed modes method. *Applied Mathematics and Computation*, 218(8):4483–4493, 2011.

23 Wei-Der Chang and Shun-Peng Shih. PID controller design of nonlinear systems using an improved particle swarm optimization approach. *Communications in Nonlinear Science and Numerical Simulation*, 15(11):3632–3639, 2010.

24 Ching-Chen Chen and Jen-Yuan Chang. Vision-assisted vibration analysis of inhomogeneous flexible cables in hard disk drives. *IEEE Transactions on Magnetics*, 49(6):2628–2633, 2013.

25 Mou Chen and Shuzhi Sam Ge. Direct adaptive neural control for a class of uncertain nonaffine nonlinear systems based on disturbance observer. *IEEE Transactions on Cybernetics*, 43(4):1213–1225, 2013.

26 Mou Chen and Shuzhi Sam Ge. Adaptive neural output feedback control of uncertain nonlinear systems with unknown hysteresis using disturbance observer. *IEEE Transactions on Industrial Electronics*, 62(12):7706–7716, 2015.

27 Jiahao Chen and Hong Qiao. Muscle-synergies-based neuromuscular control for motion learning and generalization of a musculoskeletal system. *IEEE Transactions on Systems, Man, and Cybernetics: Systems*, 51(6):3993–4006, 2020.

28 Mou Chen and Gang Tao. Adaptive fault-tolerant control of uncertain nonlinear large-scale systems with unknown dead zone. *IEEE Transactions on Cybernetics*, 46(8):1851–1862, 2016.

29 Chen Chen and Tianyu Zhang. A review of design and fabrication of the bionic flapping wing micro air vehicles. *Micromachines*, 10(2):144–163, 2019.

30 Chieh-Li Chen, Chao Chung Peng, and Her-Terng Yau. High-order sliding mode controller with backstepping design for aeroelastic systems. *Communications in Nonlinear Science and Numerical Simulation*, 17(4):1813–1823, 2012.

31 Mou Chen, Shuzhi Sam Ge, Bernard Voon Ee How, and Yoo Sang Choo. Robust adaptive position mooring control for marine vessels. *IEEE Transactions on Control Systems Technology*, 21(2):395–409, 2013.

32 Chun Lung Chen, Yan-Jun Liu, and Guo-Xing Wen. Fuzzy neural network-based adaptive control for a class of uncertain nonlinear stochastic systems. *IEEE Transactions on Cybernetics*, 44(5):583–593, 2014.

33 Fu-Ming Chen, Jason Sheng-Hong Tsai, Ying-Ting Liao, Shu-Mei Guo, Ming-Chung Ho, Fu-Zen Shaw, and Leang-San Shieh. An improvement on the transient response of tracking for the sampled-data system based on an improved PD-type iterative learning control. *Journal of the Franklin Institute*, 351(2):1130–1150, 2014.

34 Mou Chen, Peng Shi, and Cheng-Chew Lim. Adaptive neural fault-tolerant control of a 3-DOF model helicopter system. *IEEE Transactions on Systems, Man, and Cybernetics: Systems*, 46(2):260–270, 2015.

35 C Chen, Z Liu, Y Zhang, CLP Chen, and S Xie. Adaptive control of MIMO mechanical systems with unknown actuator nonlinearities based on the Nussbaum gain approach. *IEEE/CAA Journal of Automatica Sinica*, 3(1):26–34, 2016. ISSN: 2329-9266.

36 Chun Lung Philip Chen, Guo-Xing Wen, Yan-Jun Liu, and Zhi Liu. Observer-based adaptive backstepping consensus tracking control for high-order nonlinear semi-strict-feedback multiagent systems. *IEEE Transactions on Cybernetics*, 46(7):1591–1601, 2016.

37 Lei Chen, Yan Yan, Chaoxu Mu, and Changyin Sun. Characteristic model-based discrete-time sliding mode control for spacecraft with variable tilt of flexible structures. *IEEE/CAA Journal of Automatica Sinica*, 3(1):42–50, 2016.

38 Chun Lung Philip Chen, Tong Zhang, Long Chen, and Sik Chung Tam. I-Ching divination evolutionary algorithm and its convergence analysis. *IEEE Transactions on Cybernetics*, 47(1):2–13, 2017.

39 Mou Chen, Shuyi Shao, and Bin Jiang. Adaptive neural control of uncertain nonlinear systems using disturbance observer. *IEEE Transactions on Cybernetics*, 47(10):3110–3123, 2017.

40 Ci Chen, Zhi Liu, Yun Zhang, Philip Chen, and Shengli Xie. Robust optimal adaptive sliding mode control with the disturbance observer for a manipulator robot system. *International Journal of Control, Automation and Systems*, 16(4):1701–1715, 2018.

41 Dae Seung Cho, Byung Hee Kim, Jin-Hyeong Kim, Nikola Vladimir, and Tae Muk Choi. Forced vibration analysis of arbitrarily constrained rectangular plates and stiffened panels using the assumed mode method. *Thin-Walled Structures*, 90:182–190, 2015.

42 Ji-Yun Choi, Keum-Shik Hong, and Kyung-Jinn Yang. Exponential stabilization of an axially moving tensioned strip by passive damping and boundary control. *Journal of Vibration and Control*, 10(5):661–682, 2004.

43 Kang Min Choi, Hyung Jo Jung, Heon Jae Lee, and Sang Won Cho. Feasibility study of an MR damper-based smart passive control system employing an electromagnetic induction device. *Smart Materials and Structures*, 16(6):2323, 2007.

44 Pierluigi Colli, Gianni Gilardi, and Jürgen Sprekels. A boundary control problem for the viscous Cahn–Hilliard equation with dynamic boundary conditions. *Applied Mathematics and Optimization*, 73(2):195–225, 2016.

45 Chaoqun Cui, Bing Wang, Yixin Zhao, Yongjin Zhang, and Liming Xue. Risk management for mine closure: A cloud model and hybrid semi-quantitative decision method. *International Journal of Minerals, Metallurgy, and Materials*, 27(8):1021–1035, 2020.

46 Shi-Lu Dai, Cong Wang, and Min Wang. Dynamic learning from adaptive neural network control of a class of nonaffine nonlinear systems. *IEEE Transactions on Neural Networks and Learning Systems*, 25(1):111–123, 2014.

47 Shi-Lu Dai, Min Wang, and Cong Wang. Neural learning control of marine surface vessels with guaranteed transient tracking performance. *IEEE Transactions on Industrial Electronics*, 63(3):1717–1727, 2016.

48 Shi-Lu Dai, Shude He, Min Wang, and Chengzhi Yuan. Adaptive neural control of underactuated surface vessels with prescribed performance guarantees. *IEEE Transactions on Neural Networks and Learning Systems*, 30(12):3686–3698, 2018.

49 Shi-Lu Dai, Shude He, Hai Lin, and Cong Wang. Platoon formation control with prescribed performance guarantees for USVs. *IEEE Transactions on Industrial Electronics*, 65(5):4237–4246, 2018.

50 Alex De Bonis, Federica Angeletti, Paolo Iannelli, and Paolo Gasbarri. Mixed Kane–Newton multi-body analysis of a dual-arm robotic system for on-orbit servicing missions. *Aerotecnica Missili & Spazio*, 100:375–386, 2021.

51 Dario De Domenico and Giuseppe Ricciardi. Optimal design and seismic performance of tuned mass damper inerter (TMDI) for structures with nonlinear base isolation systems. *Earthquake Engineering and Structural Dynamics*, 47(12):2539–2560, 2018.

52 Marcio S De Queiroz, Darren M Dawson, Siddharth P Nagarkatti, and Fumin Zhang. *Lyapunov-based control of mechanical systems*. Springer Science & Business Media, 2012.

53 Marco De Stefano, Ribin Balachandran, and Cristian Secchi. A passivity-based approach for simulating satellite dynamics with robots: Discrete-time integration and time-delay compensation. *IEEE Transactions on Robotics*, 36(1):189–203, 2019.

54 Mingdi Deng, Zhijun Li, Yu Kang, Chun Lung Philip Chen, and Xiaoli Chu. A learning-based hierarchical control scheme for an exoskeleton robot in human–robot cooperative manipulation. *IEEE Transactions on Cybernetics*, 50(1):112–125, 2020.

55 Gouri Dhatt, Emmanuel Lefran, and Gilbert Touzot. *Finite element method*. John Wiley & Sons, 2012.

56 Politecnico Di Milano and Milano. Dynamic modeling and control of flexible manipulators. *Italy*.

57 Pingan Ding, Hesheng Zhang, Xiaojin Zhu, and Li Li. Online curvature calibration for flexible structure deformation detection based on the finite element method. In *International Congress on Image and Signal Processing*, pages 1457–1461. IEEE, 2013.

58 Jiheng Ding, Luo Xin, Xuedong Chen, Bai Ou, and Bin Han. Design of active controller for low-frequency vibration isolation considering noise levels of bandwidth-extended absolute velocity sensors. *IEEE/ASME Transactions on Mechatronics*, 23(4):1832–1842, 2018.

59 Hanh T Dinh, Ramesh Kamalapurkar, Sandeep Bhasin, and William E Dixon. Dynamic neural network-based robust observers for uncertain nonlinear systems. *Neural Networks*, 60(C):44–52, 2014.

60 Viet-Phan Doan and Yoshiki Nishi. Modeling of flexible structure interaction for simulating vortex-induced vibration of flexible riser: Finite difference method combined with wake oscillator model. *Journal of Marine Science and Technology*, 20:309–321, 2015.

61 Zhi Hua Dong and Ju Yun Yuan. Vibration control device and its performance under wind load in high-rise buildings. *Applied Mechanics and Materials*, 166–169:1358–1361, 2012.

62 Mihai Dupac and Siamak Noroozi. Dynamic modeling and simulation of a rotating single link flexible robotic manipulator subject to quick stops. *Strojniski Vestnik*, 60(7-8):475–482, 2014.

63 Takahiro Endo, Minoru Sasaki, and Fumitoshi Matsuno. Contact-force control of a flexible Timoshenko arm. *IEEE Transactions on Automatic Control*, 62(2):1004–1009.

64 Takahiro Endo, Minoru Sasaki, Fumitoshi Matsuno, and Yingmin Jia. Contact-force control of a flexible Timoshenko arm in rigid/soft environment. *IEEE Transactions on Automatic Control*, 62(5):2546–2553.

65 Lingling Fan. Iterative learning and adaptive fault-tolerant control with application to high-speed trains under unknown speed delays and control input saturations. *IET Control Theory & Applications*, 8(9):675–687, 2014.

66 Wen Fan, Hugh H Liu, and Raymond Kwong. Gust load alleviation control for a flexible aircraft with loss of control effectiveness. In *AIAA Guidance, Navigation, and Control Conference*, pages 1721–1736, 2017.

67 Yunmei Fang, Juntao Fei, and Tongyue Hu. Adaptive backstepping fuzzy sliding mode vibration control of flexible structure. *Journal of Low Frequency Noise, Vibration and Active Control*, 37(4):1079–1096, 2018.

68 Mohammad Mehdi Fateh. Robust control of flexible-joint robots using voltage control strategy. *Nonlinear Dynamics*, 67(2):1525–1537, 2012.

69 Hongyinping Feng and Bao-Zhu Guo. Observer design and exponential stabilization for wave equation in energy space by boundary displacement measurement only. *IEEE Transactions on Automatic Control*, 62(3):1438–1444, 2017.

70 Naishi Feng, Qiurong Shi, Hong Wang, Jiale Gong, Chong Liu, and Zhiguo Lu. A soft robotic hand: Design, analysis, sEMG control, and experiment. *The International Journal of Advanced Manufacturing Technology*, 97:319–333, 2018.

71 Erhan Ferhatoglu, Ender Cigeroglu, and Hikmet Nevzat Özgüven. A new modal superposition method for nonlinear vibration analysis of structures using hybrid mode shapes. *Mechanical Systems and Signal Processing*, 107:317–342, 2018.

72 Amirhossein Eshaghiyeh Firoozabadi, Saeed Ebrahimi, and Josep M Font-Llagunes. A comparative study of elastic motions in trajectory tracking of flexible RPR planar manipulators moving with high speed. *Robotica*, 1:1–18, 2016.

73 Nicholas R Fisco and Hojjat Adeli. Smart structures: Part I-active and semi-active control. *Scientia Iranica*, 18(3):275–284, 2011.

74 Aleksandr L Fradkov, Iliya V Miroshnik, and Vladimir O Nikiforov. *Nonlinear and adaptive control of complex systems*, volume 491. Springer Science & Business Media, 2013.

75 Jun Fu, Ruicheng Ma, and Tianyou Chai. Adaptive finite-time stabilization of a class of uncertain nonlinear systems via logic-based switchings. *IEEE Transactions on Automatic Control*, 62(11):5998–6003, 2017.

76 Bo Fu, Huanjun Jiang, and Tao Wu. Comparative studies of vibration control effects between structures with particle dampers and tuned liquid dampers using substructure shake table testing methods. *Soil Dynamics and Earthquake Engineering*, 121:421–435, 2019.

77 Qiang Fu, Xiang Yang Chen, and Wei He. A survey on 3D visual tracking of multicopters. *International Journal of Automation and Computing*, 16(6):707–719, 2019.

78 Qiang Fu, Yuhang Yang, Xiangyang Chen, and Yalin Shang. Vision-based obstacle avoidance for flapping-wing aerial vehicles. *Science China Information Sciences*, 63(7):119–121, 2020.

79 Shiqi Gao and Jinkun Liu. Adaptive neural network vibration control of a flexible aircraft wing system with input signal quantization. *Aerospace Science and Technology*, 96:105593–105611, 2020.

80 Hejia Gao, Wei He, Yuhua Song, Zhang Shuang, and Changyin Sun. Modeling and neural network control of a flexible beam with unknown spatiotemporally varying disturbance using assumed mode method. *Neurocomputing*, 314:458–467, 2018.

81 Hejia Gao, Wei He, Chen Zhou, and Changyin Sun. Neural network control of a two-link flexible robotic manipulator using assumed mode method. *IEEE Transactions on Industrial Informatics*, 15(2):755–765, 2019.

82 Hejia Gao, Wei He, Liang Zhang, and Changyin Sun. Neural-network control of a stand-alone tall building-like structure with an eccentric load: An experimental investigation. *IEEE Transactions on Cybernetics*, 52(6):4083–4094, 2020.

83 Hejia Gao, Wei He, Youmin Zhang, and Changyin Sun. Vibration control based on reinforcement learning for a flexible building-like structure system with active mass damper against disturbance effects. In *2020 59th IEEE Conference on Decision and Control (CDC)*, pages 2380–2385. IEEE, 2020.

84 Hejia Gao, Wei He, Youmin Zhang, and Changyin Sun. Adaptive finite-time fault-tolerant control for uncertain flexible flapping wings based on rigid finite element method. *IEEE Transactions on Cybernetics*, 52(9):9036–9047, 2021.

85 Shuzhi Sam Ge, Tong Heng Lee, and Guodong Zhu. A nonlinear feedback controller for a single-link flexible manipulator based on a finite element model. *Journal of Robotic Systems*, 14(3):165–178, 1997.

86 Shuzhi Sam Ge, Shuang Zhang, and Wei He. Vibration control of an Euler–Bernoulli beam under unknown spatiotemporally varying disturbance. *International Journal of Control*, 84(5):947–960, 2011.

87 Andreas Geist and Sebastian Trimpe. Learning constrained dynamics with Gauss' principle adhering Gaussian processes. In *Learning for Dynamics and Control*, pages 225–234. PMLR, 2020.

88 Khaled Ghaedi, Zainah Ibrahim, Hojjat Adeli, and Ahad Javanmardi. Invited Review: Recent developments in vibration control of building and bridge structures. *Journal of Vibroengineering*, 19(5):3564–3580, 2017.

89 Herbert Goldstein. *Classical mechanics*. Massachusetts, USA: Addison-Wesley, 1951.

90 Martin Goubej, Jana Königsmarková, Ronald Kampinga, Jakko Nieuwenkamp, and Stéphane Paquay. Employing finite element analysis and robust control concepts in mechatronic system design-flexible manipulator case study. *Applied Sciences*, 11(8):3689, 2021.

91 Michael A Graule, Panyaphat Chirarattananon, Shelley B Fuller, Nicolas T Jafferis, Kai-Yu Ma, Michaela Spenko, Robert Kornbluh, and Robert J Wood. Perching and takeoff of a robotic insect on overhangs using switchable electrostatic adhesion. *Science*, 352(6288):978–982, 2016.

92 Bao-Zhu Guo and Wei Guo. The strong stabilization of a one-dimensional wave equation by non-collocated dynamic boundary feedback control. *Automatica*, 45(3):790–797, 2009.

93 Wei Guo and Bao-Zhu Guo. Parameter estimation and non-collocated adaptive stabilization for a wave equation subject to general boundary harmonic disturbance. *IEEE Transactions on Automatic Control*, 58(7):1631–1643, 2013.

94 Bao-Zhu Guo and Feng-Fei Jin. The active disturbance rejection and sliding mode control approach to the stabilization of the Euler–Bernoulli beam equation with boundary input disturbance. *Automatica*, 49(9):2911–2918, 2013.

95 Bao-Zhu Guo and Feng-Fei Jin. Output feedback stabilization for one-dimensional wave equation subject to boundary disturbance. *IEEE Transactions on Automatic Control*, 60(3):824–830, 2015.

96 Bao-Zhu Guo and Hua-Cheng Zhou. The active disturbance rejection control to stabilization for multi-dimensional wave equation with boundary control matched disturbance. *IEEE Transactions on Automatic Control*, 60(1):143–157, 2014.

97 Tamas Haidegger, Levente Kovács, Stefan Preitl, Radu-Emil Precup, Balázs Benyó, and Zoltán Benyó. Controller design solutions for long distance telesurgical applications. *International Journal of Artificial Intelligence*, 6(S11):48–71, 2011.

98 Martin B Hamper, Khaled E Zaazaa, and Ahmed A Shabana. Modeling railroad track structures using the finite segment method. *Acta Mechanica*, 223(8):1707–1721, 2012.

99 Godfrey Harold Hardy, John Edensor Littlewood, and George Pólya. *Inequalities*. Cambridge University Press, 1959.

100 Wei He and Shuzhi Sam Ge. Robust adaptive boundary control of a vibrating string under unknown time-varying disturbance. *IEEE Transactions on Control Systems Technology*, 20(1):48–58, 2012.

101 Wei He and Shuzhi Sam Ge. Vibration control of a nonuniform wind turbine tower via disturbance observer. *IEEE/ASME Transactions on Mechatronics*, 20(1):237–244, 2015.

102 Wei He, Shuzhi Sam Ge, Bernard Voon Ee How, Yoo Sang Choo, and Keum-Shik Hong. Robust adaptive boundary control of a flexible marine riser with vessel dynamics. *Automatica*, 47(4):722–732, 2011.

103 Wei He, Shuang Zhang, and Shuzhi Sam Ge. Robust adaptive control of a thruster assisted position mooring system. *Automatica*, 50(7):1843–1851, 2014.

104 Wei He, Shuzhi Sam Ge, Bernard Voon Ee How, and Yoo Sang Choo. *Dynamics and control of mechanical systems in offshore engineering*. Springer, 2014.

105 Wei He, Shuang Zhang, and Shuzhi Sam Ge. Adaptive boundary control of a nonlinear flexible string system. *IEEE Transactions on Control Systems Technology*, 22(3):1088–1093, 2014.

106 Wei He, Shuang Zhang, and Shuzhi Sam Ge. Adaptive control of a flexible crane system with the boundary output constraint. *IEEE Transactions on Industrial Electronics*, 61(8):4126–4133, 2014.

107 Wei He, Shuzhi Sam Ge, and D Huang. Modeling and vibration control for a nonlinear moving string with output constraint. *IEEE/ASME Transactions on Mechatronics*, 20(4):1886–1897, 2015.

108 Wei He, Shuzhi Sam Ge, Yanan Li, Effie Chew, and Yee Sien Ng. Neural network control of a rehabilitation robot by state and output feedback. *Journal of Intelligent and Robotic Systems*, 80(1):15–31, 2015.

109 Wei He, Amoateng Ofosu David, Zhao Yin, and Changyin Sun. Neural network control of a robotic manipulator with input deadzone and output constraint. *IEEE Transactions on Systems, Man, and Cybernetics: Systems*, 46(6):759–770, 2016.

110 Wei He, Yiting Dong, and Changyin Sun. Adaptive neural impedance control of a robotic manipulator with input saturation. *IEEE Transactions on Systems, Man, and Cybernetics: Systems*, 46(3):334–344, 2016.

111 Yuanbin He, Henry Shu-Hung Chung, Carl Ngai-Man Ho, and Weimin Wu. Use of boundary control with second-order switching surface to reduce the system order for deadbeat controller in grid-connected inverter. *IEEE Transactions on Power Electronics*, 31(3):2638–2653, 2016.

112 Wei He, Yuncheng Ouyang, and Jie Hong. Vibration control of a flexible robotic manipulator in the presence of input deadzone. *IEEE Transactions on Industrial Informatics*, 13(1):48–59, 2017.

113 Wei He, Weiliang Ge, Yunchuan Li, Yanjun Liu, and Changyin Sun. Model identification and control design for a humanoid robot. *IEEE Transactions on Systems, Man, and Cybernetics: Systems*, 47(1):45–57, 2017.

114 Wei He, Xiuyu He, and Changyin Sun. Vibration control of an industrial moving strip in the presence of input deadzone. *IEEE Transactions on Industrial Electronics*, 64(6):4680–4689, 2017.

115 Wei He, Haifeng Huang, Yunan Chen, Wenzhen Xie, Fusen Feng, Yemeng Kang, and Changyin Sun. Development of an autonomous flapping-wing aerial vehicle. *Science China Information Sciences*, 60:1–8, 2017.

116 Wei He, Zhijun Li, and Chun Lung Philip Chen. Guest editorial for special issue on human-centered intelligent robots: Issues and challenges. *IEEE/CAA Journal of Automatica Sinica*, 4(4):602–609, 2017.

117 Wei He, Zichen Yan, Changyin Sun, and Yunan Chen. Adaptive neural network control of a flapping wing micro aerial vehicle with disturbance observer. *IEEE Transactions on Cybernetics*, 47(10):3452–3465, 2017.

118 Wei He, Zhao Yin, and Changyin Sun. Adaptive neural network control of a marine vessel with constraints using the asymmetric barrier Lyapunov function. *IEEE Transactions on Cybernetics*, 47(7):1641–1651, 2017.

119 Xiuyu He, Wei He, Jing Shi, and Changyin Sun. Boundary vibration control of variable length crane systems in two-dimensional space with output constraints. *IEEE/ASME Transactions on Mechatronics*, 22(5):1952–1962, 2017.

120 Wei He, Tingting Meng, Xiuyu He, and Shuzhi Sam Ge. Unified iterative learning control for flexible structures with input constraints. *Automatica*, 86:326–336, 2018.

121 Wei He, Tingting Meng, Shuang Zhang, Quanbo Ge, and Changyin Sun. Trajectory tracking control for the flexible wings of a micro aerial vehicle. *IEEE Transactions on Systems, Man, and Cybernetics: Systems*, 48(12):2431–2441, 2018.

122 Xiuyu He, Wei He, Yingru Liu, Yiheng Wang, Guang Li, and Yu Wang. Robust adaptive control of an offshore ocean thermal energy conversion system. *IEEE Transactions on Systems, Man, and Cybernetics: Systems*, 50(12):5285–5295, 2018.

123 Wei He, Tingting Wang, Xiuyu He, Lung-Jieh Yang, and Okyay Kaynak. Dynamical modeling and boundary vibration control of a rigid-flexible wing system. *IEEE/ASME Transactions on Mechatronics*, 25(6):2711–2721, 2020.

124 Mitsuhito Hirose and Jerome P Lynch. Adaptive feedback control of time varying structures utilizing dual-core wireless sensing and actuation nodes. In *2015 American Control Conference (ACC)*, pages 1561–1566, 2015.

125 Leo H Holthuijsen. *Waves in oceanic and coastal waters*. Cambridge University Press, 2007.

126 Shengbin Hu, Jinyuan Xu, Xuan Wu, Chi Zhang, and Yihao He. Fast terminal fuzzy sliding mode control for attitude of flapping wing micro aerial vehicle. *Applied Mechanics and Materials*, 427–429:1179–1182, 2013.

127 Rui Hu, Shiping Wen, Zhigang Zeng, and Tingwen Huang. A short-term power load forecasting model based on the generalized regression neural network with decreasing step fruit fly optimization algorithm. *Neurocomputing*, 221:24–31, 2017.

128 Qinglei Hu, Boyan Jiang, and Youmin Zhang. Observer-based output feedback attitude stabilization for spacecraft with finite-time convergence. *IEEE Transactions on Control Systems Technology*, 27(2):781–789, 2019.

129 Mingyang Huang. Optimization of flapping wing mechanism of bionic eagle. *Proceedings of the Institution of Mechanical Engineers, Part G: Journal of Aerospace Engineering*, 233(9):3260–3272, 2019.

130 Deqing Huang, Jian-Xin Xu, Xuefang Li, Chao Xu, and Miao Yu. D-type anticipatory iterative learning control for a class of inhomogeneous heat equations. *Automatica*, 49(8):2397–2408, 2013.

131 Yiqing Huang, Changyin Sun, Chengshan Qian, and Li Wang. Non-fragile switching tracking control for a flexible air-breathing hypersonic vehicle based on polytopic LPV model. *Chinese Journal of Aeronautics*, 26(4):948–959, 2013.

132 Panfeng Huang, Dongke Wang, Zhongjie Meng, Fan Zhang, and Zhengxiong Liu. Impact dynamic modeling and adaptive target capturing control for tethered space robots with uncertainties. *IEEE/ASME Transactions on Mechatronics*, 21(5):2260–2271, 2016.

133 Panfeng Huang, Fan Zhang, Jia Cai, Dongke Wang, Zhongjie Meng, and Jian Guo. Dexterous tethered space robot: Design, measurement, control, and experiment. *IEEE Transactions on Aerospace and Electronic Systems*, 53(3):1452–1468, 2017.

134 Jagmohan Humar. *Dynamics of structures*. CRC Press, 2012.

135 Thang Pham Huu, Nanako Miura, and Daisuke Iba. Multi active tuned mass dampers for earthquake-induced vibration response control of high rise building. *Journal of Mechanical Science and Technology*, 36(4):1655–1666, 2022.

136 Mohammad Jafari and Alice Alipour. Methodologies to mitigate wind-induced vibration of tall buildings: A state-of-the-art review. *Journal of Building Engineering*, 33:101582, 2021.

137 Hamed Jamshidifar, Saeid Khosravani, Baris Fidan, and Amir Khajepour. Vibration decoupled modeling and robust control of redundant cable-driven parallel robots. *IEEE/ASME Transactions on Mechatronics*, 23(2):690–701, 2018.

138 Dong-Doo Jang, Jeongsu Park, and Hyung-Jo Jung. Active mass damper system employing time delay control algorithm for vibration mitigation of building structure. In *Sensors and Smart Structures Technologies for Civil, Mechanical, and Aerospace Systems 2013*, volume 8692, page 86923Z, 2013.

139 Wenqiang Ji, Jianbin Qiu, Ligang Wu, and Hak-Keung Lam. Fuzzy-affine model-based output feedback dynamic sliding mode controller design of nonlinear systems. *IEEE Transactions on Systems, Man, and Cybernetics: Systems*, 51(3):1652–1661, 2019.

140 Han Jiakun, Hui Zhe, Tian Fangbao, and Chen Gang. Review on bio-inspired flight systems and bionic aerodynamics. *Chinese Journal of Aeronautics*, 34(7):170–186, 2021.

141 Tingting Jiang, Jinkun Liu, and Wei He. Boundary control for a flexible manipulator based on infinite dimensional disturbance observer. *Journal of Sound and Vibration*, 348:1–14, 2015.

142 Wanyue Jiang, Dongyu Li, and Shuzhi Sam Ge. Micro flapping-wing vehicles formation control with attitude estimation. *IEEE Transactions on Cybernetics*, 52(2):1035–1047, 2020.

143 Fengfei Jin and Baozhu Guo. Lyapunov approach to output feedback stabilization for the Euler–Bernoulli beam equation with boundary input disturbance. *Automatica*, 52:95–102, 2015.

144 Maolin Jin, Jinoh Lee, and Nikos G Tsagarakis. Model-free robust adaptive control of humanoid robots with flexible joints. *IEEE Transactions on Industrial Electronics*, 64(2):1706–1715, 2017.

145 Zhe Jing, Yonghao Ma, Xiaoyang Wu, Xiuyu He, and Yongbin Sun. Backstepping control for vibration suppression of 2-D Euler–Bernoulli beam based on nonlinear saturation compensator. *IEEE Transactions on Systems, Man, and Cybernetics: Systems*, 53(5):2562–2571, 2022.

146 Karam Dad Kallu, Wang Jie, and Min Cheol Lee. Sensorless reaction force estimation of the end effector of a dual-arm robot manipulator using sliding mode control with a sliding perturbation observer. *International Journal of Control, Automation and Systems*, 16(3):1367–1378, 2018.

147 Iasson Karafyllis and Miroslav Krstic. Sampled-data boundary feedback control of 1-D parabolic PDEs. *Automatica*, 87:226–237, 2018.

148 Parham M Kebria, Abbas Khosravi, Saeid Nahavandi, Peng Shi, and Roohallah Alizadehsani. Robust adaptive control scheme for teleoperation systems with delay and uncertainties. *IEEE Transactions on Cybernetics*, 50(7):3243–3253, 2019.

149 Said G Khan, Guido Herrmann, Frank L Lewis, Tony Pipe, and Chris Melhuish. Reinforcement learning and optimal adaptive control: An overview and implementation examples. *Annual Reviews in Control*, 36(1):42–59, 2012.

150 Fouzia Khan, Abdulhamit Donder, Stefano Galvan, Ferdinando Rodriguez y Baena, and Sarthak Misra. Pose measurement of flexible medical instruments using fiber Bragg gratings in multi-core fiber. *IEEE Sensors Journal*, 20(18):10955–10962, 2020.

151 Oussama Khatib, Xiyang Yeh, Gerald Brantner, Brian Soe, Boyeon Kim, Shameek Ganguly, Hannah Stuart, Shiquan Wang, Mark Cutkosky, Aaron Edsinger, et al. Ocean one: A robotic avatar for oceanic discovery. *IEEE Robotics and Automation Magazine*, 23(4):20–29, 2016.

152 Joonyoung Kim and Elizabeth A Croft. Full-state tracking control for flexible joint robots with singular perturbation techniques. *IEEE Transactions on Control Systems Technology*, 27(1):63–73, 2017.

153 Young-moon Kim, Ki-pyo You, and Jang-youl You. Passive control of along-wind response of tall building. *Journal of Central South University*, 21(10):4002–4006, 2014.

154 Mojtaba Kordestani, Karim Salahshoor, Ali Akbar Safavi, and Mehrdad Saif. An adaptive passive fault tolerant control system for a steam turbine using a PCA based inverse neural network control strategy. In *2018 World Automation Congress (WAC)*, pages 1–6. IEEE, 2018.

155 Vadim A Krysko Jr, Jacek Awrejcewicz, and Irina V Papkova. Chaotic vibrations of size-dependent flexible rectangular plates. *Chaos: An Interdisciplinary Journal of Nonlinear Science*, 31(4):043119, 2021.

156 Manoj Kumar, Karthikeyan Rajagopal, Sivasubramanya Nadar Balakrishnan, and Nhan T Nguyen. Reinforcement learning based controller synthesis for flexible aircraft wings. *IEEE/CAA Journal of Automatica Sinica*, 1(4):435–448, 2014.

157 Shailaja Kurode and Prashant Dixit. Control of tip position of flexible link manipulator using sliding modes. *System*, 3:5, 2012.

158 Pardo Miguel Lamas and Rodrigo Perez Fernandez. Floating offshore wind farms: The seed of artificial. *Journal of Marine Technology and Environment*, 2:95–106, 2011.

159 Fabian Andres Lara-Molina, Karina Assolari Takano, and Edson Hideki Koroishi. Set-point regulation of a robot manipulator with flexible joints using fuzzy control. In *2015 12th Latin American Robotics Symposium and 2015 3rd Brazilian Symposium on Robotics (LARS-SBR)*, pages 103–108. IEEE, 2015.

160 Fabian Andres Lara-Molina, Edson Hideki Koroishi, and Didier Dumur. Design criteria of 2-DOF planar parallel manipulator with flexible joints. In *2016 12th IEEE International Conference on Industry Applications (INDUSCON)*, pages 1–6. IEEE, 2016.

161 Younbaek Lee, Yong Jae Kim, Jongwon Lee, Minhyung Lee, Byungjune Choi, Jeonghun Kim, Young Jin Park, and Jungyun Choi. Biomechanical design of a novel flexible exoskeleton for lower extremities. *IEEE/ASME Transactions on Mechatronics*, 22(5):2058–2069, 2017.

162 Frank L Lewis, Draguna Vrabie, and Kyriakos G Vamvoudakis. Reinforcement learning and feedback control: Using natural decision methods to design optimal adaptive controllers. *IEEE Control Systems Magazine*, 32(6):76–105, 2012.

163 Xianwei Li and Huijun Gao. Load mitigation for a floating wind turbine via generalized H_∞ structural control. *IEEE Transactions on Industrial Electronics*, 63(1):332–342, 2016.

164 Jian Li and Yungang Liu. Adaptive stabilization for ODE systems via boundary measurement of uncertain diffusion-dominated actuator dynamics. *International Journal of Robust and Nonlinear Control*, 24(18):3214–3238, 2014.

165 Haijie Li and Xuping Zhang. A method for modeling flexible manipulators: Transfer matrix method with finite segments. *International Journal of Computer, Electrical, Automation, Control and Information Engineering*, 10(6):1086–1093, 2016.

166 Hua Jun Li, Sau Lon James Hu, and Christopher Jakubiak. H_2 active vibration control for offshore platform subjected to wave loading. *Journal of Sound and Vibration*, 263(4):709–724, 2003.

167 Zhijun Li, Jianxun Li, and Yu Kang. Adaptive robust coordinated control of multiple mobile manipulators interacting with rigid environments. *Automatica*, 46(12):2028–2034, 2010.

168 Luyu Li, Gangbing Song, and Jinping Ou. Hybrid active mass damper (AMD) vibration suppression of nonlinear high-rise structure using fuzzy logic control algorithm under earthquake excitations. *Structural Control & Health Monitoring*, 18(6):698–709, 2011.

169 Zhijun Li, Chun-Yi Su, Liangyong Wang, Ziting Chen, and Tianyou Chai. Nonlinear disturbance observer-based control design for a robotic exoskeleton incorporating fuzzy approximation. *IEEE Transactions on Industrial Electronics*, 62(9):5763–5775, 2015.

170 Zhijun Li, Ziting Chen, Jun Fu, and Changyin Sun. Direct adaptive controller for uncertain MIMO dynamic systems with time-varying delay and dead-zone inputs. *Automatica*, 63:287–291, 2016.

171 Zhijun Li, Quanbo Ge, Wenjun Ye, and Peijian Yuan. Constrained multilegged robot system modeling and fuzzy control with uncertain kinematics and dynamics incorporating foot force optimization. *IEEE Transactions on Systems, Man, and Cybernetics: Systems*, 46(10):1338–1351, 2016.

172 Da-Quan Li, Hua-Jie Hong, and Xian-Liang Jiang. Dynamics modeling, control system design and simulation of manipulator based on Lagrange equation. In *Mechanism and Machine Science*. page 1129–1141. Springer Singapore, 2017.

173 Hongyi Li, Lijie Wang, Du Haiping, and Abdesselem Boulkroune. Adaptive fuzzy backstepping tracking control for strict-feedback systems with input delay. *IEEE Transactions on Fuzzy Systems*, 25(3):642–652, 2017.

174 Zhijun Li, Huang Bo, Arash Ajoudani, Chenguang Yang, Chun Yi Su, and Antonio Bicchi. Asymmetric bimanual control of dual-arm exoskeletons for human-cooperative manipulations. *IEEE Transactions on Robotics*, 34(1):264–271, 2017.

175 Zhijun Li, Zhicong Huang, Wei He, and Chun-Yi Su. Adaptive impedance control for an upper limb robotic exoskeleton using biological signals. *IEEE Transactions on Industrial Electronics*, 64(2):1664–1674, 2017.

176 Jian Li, Xin Li, and Qingyu Su. Fault detection filter design for discrete-time switched time-delay systems with static quantizer and quantized feedback. *International Journal of Innovative Computing, Information and Control*, 14(4):1465–1479, 2018.

177 Zhijun Li, Huang Bo, Zhifeng Ye, Mingdi Deng, and Chenguang Yang. Physical human–robot interaction of a robotic exoskeleton by admittance control. *IEEE Transactions on Industrial Electronics*, 65(12):9614–9624, 2018.

178 Zhijun Li, Bo Huang, Arash Ajoudani, Chenguang Yang, Chun-Yi Su, and Antonio Bicchi. Asymmetric bimanual control of dual-arm exoskeletons for human-cooperative manipulations. *IEEE Transactions on Robotics*, 34(1):264–271, 2018.

179 Hongyi Li, Shiyi Zhao, Wei He, and Renquan Lu. Adaptive finite-time tracking control of full state constrained nonlinear systems with dead-zone. *Automatica*, 100:99–107, 2019.

180 Liang Li, Ziyu Chen, Yaobing Wang, Xiaodong Zhang, and Ningfei Wang. Robust task-space tracking for free-floating space manipulators by cerebellar model articulation controller. *Assembly Automation*, 39(1):26–33, 2019.

181 Haihong Li, Lingxiao Xun, Gang Zheng, and Federico Renda. Discrete Cosserat static model-based control of soft manipulator. *IEEE Robotics and Automation Letters*, 8(3):1739–1746, 2023.

182 Yang Yin Lin, Jyh Yeong Chang, and Chin Teng Lin. Identification and prediction of dynamic systems using an interactively recurrent self-evolving fuzzy neural network. *IEEE Transactions on Neural Networks and Learning Systems*, 24(2):310–321, 2013.

183 Haowei Lin, Bo Zhao, Derong Liu, and Cesare Alippi. Data-based fault tolerant control for affine nonlinear systems through particle swarm optimized neural networks. *IEEE/CAA Journal of Automatica Sinica*, 7(4):954–964, 2020.

184 Yan-Jun Liu and Shaocheng Tong. Barrier Lyapunov functions for Nussbaum gain adaptive control of full state constrained nonlinear systems. *Automatica*, 76:143–152, 2017.

185 Derong Liu, Ding Wang, and Hongliang Li. Decentralized stabilization for a class of continuous-time nonlinear interconnected systems using online learning optimal control approach. *IEEE Transactions on Neural Networks and Learning Systems*, 25(2):418–428, 2014.

186 Derong Liu, Ding Wang, Fei-Yue Wang, Hongliang Li, and Xiong Yang. Neural-network-based online HJB solution for optimal robust guaranteed cost control of continuous-time uncertain nonlinear systems. *IEEE Transactions on Cybernetics*, 44(12):2834–2847, 2014.

187 Dongyi Liu, Liping Zhang, and Genqi Xu. Stabilisation of Timoshenko beam system with a tip payload under the unknown boundary external disturbances. *International Journal of Control*, 88(9):1830–1840, 2015.

188 Yan-Jun Liu, Jing Li, Shaocheng Tong, and Chun Lung Philip Chen. Neural network control-based adaptive learning design for nonlinear systems with full-state constraints. *IEEE Transactions on Neural Networks and Learning Systems*, 27(7):1562–1571, 2016.

189 Yanjun Liu, Shaocheng Tong, Dongjuan Li, and Ying Gao. Fuzzy adaptive control with state observer for a class of nonlinear discrete-time systems with input constraint. *IEEE Transactions on Fuzzy Systems*, 24(5):1147–1158, 2016.

190 Zhijie Liu, Jin-Kun Liu, and Wei He. Adaptive boundary control of a flexible manipulator with input saturation. *International Journal of Control*, 89(6):1191–1202, 2016.

191 Peng Liu, Zhigang Zeng, and Jun Wang. Multiple Mittag–Leffler stability of fractional-order recurrent neural networks. *IEEE Transactions on Systems, Man, and Cybernetics: Systems*, 47(8):2279–2288, 2017.

192 Zhijie Liu, Jinkun Liu, and Wei He. *Dynamic modeling and vibration control for a nonlinear 3-dimensional flexible manipulator*, volume 28. Wiley Online Library, 2018.

193 Yu Liu, Fang Guo, Xiuyu He, and Qing Hui. Boundary control for an axially moving system with input restriction based on disturbance observers. *IEEE Transactions on Systems, Man, and Cybernetics: Systems*, 49(11):2242–2253, 2019.

194 Zhijie Liu, Xiuyu He, Zhijia Zhao, Choon Ki Ahn, and Han-Xiong Li. Vibration control for spatial aerial refueling hoses with bounded actuators. *IEEE Transactions on Industrial Electronics*, 2020. doi: 10.1109/TIE.2020.2984442.

195 Yu Liu, Xiongbin Chen, Yilin Wu, He Cai, and Hiroshi Yokoi. Adaptive neural network control of a flexible spacecraft subject to input nonlinearity and asymmetric output constraint. *IEEE Transactions on Neural Networks and Learning Systems*, 33(11):6226–6234, 2021.

196 Yu Liu, Yanfang Mei, Changran He, Tao Liu, and Guoqiang Hu. Asymmetric input–output constraint control of a flexible variable-length rotary crane arm. *IEEE Transactions on Cybernetics*, 52(10):10582–10591, 2021.

197 Yu Liu, Xiongbin Chen, Yanfang Mei, and Yilin Wu. Observer-based boundary control for an asymmetric output-constrained flexible robotic manipulator. *Science China Information Sciences*, 65(3):139203:1–139203:3, 2022.

198 Chin Hsiung Loh, Pay Yang Lin, and Nan Hau Chung. Experimental verification of building control using active bracing system. *Earthquake Engineering and Structural Dynamics*, 28(10):1099–1119, 2010.

199 Xilin Lu, Lu Zheng, Dianchao Wang, and Sami F Masri. An experimental study of vibration control of wind-excited high-rise buildings using particle tuned mass dampers. *Smart Structures and Systems*, 18(1):93–115, 2016.

200 Biao Luo, Tingwen Huang, Huai Ning Wu, and Xiong Yang. Data-driven H_∞ control for nonlinear distributed parameter systems. *IEEE Transactions on Neural Networks and Learning Systems*, 26(11):2949–2961, 2015.

201 Guoqin Lyu, Huiqin Zhang, and Jiachun Li. Effects of incident wind/wave directions on dynamic response of a spar-type floating offshore wind turbine system. *Acta Mechanica Sinica*, 35(5):954–963, 2019.

202 Kevin Y Ma, Pakpong Chirarattananon, Sawyer B Fuller, and Robert J Wood. Controlled flight of a biologically inspired, insect-scale robot. *Science*, 340(6132):603–607, 2013.

203 Chris JB Macnab, Gabriele MT D'Eleuterio, and Max Meng. CMAC adaptive control of flexible-joint robots using backstepping with tuning functions. In *IEEE International Conference on Robotics and Automation*, volume 3, pages 2679–2686, 2004.

204 Leonard Meirovitch and Hans Baruh. On the problem of observation spillover in self-adjoint distributed parameter systems. *Journal of Optimization Theory and Applications*, 39(2):269–291, 1983.

205 Tingting Meng, Wei He, and Jin-Kun Liu. Boundary control for a vibrating string with asymmetrically constrained outputs. *International Journal of Robust and Nonlinear Control*, 28(3):798–807, 2018.

206 Jun Kim Min and Kyun Chung Wan. Disturbance-observer-based PD control of flexible joint robots for asymptotic convergence. *IEEE Transactions on Robotics*, 31(6):1508–1516, 2015.

207 Wang Ming, Zhang Yanlu, Dong Huifang, and Yu Junzhi. Trajectory tracking control of a bionic robotic fish based on iterative learning. *Science China Information Sciences*, 63:1–9, 2020.

208 Ranjan Kumar Mitra, Atul Krishna Banik, and Shyamal Chatterjee. State feedback control of surge oscillations of two-point mooring system. *Journal of Sound and Vibration*, 386:1–20, 2017.

209 Rajesh Kumar Moolam. *Dynamic modeling and control of flexible manipulators*. PhD thesis, Italy, [D]. POLITECNICO DI MILANO, 2014.

210 Chaoxu Mu, Zhen Ni, Changyin Sun, and Haibo He. Air-breathing hypersonic vehicle tracking control based on adaptive dynamic programming. *IEEE Transactions on Neural Networks and Learning Systems*, 47(6):1460–1470, 2017.

211 Imad H Mualla and Borislav Belev. Performance of steel frames with a new friction damper device under earthquake excitation. *Engineering Structures*, 24(3):365–371, 2002.

212 Walter Musial, Sandy Butterfield, and Andrew Boone. Feasibility of floating platform systems for wind turbines. In *42nd AIAA aerospace sciences meeting and exhibit*, page 1007, 2004.

213 Ichiro Nagashima and Yuzo Shinozaki. Variable gain feedback control technique of active mass damper and its application to hybrid structural control. *Earthquake Engineering and Structural Dynamics*, 26(8):815–838, 2015.

214 Noel Naughton, Jiarui Sun, Arman Tekinalp, Tejaswin Parthasarathy, Girish Chowdhary, and Mattia Gazzola. Elastica: A compliant mechanics environment for soft robotic control. *IEEE Robotics and Automation Letters*, 6(2):3389–3396, 2021.

215 Quoc Chi Nguyen and Keum-Shik Hong. Asymptotic stabilization of a nonlinear axially moving string by adaptive boundary control. *Journal of Sound and Vibration*, 329(22):4588–4603, 2010.

216 Quoc Chi Nguyen and Keum-Shik Hong. Simultaneous control of longitudinal and transverse vibrations of an axially moving string with velocity tracking. *Journal of Sound and Vibration*, 331(13):3006–3019, 2012.

217 Germán G Oggier, Martin Ordonez, Juan M Galvez, and Federico Luchino. Fast transient boundary control and steady-state operation of the dual active bridge converter using the natural switching surface. *IEEE Transactions on Power Electronics*, 29(2):946–957, 2014.

218 Giuseppe Oliveto, Anastasia Athanasiou, and Nicholas D Oliveto. Analytical earthquake response of 1D hybrid base isolation systems. *Soil Dynamics and Earthquake Engineering*, 43:1–15, 2012.

219 Gina Olson, Ross L Hatton, Julie A Adams, and Yiğit Mengüç. An Euler–Bernoulli beam model for soft robot arms bent through self-stress and external loads. *International Journal of Solids and Structures*, 207:113–131, 2020.

220 Daniel Ossmann and Manuel Pusch. Fault tolerant control of an experimental flexible wing. *Aerospace*, 6(7):76–94, 2019.

221 Yuncheng Ouyang, Wei He, and Xiajing Li. Reinforcement learning control of a single-link flexible robotic manipulator. *IET Control Theory and Applications*, 11(9):1426–1433, 2017.

222 Erzhen Pan, Lianrui Chen, Bing Zhang, and Wenfu Xu. A kind of large-sized flapping wing robotic bird: Design and experiments. In *Intelligent Robotics and Applications: 10th International Conference, ICIRA 2017, Wuhan, China, August 16–18, 2017, Proceedings, Part III 10*, pages 538–550. Springer, 2017.

223 Aditya A Paranjape, Soon-Jo Chung, and Joseph Kim. Novel dihedral-based control of flapping-wing aircraft with application to perching. *IEEE Transactions on Robotics*, 29(5):1071–1084, 2013.

224 Aditya A Paranjape, Jinyu Guan, Soon-Jo Chung, and Miroslav Krstic. PDE boundary control for flexible articulated wings on a robotic aircraft. *IEEE Transactions on Robotics*, 29(3):625–640, 2013.

225 Liam Paull, Carl Thibault, Amr Nagaty, Mae Seto, and Howard Li. Sensor-driven area coverage for an autonomous fixed-wing unmanned aerial vehicle. *IEEE Transactions on Cybernetics*, 44(9):1605–1618, 2014.

226 Vincenzo Pesce, Stefano Silvestrini, and Michèle Lavagna. Radial basis function neural network aided adaptive extended Kalman filter for spacecraft relative navigation. *Aerospace Science and Technology*, 96:105527, 2020.

227 Francesc Pozo, Pere Marc Montserrat, José Rodellar, and Leonardo Acho. Robust active control of hysteretic base-isolated structures: Application to the benchmark smart base-isolated building. *Structural Control and Health Monitoring*, 15(5):720–736, 2008.

228 Santanu Kumar Pradhan and Bidyadhar Subudhi. Real-time adaptive control of a flexible manipulator using reinforcement learning. *IEEE Transactions on Automation Science and Engineering*, 9(2):237–249, 2012.

229 Radu Emil Precup and Marius L Tomescu. Stable fuzzy logic control of a general class of chaotic systems. *Neural Computing and Applications*, 26(3):541–550, 2015.

230 Chen Qian and Yongchun Fang. Adaptive tracking control of flapping wing micro-air vehicles with averaging theory. *CAAI Transactions on Intelligence Technology*, 3(1):18–27, 2018.

231 Hong Qiao, Min Wang, Jianhua Su, Shengxin Jia, and Rui Li. The concept of "attractive region in environment" and its application in high-precision tasks with low-precision systems. *IEEE/ASME Transactions on Mechatronics*, 20(5):2311–2327, 2015.

232 Zhicheng Qiu and Zhili Zhao. Pneumatic drive active vibration control for flexible manipulator using an adaptive interactive PD controller. In *IEEE International Conference on Robotics and Biomimetics*, pages 1956–1961. IEEE, 2012.

233 Shiyuan Qiu, Zhijun Li, Wei He, Longbin Zhang, Chenguang Yang, and Chun-Yi Su. Brain–machine interface and visual compressive sensing-based teleoperation control of an exoskeleton robot. *IEEE Transactions on Fuzzy Systems*, 25(1):58–69, 2017.

234 Quanser. Smart structure user manual. Available: https://www.quanser.com/products/.

235 Marcio S Queiroz, Darren M Dawson, Siddharth P Nagarkatti, and Fumin Zhang. *Lyapunov based control of mechanical systems*. Boston, USA: Birkhauser, 2000.

236 Fatemeh Rahimi, Reza Aghayari, and Bijan Samali. Application of tuned mass dampers for structural vibration control: A state-of-the-art review. *Civil Engineering Journal*, 6(8):1622–1651, 2020.

237 Alireza Ramezani, Soon-Jo Chung, and Seth Hutchinson. A biomimetic robotic platform to study flight specializations of bats. *Science Robotics*, 2(3):eaal2505, 2017.

238 Mule Pala Prasad Reddy, Shihab Kv, and Jeevamma Jacob. Precise non linear modeling of flexible link flexible joint manipulator. *International Review on Modelling and Simulations*, 5(3):1368–1374, 2012.

239 Beibei Ren, Shuzhi Sam Ge, Keng Peng Tee, and Tong Heng Lee. Adaptive neural control for output feedback nonlinear systems using a barrier Lyapunov function. *IEEE Transactions on Neural Networks*, 21(8):1339–1345, 2010.

240 Beibei Ren, Jun-Min Wang, and Miroslav Krstic. Stabilization of an ODE–Schrödinger cascade. *Systems & Control Letters*, 62(6):503–510, 2013.

241 Yong Ren, Mou Chen, and Jianye Liu. Bilateral coordinate boundary adaptive control for a helicopter lifting system with backlash-like hysteresis. *Science China Information Sciences*, 63(119203):1–3, 2020.

242 Bao Rong, Xiaoting Rui, Ling Tao, and Guoping Wang. Theoretical modeling and numerical solution methods for flexible multibody system dynamics. *Nonlinear Dynamics*, 98:1519–1553, 2019.

243 Carlo Ruzzo, Nilanjan Saha, and Felice Arena. Wave spectral analysis for design of a spar floating wind turbine in Mediterranean Sea. *Ocean Engineering*, 184:255–272, 2019.

244 Nirmal Sadasivan. Design and realization of an unmanned aerial rotorcraft vehicle using pressurized inflatable structure. *International Journal of Aviation, Aeronautics, and Aerospace*, 6(4):3, 2019.

245 Avimanyu Sahoo, Hao Xu, and Sarangapani Jagannathan. Neural network-based event-triggered state feedback control of nonlinear continuous-time systems. *IEEE Transactions on Neural Networks and Learning Systems*, 27(3):497–509, 2015.

246 Yoshiyuki Sakawa, Fumitoshi Matsuno, and Shigenobu Fukushima. Modeling and feedback control of a flexible arm. *Journal of Robotic Systems*, 2(4):453–472, 2007.

247 Cosimo Della Santina and Daniela Rus. Control oriented modeling of soft robots: The polynomial curvature case. *IEEE Robotics and Automation Letters*, 5(2):290–298, 2019.

248 Ke Shao, Zhigang Wu, Chao Yang, and Lei Chen. Theoretical and experimental study of gust response alleviation using neuro-fuzzy control law for a flexible wing model. *Chinese Journal of Aeronautics*, 23(3):290–297, 2010.

249 Shuyi Shao, Mou Chen, and Youmin Zhang. Adaptive discrete-time flight control using disturbance observer and neural networks. *IEEE Transactions on Neural Networks and Learning Systems*, 30(12):3708–3721, 2019.

250 Weiping Shao, Hui Zhang, and Pengfei Zeng. Design and aerodynamic characteristic analysis of unfolding-bending bionic flapping-wing aircraft. *Journal of Mechanical Science and Technology*, 36(6):2981–2991, 2022.

251 Tong Shaocheng, Tong Wang, Yongming Li, and Huaguang Zhang. Adaptive neural network output feedback control for stochastic nonlinear systems with unknown dead-zone and unmodeled dynamics. *IEEE Transactions on Cybernetics*, 44(6):910–921, 2017.

252 Qing Shi, Chang Li, Kang Li, Qiang Huang, Hiroyuki Ishii, Atsuo Takanishi, and Toshio Fukuda. A modified robotic rat to study rat-like pitch and yaw movements. *IEEE/ASME Transactions on Mechatronics*, 23(5):2448–2458, 2018.

253 Kei-Man Shum, Simon Leung, Youlin Xu, and Michael Hui. Vibration control of a long span cable-stayed bridge tower using pendulum mass damper. *HKIE Transactions*, 13(4):44–56, 2013.

254 Yulin Si, Hamid Reza Karimi, and Huijun Gao. Modeling and parameter analysis of the OC3-hywind floating wind turbine with a tuned mass damper in nacelle. *Journal of Applied Mathematics*, 2013(4):1–10, 2013.

255 Jakub Sikorski, Sumit Mohanty, and Sarthak Misra. MILiMAC: Flexible catheter with miniaturized electromagnets as a small-footprint system for microrobotic tasks. *IEEE Robotics and Automation Letters*, 5(4):5260–5267, 2020.

256 Andrey Smyshlyaev and Miroslav Krstic. *Adaptive control of parabolic PDEs*. New Jersey, USA: Princeton University Press, 2010.

257 Ruizhuo Song, Wendong Xiao, Huaguang Zhang, and Changyin Sun. Adaptive dynamic programming for a class of complex-valued nonlinear systems. *IEEE Transactions on Neural Networks and Learning Systems*, 25(9):1733–1739, 2014.

258 Yuhua Song, Xiuyu He, Zhijie Liu, Wei He, Changyin Sun, and Fei-Yue Wang. Parallel control of distributed parameter systems. *IEEE Transactions on Cybernetics*, 48(12):3291–3301, 2018.

259 Mohammad H Stanikzai, Said Elias, and Yunbyeong Chae. Recent advances in hybrid vibration-control systems. *Practice Periodical on Structural Design and Construction*, 27(3):03122003, 2022.

260 Gordon Stewart and Matthew Lackner. Offshore wind turbine load reduction employing optimal passive tuned mass damping systems. *IEEE Transactions on Control Systems Technology*, 21(4):1090–1104, 2013.

261 Shanwei Su. Output-feedback dynamic surface control for a class of nonlinear non-minimum phase systems. *IEEE/CAA Journal of Automatica Sinica*, 3(1):96–104, 2016. ISSN: 2329-9266.

262 Changyin Sun and Youshen Xia. An analysis of a neural dynamical approach to solving optimization problems. *IEEE Transactions on Automatic Control*, 54(8):1972–1977, 2009.

263 Changyin Sun, Wei He, and Jie Hong. Neural network control of a flexible robotic manipulator using the lumped spring-mass model. *IEEE Transactions on Systems, Man, and Cybernetics: Systems*, 47(8):1863–1874, 2017.

264 Guanghui Sun, Shidong Xu, and Li Zhan. Finite-time fuzzy sampled-data control for nonlinear flexible spacecraft with stochastic actuator failures. *IEEE Transactions on Industrial Electronics*, 64(5):3851–3861, 2017.

265 Changyin Sun, Hejia Gao, Wei He, and Yao Yu. Fuzzy neural network control of a flexible robotic manipulator using assumed mode method. *IEEE Transactions on Neural Networks and Learning Systems*, 29(11):5214–5227, 2018.

266 Nagender Kumar Suryadevara, Subhas Chandra Mukhopadhyay, Sean Dieter Tebje Kelly, and Satinder Pal Singh Gill. WSN-based smart sensors and actuator for power management in intelligent buildings. *IEEE/ASME Transactions on Mechatronics*, 20(2):564–571, 2014.

267 Mahmoud Tahmasebi and Seyed Majid Esmailzadeh. Modeling and co-simulating of a large flexible satellites with three reaction wheels in ADAMS and MATLAB. *International Journal of Dynamics and Control*, 6(1):79–88, 2017.

268 Yufei Tang, Haibo He, Jinyu Wen, and Ju Liu. Power system stability control for a wind farm based on adaptive dynamic programming. *IEEE Transactions on Smart Grid*, 6(1):166–177, 2017.

269 Ali Reza Tavakolpour, Intan Z Mat Darus, and Musa Mailah. Modeling and simulation of an active vibration control system for a flexible structure using finite difference method. In *2009 3rd Asia International Conference on Modelling and Simulation*, pages 448–453, 2009.

270 Keng Peng Tee and Shuzhi Sam Ge. Control of fully actuated ocean surface vessels using a class of feedforward approximators. *IEEE Transactions on Control Systems Technology*, 14(4):750–756, 2006.

271 Keng Peng Tee, Shuzhi Sam Ge, and Eng Hock Tay. Barrier Lyapunov functions for the control of output-constrained nonlinear systems. *Automatica*, 45(4):918–927, 2009.

272 Keng Peng Tee, Beibei Ren, and Shuzhi Sam Ge. Control of nonlinear systems with time-varying output constraints. *Automatica*, 47(11):2511–2516, 2011.

273 Suresh Thenozhi and Yu Wen. Stability analysis of active vibration control of building structures using PD/PID control. *Engineering Structures*, 81:208–218, 2014.

274 Thomas George Thuruthel, Egidio Falotico, Mariangela Manti, Andrea Pratesi, Matteo Cianchetti, and Cecilia Laschi. Learning closed loop kinematic controllers for continuum manipulators in unstructured environments. *Soft Robotics*, 4(3):285–296, 2017.

275 Mustafa Tinkir, Mete Kalyoncu, and Yusuf ŞLahin. Modelling and controller design for a flexible structure system against disturbance effects. *Journal of Low Frequency Noise, Vibration and Active Control*, 34(4):525–547, 2015.

276 Shao Cheng Tong, Yong Ming Li, Gang Feng, and Tie Shan Li. Observer-based adaptive fuzzy backstepping dynamic surface control for a class of MIMO nonlinear systems. *IEEE Transactions on Systems Man and Cybernetics Part B*, 41(4):1124–1135, 2011.

277 Shaocheng Tong, Tong Wang, and Yongming Li. Fuzzy adaptive actuator failure compensation control of uncertain stochastic nonlinear systems with unmodeled dynamics. *IEEE Transactions on Fuzzy Systems*, 22(3):563–574, 2014.

278 Xin Tong, Xiaowei Zhao, and Shi Zhao. Load reduction of a monopile wind turbine tower using optimal tuned mass dampers. *International Journal of Control*, 90(7):1283–1298, 2017.

279 Natsuki Tsushima and Weihua Su. Flutter suppression for highly flexible wings using passive and active piezoelectric effects. *Aerospace Science and Technology*, 65:78–89, 2017.

280 Mohammad Vakil, Reza Fotouhi, and Peter N Nikiforuk. A new method for dynamic modeling of flexible-link flexible-joint manipulators. *Journal of Vibration and Acoustics*, 134(1), 2012.

281 Chad Van der Woude. *A study on vibration isolation in a wind turbine subjected to wind and seismic loading*. Master thesis, University of Waterloo, 2011.

282 Rong-Jong Wai and Rajkumar Muthusamy. Fuzzy-neural-network inherited sliding-mode control for robot manipulator including actuator dynamics. *IEEE Transactions on Neural Networks and Learning Systems*, 24(2):274–287, 2013.

283 Nengmou Wang and Hojjat Adeli. Robust vibration control of wind-excited highrise building structures. *Statyba*, 21(8):967–976, 2015.

284 Fei Yue Wang and Yanqing Gao. On frequency sensitivity and mode orthogonality of flexible robotic manipulators. *IEEE/CAA Journal of Automatica Sinica*, 3(4):394–397, 2016.

285 Mingliang Wang and Han-Xiong Li. Real-time estimation of temperature distribution for cylindrical lithium-ion batteries under boundary cooling. *IEEE Transactions on Industrial Electronics*, 64(3):2316–2324, 2017.

286 Yafei Wang and Changguo Wang. Mechanics of strain-limiting wrinkled kirigami for flexible devices: High flexibility, stretchability and compressibility. *International Journal of Solids and Structures*, 238:111382, 2022.

287 Zhongfei Wang, Shiming Ji, Yanbiao Li, and Yuehua Wan. A unified algorithm to determine the reachable and dexterous workspace of parallel manipulators. *Robotics and Computer-Integrated Manufacturing*, 26(5):454–460, 2010.

288 Jun-Min Wang, Beibei Ren, and Miroslav Krstic. Stabilization and Gevrey regularity of a Schrödinger equation in boundary feedback with a heat equation. *IEEE Transactions on Automatic Control*, 57(1):179–185, 2012.

289 Na Wang, Huai-Ning Wu, and Lei Guo. Coupling-observer-based nonlinear control for flexible air-breathing hypersonic vehicles. *Nonlinear Dynamics*, 78(3):2141–2159, 2014.

290 Jun Min Wang, Jun Jun Liu, Beibei Ren, and Jinhao Chen. Sliding mode control to stabilization of cascaded heat PDE–ODE systems subject to boundary control matched disturbance. *Automatica*, 52:23–34, 2015.

291 Ding Wang, Derong Liu, Hongliang Li, Biao Luo, and Hongwen Ma. An approximate optimal control approach for robust stabilization of a class of discrete-time nonlinear systems with uncertainties. *IEEE Transactions on Systems, Man, and Cybernetics: Systems*, 46(5):713–717, 2016.

292 Huanqing Wang, Wanjing Sun, and Peter Xiaoping Liu. Adaptive intelligent control of nonaffine nonlinear time-delay systems with dynamic uncertainties. *IEEE Transactions on Systems, Man, and Cybernetics: Systems*, 47(7):1474–1485, 2016.

293 Ding Wang, Haibo He, and Derong Liu. Adaptive critic nonlinear robust control: A survey. *IEEE Transactions on Cybernetics*, 47(10):3429–3451, 2017.

294 Fei-Yue Wang, Jie Zhang, Qinglai Wei, Xinhu Zheng, and Li Li. PDP: Parallel dynamic programming. *IEEE/CAA Journal of Automatica Sinica*, 4(1):1–5, 2017.

295 Fei-Yue Wang, Nan-Ning Zheng, Dongpu Cao, Clara Marina Martinez, Li Li, and Teng Liu. Parallel driving in CPSS: A unified approach for transport automation and vehicle intelligence. *IEEE/CAA Journal of Automatica Sinica*, 4(4):577–587, 2017.

296 Lijie Wang, Michael V Basin, Hongyi Li, and Renquan Lu. Observer-based composite adaptive fuzzy control for nonstrict-feedback systems with actuator failures. *IEEE Transactions on Fuzzy Systems*, 26(4):2336–2347, 2017.

297 Han Wang, Zhi-Qiang Hu, and Xiang-Yin Meng. Dynamic performance investigation of a spar-type floating wind turbine under different sea conditions. *China Ocean Engineering*, 32(3):256–265, 2018.

298 Jianhui Wang, Houyao Zhu, Chun Liang Zhang, Zicong Chen, Yunchang Huang, Wenli Chen, Xing Huang, and Fang Wang. Adaptive hyperbolic tangent sliding-mode control for building structural vibration systems for uncertain earthquakes. *IEEE Access*, 6:74728–74736, 2018.

299 Qian Wang, Chenkun Qi, Feng Gao, Xianchao Zhao, Anye Ren, and Qiao Sun. Force-based delay compensation for hardware-in-the-loop simulation divergence of 6-DOF space contact.

Proceedings of the Institution of Mechanical Engineers, Part G: Journal of Aerospace Engineering, 233(1):151–165, 2019.

300 Yuejiao Wang, Mukhtar Lawan Adam, Yunlong Zhao, Weihao Zheng, Libo Gao, Zongyou Yin, and Haitao Zhao. Machine learning-enhanced flexible mechanical sensing. *Nano-Micro Letters*, 15(1):1–33, 2023.

301 Pennung Warnitchai, Yozo Fujino, Benito M Pacheco, and Remi Agret. An experimental study on active tendon control of cable-stayed bridges. *Earthquake Engineering and Structural Dynamics*, 22(2):93–111, 2010.

302 Xiaojun Wei and John E Mottershead. Robust passivity-based continuous sliding-mode control for under-actuated nonlinear wing sections. *Aerospace Science and Technology*, 60:9–19, 2017.

303 Daehee Won, Wonhee Kim, Donghoon Shin, and Chung Choo Chung. High-gain disturbance observer-based backstepping control with output tracking error constraint for electro-hydraulic systems. *IEEE Transactions on Control Systems Technology*, 23(2):787–795, 2015.

304 Xiaojuan Wu and Danhui Gao. Fault tolerance control of SOFC systems based on nonlinear model predictive control. *International Journal of Hydrogen Energy*, 42(4):2288–2308, 2017.

305 Ruo-Yang Wu and Chris P Pantelides. Seismic evaluation of repaired multi-column bridge bent using static and dynamic analysis. *Construction and Building Materials*, 208:792–807, 2019.

306 Huaining Wu and Junwei Wang. Observer design and output feedback stabilization for nonlinear multivariable systems with diffusion PDE-governed sensor dynamics. *Nonlinear Dynamics*, 72(3):615–628, 2013.

307 Huai-Ning Wu and Jun-Wei Wang. Static output feedback control via PDE boundary and ODE measurements in linear cascaded ODE–beam systems. *Automatica*, 50(11):2787–2798, 2014.

308 Huai-Ning Wu and Zi-Peng Wang. Observer-based H_∞ sampled-data fuzzy control for a class of nonlinear parabolic PDE systems. *IEEE Transactions on Fuzzy Systems*, 26(2):454–473, 2018.

309 Ailong Wu and Zhigang Zeng. Lagrange stability of memristive neural networks with discrete and distributed delays. *IEEE Transactions on Neural Networks and Learning Systems*, 25(4):690–703, 2014.

310 Huai-Ning Wu, Jun-Wei Wang, and Han-Xiong Li. Design of distributed H_∞ fuzzy controllers with constraint for nonlinear hyperbolic PDE systems. *Automatica*, 48(10):2535–2543, 2012.

311 Huai-Ning Wu, Jun-Wei Wang, and Han-Xiong Li. Fuzzy boundary control design for a class of nonlinear parabolic distributed parameter systems. *IEEE Transactions on Fuzzy Systems*, 22(3):642–652, 2014.

312 Guangyu Wu, Jian Sun, and Jie Chen. Optimal linear quadratic regulator of switched systems. *IEEE Transactions on Automatic Control*, 64(7):2898–2904, 2019.

313 Honghui Wu, Zhi Liu, Yun Zhang, and Chun Lung Philip Chen. Adaptive fuzzy quantized control for nonlinear systems with hysteretic actuator using a new filter-connected quantizer. *IEEE Transactions on Cybernetics*, 50(3):876–889, 2020.

314 Shiqing Wu, Zhonghou Wang, Bin Shen, Jia-Hai Wang, and Li Dongdong. Human-computer interaction based on machine vision of a smart assembly workbench. *Assembly Automation*, 40(3):475–482, 2020.

315 Si-wei Wu, Jian Yang, and Guang-ming Cao. Prediction of the Charpy V-notch impact energy of low carbon steel using a shallow neural network and deep learning. *International Journal of Minerals, Metallurgy, and Materials*, 28(8):1309–1320, 2021.

316 Shengjie Xiao, Kai Hu, Binxiao Huang, Huichao Deng, and Xilun Ding. A review of research on the mechanical design of hoverable flapping wing micro-air vehicles. *Journal of Bionic Engineering*, 18:1235–1254, 2021.

317 Guo Xie, Lanlan Sun, Tao Wen, Xinhong Hei, and Fucai Qian. Adaptive transition probability matrix-based parallel IMM algorithm. *IEEE Transactions on Systems, Man, and Cybernetics: Systems*, 51(5):2980–2989, 2019.

318 Guo Xie, Anqi Shangguan, Rong Fei, Wenjiang Ji, Weigang Ma, and Xinhong Hei. Motion trajectory prediction based on a CNN-LSTM sequential model. *Science China Information Sciences*, 63:1–21, 2020.

319 Tang Xin, Haiping Du, Shuaishuai Sun, Donghong Ning, Zhiwei Xing, and Weihua Li. Takagi–Sugeno fuzzy control for semi-active vehicle suspension with a magnetorheological damper and experimental validation. *IEEE/ASME Transactions on Mechatronics*, 22(1):291–300, 2017.

320 Xueyan Xing and Jinkun Liu. Vibration and position control of overhead crane with three-dimensional variable length cable subject to input amplitude and rate constraints. *IEEE Transactions on Systems, Man, and Cybernetics: Systems*, 51(7):4127–4138, 2019.

321 Bin Xu, Chenguang Yang, and Zhongke Shi. Reinforcement learning output feedback NN control using deterministic learning technique. *IEEE Transactions on Neural Networks and Learning Systems*, 25(3):635–641, 2014.

322 Bin Xu, Chenguang Yang, and Yongping Pan. Global neural dynamic surface tracking control of strict-feedback systems with application to hypersonic flight vehicle. *IEEE Transactions on Neural Networks and Learning Systems*, 26(10):2563–2575, 2015.

323 Bin Xu, Danwei Wang, Youmin Zhang, and Zhongke Shi. DOB-based neural control of flexible hypersonic flight vehicle considering wind effects. *IEEE Transactions on Industrial Electronics*, 64(11):8676–8685, 2017.

324 Bin Xu, Zhongke Shi, Fuchun Sun, and Wei He. Barrier Lyapunov function based learning control of hypersonic flight vehicle with AOA constraint and actuator faults. *IEEE Transactions on Cybernetics*, 49(3):1047–1057, 2018.

325 Shidong Xu, Guanghui Sun, and Li Zhan. Finite frequency vibration suppression for space flexible structures in tip position control. *International Journal of Control, Automation and Systems*, 16(2):1–9, 2018.

326 Bin Xu, Yingxin Shou, Jun Luo, Huayan Pu, and Zhongke Shi. Neural learning control of strict-feedback systems using disturbance observer. *IEEE Transactions on Neural Networks and Learning Systems*, 30(5):1296–1307, 2019.

327 Zhihao Xu, Shuai Li, Xuefeng Zhou, Songbin Zhou, Taobo Cheng, and Yisheng Guan. Dynamic neural networks for motion-force control of redundant manipulators: An optimization perspective. *IEEE Transactions on Industrial Electronics*, 68(2):1525–1536, 2020.

328 Gao Yan, Huai-Ning Wu, Jun-Wei Wang, and Guo Lei. Feedback control design with vibration suppression for flexible air-breathing hypersonic vehicles. *Science China Information Sciences*, 57(3):1–14, 2014.

329 Guiyun Yan, Fuquan Chen, and Yingxiong Wu. A semi-active H_∞ control strategy with application to the vibration suppression of nonlinear high-rise building under earthquake excitations. *Springerplus*, 5(1):1–17, 2016.

330 Hongjun Yang and Jinkun Liu. An adaptive RBF neural network control method for a class of nonlinear systems. *IEEE/CAA Journal of Automatica Sinica*, 5(2):457–462, 2018.

331 Liang Yang and Jianying Yang. Nonsingular fast terminal sliding-mode control for nonlinear dynamical systems. *International Journal of Robust and Nonlinear Control*, 21(16):1865–1879, 2011.

332 Kyung-Jinn Yang, Keum-Shik Hong, and Fumitoshi Matsuno. Boundary control of an axially moving steel strip under a spatiotemporally varying tension. *JSME International Journal Series*, 47(2):665–674, 2004.

333 Jun Yang, Shihua Li, and Xinghuo Yu. Sliding-mode control for systems with mismatched uncertainties via a disturbance observer. *IEEE Transactions on Industrial Electronics*, 60(1):160–169, 2012.

334 Chenguang Yang, Zhijun Li, Rongxin Cui, and Bugong Xu. Neural network-based motion control of an underactuated wheeled inverted pendulum model. *IEEE Transactions on Neural Networks and Learning Systems*, 25(11):2004–2016, 2014.

335 Yonggang Yang, Youzeng Xie, and Lipeng Huo. Design and research on mechanism of bionic flapping-wing air vehicle. *International Journal of Hybrid Information Technology*, 8(4):299–306, 2015.

336 Chenguang Yang, Yiming Jiang, Zhijun Li, Wei He, and Chun-Yi Su. Neural control of bimanual robots with guaranteed global stability and motion precision. *IEEE Transactions on Industrial Informatics*, 13(3):1162–1171, 2017.

337 Chenguang Yang, Xinyu Wang, Long Cheng, and Hongbin Ma. Neural-learning-based telerobot control with guaranteed performance. *IEEE Transactions on Cybernetics*, 47(10):3148–3159, 2017.

338 Chenguang Yang, Chao Zeng, Peidong Liang, Zhijun Li, Ruifeng Li, and Chun-Yi Su. Interface design of a physical human–robot interaction system for human impedance adaptive skill transfer. *IEEE Transactions on Automation Science and Engineering*, 15(1):329–340, 2018.

339 Yongliang Yang, Hamidreza Modares, Donald C Wunsch, and Yixin Yin. Leader-follower output synchronization of linear heterogeneous systems with active leader using reinforcement learning. *IEEE Transactions on Neural Networks and Learning Systems*, 29(6):2139–2153, 2018.

340 Deyin Yao, Hongyi Li, Renquan Lu, and Yang Shi. Distributed sliding-mode tracking control of second-order nonlinear multiagent systems: An event-triggered approach. *IEEE Transactions on Cybernetics*, 50(9):3892–3902, 2020.

341 Yao Yu, Chang-Yin Sun, and Zhuang Jiao. Robust-decentralized tracking control for a class of uncertain MIMO nonlinear systems with time-varying delays. *International Journal of Robust and Nonlinear Control*, 24(18):3474–3490, 2015.

342 Xiang Yu, Yu Fu, and Youmin Zhang. Aircraft fault accommodation with consideration of actuator control authority and gyro availability. *IEEE Transactions on Control Systems Technology*, 26(4):1285–1299, 2018.

343 Ju Yun Yuan and Zhi Hua Dong. Modeling and application of TMD for vibration control under wind load in high-rise buildings. *Advanced Materials Research*, 446–449:3490–3494, 2012.

344 Melak Zebenay, Toralf Boge, Rainer Krenn, and Daniel Choukroun. Analytical and experimental stability investigation of a hardware-in-the-loop satellite docking simulator. *Proceedings of the Institution of Mechanical Engineers, Part G: Journal of Aerospace Engineering*, 229(4):666–681, 2015.

345 Tahmina Zebin and Mohammad Shafiul Alam. Modeling and control of a two-link flexible manipulator using fuzzy logic and genetic optimization techniques. *Journal of Computers*, 7(3):578–585, 2012.

346 Chao Zeng, Chenguang Yang, and Zhaopeng Chen. Bio-inspired robotic impedance adaptation for human–robot collaborative tasks. *Science China Information Sciences*, 63(170201):1–10, 2020.

347 Shuang Zhang and Deqing Huang. End-point regulation and vibration suppression of a flexible robotic manipulator. *Asian Journal of Control*, 19(1):245–254, 2017.

348 Leiqi Zhang and Jiangchuan Liu. Observer-based partial differential equation boundary control for a flexible two-link manipulator in task space. *IET Control Theory and Applications*, 6(13):2120–2133, 2012.

349 Shuang Zhang, Hui Qin, Xiuyu He, and Wei He. Modeling and vibration control of a coupled floating wind turbine system. In *Proceedings of the 2015 15th International Conference on Control, Automation and Systems, Busan, Korea*, pages 1989–1994, 2015.

350 Peixi Zhang, Jianzhong Qiao, Lei Guo, and Wenshuo Li. Sliding mode friction observer based control for flexible spacecraft with reaction wheel. *IET Control Theory and Applications*, 11(8):1274–1281, 2016.

351 Shuang Zhang, Wei He, and Deqing Huang. Active vibration control for a flexible string system with input backlash. *IET Control Theory & Applications*, 10(7):800–805, 2016.

352 Shuang Zhang, Yiting Dong, Yuncheng Ouyang, Zhao Yin, and Kaixiang Peng. Adaptive neural control for robotic manipulators with output constraints and uncertainties. *IEEE Transactions on Neural Networks and Learning Systems*, 29(11):5554–5564, 2018.

353 Tong Zhang, Chun Lung Philip Chen, Long Chen, Xiangmin Xu, and Bin Hu. Design of highly nonlinear substitution boxes based on I-Ching operators. *IEEE Transactions on Cybernetics*, 48(12):3349–3358, 2018.

354 Liguo Zhang, Jianru Hao, and Junfei Qiao. Input-to-state stability of coupled hyperbolic PDE-ODE systems via boundary feedback control. *Science China Information Sciences*, 62(4):42201, 2019.

355 Tong Zhang, Guoxi Su, Chunmei Qing, Xiangmin Xu, Bolun Cai, and Xiaofen Xing. Hierarchical lifelong learning by sharing representations and integrating hypothesis. *IEEE Transactions on Systems, Man, and Cybernetics: Systems*, 51(2):1004–1014, 2019.

356 Tong Zhang, Xuehan Wang, Xiangmin Xu, and Chun Lung Philip Chen. GCB-Net: Graph convolutional broad network and its application in emotion recognition. *IEEE Transactions on Affective Computing*, 13(1):379–388, 2019.

357 Yuanyuan Zhang, Jinkun Liu, and Wei He. Adaptive fault-tolerant control for a nonlinear flexible aircraft wing system. *Asian Journal of Control*, 21(5):2340–2351, 2019.

358 Zili Zhang, Biswajit Basu, and Søren RK Nielsen. Real-time hybrid aeroelastic simulation of wind turbines with various types of full-scale tuned liquid dampers. *Wind Energy*, 22(2):239–256, 2019.

359 Yanhui Zhang, Jian Sun, Hongjing Liang, and Hongyi Li. Event-triggered adaptive tracking control for multiagent systems with unknown disturbances. *IEEE Transactions on Cybernetics*, 50(3):890–901, 2020.

360 Shuang Zhang, Pengxin Yang, Linghuan Kong, Wenshi Chen, Qiang Fu, and Kaixiang Peng. Neural networks-based fault tolerant control of a robot via fast terminal sliding mode. *IEEE Transactions on Systems, Man, and Cybernetics: Systems*, 51(7):4091–4101, 2021.

361 Xiaolei Zhang, Guangwei Yu, Yanbo Wang, Shanda Wang, and Jing Zhang. Dynamic characteristic analysis of flexible-joint flexible-link space manipulator with joint friction. In *2022 8th International Conference on Mechanical Engineering and Automation Science (ICMEAS)*, pages 94–98. IEEE, 2022.

362 Zhijia Zhao, Xiuyu He, Zhigang Ren, and Guilin Wen. Boundary adaptive robust control of a flexible riser system with input nonlinearities. *IEEE Transactions on Systems, Man, and Cybernetics: Systems*, 49(10):1971–1980, 2018.

363 Lingxiao Zheng, Tyson Hedrick, and Rajat Mittal. A comparative study of the hovering efficiency of flapping and revolving wings. *Bioinspiration & Biomimetics*, 8(3):036001, 2013.

364 Lu Zheng, Dianchao Wang, and Zhou Ying. Experimental parametric study on wind-induced vibration control of particle tuned mass damper on a benchmark high-rise building. *Structural Design of Tall and Special Buildings*, 26(8), 2017.

365 Qi Zhou, Hongyi Li, Lijie Wang, and Renquan Lu. Prescribed performance observer-based adaptive fuzzy control for nonstrict-feedback stochastic nonlinear systems. *IEEE Transactions on Systems, Man, and Cybernetics: Systems*, 48(10):1747–1758, 2017.

366 Shui-Ting Zhou, Yi-Jui Chiu, Guo-Fei Yu, Chia-Hao Yang, Hong-Wu Huang, and Sheng-Rui Jian. An assumed mode method and finite element method investigation of the coupled vibration in a flexible-disk rotor system with lacing wires. *Journal of Mechanical Science and Technology*, 31(2):577–586, 2017.

367 Bin Zhu, Jianzhong Zhu, and Qingwei Chen. A bio-inspired flight control strategy for a tail-sitter unmanned aerial vehicle. *Science China Information Sciences*, 63(170203):1–10, 2020.

368 Shiqiang Zhu, Ting Yu, Tao Xu, Hongyang Chen, Schahram Dustdar, Sylvain Gigan, Deniz Gunduz, Ekram Hossain, Yaochu Jin, Feng Lin, et al. Intelligent computing: The latest advances, challenges, and future. *Intelligent Computing*, 2:0006, 2023.

369 Nico Zimmert, Alexander Pertsch, and Oliver Sawodny. 2-DOF control of a fire-rescue turntable ladder. *IEEE Transactions on Control Systems Technology*, 20(2):438–452, 2012.

370 Xiuqing Zou, Erbao Dong, Shiwu Zhang, Xu Min, and Yang Jie. Dynamics analysis of SMA flexible torsion actuator based on finite segment method. *China Mechanical Engineering*, 24(23):3205–3210, 2013.

Index

Dynamic Modeling and Neural Network-Based Intelligent Control of Flexible Systems, First Edition. Hejia Gao, Wei He, and Changyin Sun.

Books in the IEEE Press Series on Control Systems Theory and Applications

Series Editor: Maria Domenica Di Benedetto, University of l'Aquila, Italy

The series publishes monographs, edited volumes, and textbooks which are geared for control scientists and engineers, as well as those working in various areas of applied mathematics such as optimization, game theory, and operations.

1. *Autonomous Road Vehicle Path Planning and Tracking Control*
   Levent Güvenç, Bilin Aksun-Güvenç, Sheng Zhu, and Sükrü Yaren Gelbal
2. *Embedded Control for Mobile Robotic Applications*
   Leena Vachhani, Pranjal Vyas, and Arunkumar G. K.
3. *Merging Optimization and Control in Power Systems: Physical and Cyber Restrictions in Distributed Frequency Control and Beyond*
   Feng Liu, Zhaojian Wang, Changhong Zhao, and Peng Yang
4. *Dynamic System Modeling and Analysis with MATLAB and Python: For Control Engineers*
   Jongrae Kim
5. *Model-Based Reinforcement Learning: From Data to Continuous Actions with a Python-based Toolbox*
   Milad Farsi and Jun Liu
6. *Disturbance Observer for Advanced Motion Control with MATLAB/Simulink*
   Akira Shimada
7. *Control over Communication Networks: Modeling, Analysis, and Design of Networked Control Systems and Multi-Agent Systems over Imperfect Communication Channels*
   Jianying Zheng, Liang Xu, Qinglei Hu, and Lihua Xie
8. *Advanced Control of Power Converters: Techniques and Matlab/Simulink Implementation*
   Hasan Komurcugil, Sertac Bayhan, Ramon Guzman, Mariusz Malinowski, and Haitham Abu-Rub
9. *Sensorless Control of Permanent Magnet Synchronous Machine Drives*
   Zi Qiang Zhu and Xi Meng Wu
10. *The Impact of Automatic Control Research on Industrial Innovation: Enabling a Sustainable Future*
    Silvia Mastellone and Alex van Delft
11. *Interval Analysis: Application in the Optimal Control Problems*
    Navid Razmjooy
12. *Dynamic Modeling and Neural Network-Based Intelligent Control of Flexible Systems*
    Hejia Gao, Wei He, and Changyin Sun

Printed and bound by CPI Group (UK) Ltd, Croydon, CR0 4YY
21/09/2026
14965828-0001